T0176707

An Introduction to Discrete-Valued Time Series

An Introduction to Discrete-Valued Time Series

Christian H. Weiss

Registered Offices
John Wiley & Sons, Inc., 111 River Street, Hoboken, NJ 07030, USA
John Wiley & Sons Ltd, The Atrium, Southern Gate, Chichester, West Sussex, PO19 8SQ, UK

Editorial Office
9600 Garsington Road, Oxford, OX4 2DQ, UK

For details of our global editorial offices, customer services, and more information about Wiley products visit us at www.wiley.com

Wiley also publishes its books in a variety of electronic formats and by print-on-demand. Some content that appears in standard print versions of this book may not be available in other formats.

Library of Congress Cataloging-in-Publication Data:

Name: Weiss, Christian H., 1977– author.
Title: An introduction to discrete-valued time series / by Christian H. Weiss.
Description: Hoboken, NJ : John Wiley & Sons, 2017. | Includes
    bibliographical references and index. |
Identifiers: LCCN 2017040687 (print) | LCCN 2017046480 (ebook) | ISBN
    9781119096986 (pdf) | ISBN 9781119096993 (epub) | ISBN 9781119096962
    (cloth)
Subjects: LCSH: Time-series analysis. | Discrete-time systems–Mathematical
    models.
Classification: LCC QA280 (ebook) | LCC QA280 .W456 2017 (print) | DDC
    519.5/5–dc23
LC record available at https://lccn.loc.gov/2017040687

Cover images: (Background) © RKaulitzki/Gettyimages; (Foreground) Courtesy of Christian H. Weiss
Cover design: Wiley

Set in 10/12pt WarnockPro by SPi Global, Chennai, India
Printed and bound in Malaysia by Vivar Printing Sdn Bhd

10 9 8 7 6 5 4 3 2 1

To Miia,
Maximilian, Tilman and Amalia

Contents

Preface

People have long been interested in time series: data observed sequentially in time. See Klein (1997) for a historical overview. Nowadays, such time series are collected in diverse fields of science and practice, such as business, computer science, epidemiology, finance, manufacturing or meteorology. In line with the increasing potential for applications, more and more textbooks on time series analysis have become available; see for example the recent ones by Box et al. (2015), Brockwell & Davis (2016), Cryer & Chan (2008), Falk et al. (2012), Shumway & Stoffer (2011) and Wei (2006). These textbooks nearly exclusively concentrate on continuous-valued time series, where real numbers or vectors are the possible outcomes. During the last few decades, however, discrete-valued time series have also become increasingly important in research and applications. These are time series arising from counting certain objects or events at specified times, but they are usually neglected in the textbook literature. Among the few introductory or overview texts on discrete-valued time series are

- the books (or parts thereof) by Fahrmeir & Tutz (2001), Kedem & Fokianos (2002) and Cameron & Trivedi (2013) about regression models
- the book by Zucchini & MacDonald (2009) about hidden-Markov models
- the survey article by McKenzie (2003) in the *Handbook of Statistics*
- the textbook by Turkman et al. (2014), which includes a chapter about models for integer-valued time series
- the book by Davis et al. (2016), which provides a collection of essays about discrete-valued time series.

The present book intends to be an introductory text to the field of discrete-valued time series, and to present the subject with a good balance between theory and application. It covers common models for time series of counts as well as for categorical time series, and it works out their most important stochastic properties. It provides statistical approaches for analyzing discrete-valued time series, and it exemplifies their practical implementation in a number of data examples. It does not constitute a purely mathematical treatment of the considered topics, but tries to be accessible to users from all those areas where discrete-valued time series arise and need to be analyzed. Inspired by the seminal time series book by Box & Jenkins (1970), there is a strong emphasis on models and methods "possessing maximum simplicity", but it also provides background and references on more sophisticated approaches. Furthermore, following again the example of Box & Jenkins, the book also includes a part on methods from statistical process control, for the monitoring of a discrete-valued process.

The book is aimed at academics at graduate level having a basic knowledge of mathematics (calculus, linear algebra) and statistics. In addition, elementary facts about time series and stochastic processes are assumed, as they are typically taught in basic courses on time series analysis (also see the textbooks listed above on time series analysis). To allow the reader to refresh their knowledge and to make this book more self-contained, Appendix B contains background information on, for example, Markov chains and ARMA models. Besides putting the reader in a position to analyze and model the discrete-valued time series occurring in practice, the book can also be used as a textbook for a lecture on this topic. The author has already used parts of the book in courses about discrete-valued time series. To support both its application in practice and its use in teaching, ready-made software implementations for the data examples and numerical examples are available to accompany the book. Although such implementations are generally not restricted to a particular software package, the program codes are written in the R language (R Core Team, 2016), since R is freely available to everyone. But each of the examples in this book could have been done with another computational software like Matlab or Mathematica as well. All the R codes, and most of the datasets, are provided on a companion website, see Appendix C for details.

I am very grateful to Prof. Dr. Konstantinos Fokianos (University of Cyprus), Prof. Dr. Robert Jung (University of Hohenheim), Prof. Dr. Dimitris Karlis (Athens University of Economics and Business) and to M. Sc. Tobias Möller (Helmut Schmidt University Hamburg) for reading the entire manuscript and for many valuable comments. I also wish to thank Prof. Dr. Sven Knoth (Helmut Schmidt University Hamburg) for useful feedback on Part III of this book, as well as M. Sc. Boris Aleksandrov and M. Sc. Sebastian Ottenstreuer (ibid.) for making me aware of some typographical errors. I want to thank Prof. Dr. Kurt Brännäs (Umeå University) for allowing me to share the transactions counts data in Example 4.1.5, Alexander Jonen (Helmut Schmidt University Hamburg) for making me aware of the rig counts data in Example 2.6.2, and again Prof. Dr. Dimitris Karlis for contributing the accidents counts data in Example 3.4.2. Thanks go to the Helmut Schmidt University in Hamburg, to the editorial staff of Wiley, especially to Blesy Regulas and Shyamala Venkateswaran for the production of this book, and to Andrew Montford

(Anglosphere Editing Limited) for the copyediting of the book. Finally, I wish to thank my wife Miia and my children Maximilian, Tilman and Amalia for their encouragement and welcome distraction during this work.

Christian H. Weiss
Hamburg
February 2017

About the Companion Website

Don't forget to visit the companion website for this book:

www.wiley.com/go/weiss/discrete-valuedtimeseries

There you will find valuable material designed to enhance your learning, including:

- codes and datasets

Scan this QR code to visit the companion website

1

Introduction

A (discrete-time) *time series* is a set of observations x_t, which are recorded at times t stemming from a discrete and linearly ordered set $\mathcal{T}_0$. An example of such a time series is plotted in Figure 1.1. This is the annual number of lynx fur returns for the MacKenzie River district in north-west Canada. The source is the Hudson's Bay Company, 1821–1934; see Elton & Nicholson (1942). These lynx data are discussed in many textbooks about time series analysis, to illustrate that real time series may exhibit quite complex seasonal patterns. Another famous example from the time series literature is the passenger data of Box & Jenkins (1970), which gives the monthly totals of international airline passengers (in thousands) for the period 1949–1960. These data (see Figure 1.2 for a plot) are often used to demonstrate the possible need for variance-stabilizing transformations.

Looking at the date of origin of the lynx data, it becomes clear that people have long been interested in data collected sequentially in time; see also the historical examples of time series in the books by Klein (1997) and Aigner et al. (2011). But even basic methods of analyzing such time series, as taught in any time series course these days, are rather new, mainly stemming from the last century. As shown by Klein (1997), the classical decomposition of time series into a trend component, a seasonal component and an "irregular component" was mostly developed in the first quarter of the 20th century. The periodogram, nowadays a standard tool to uncover seasonality, dates back to the work of A. Schuster in 1906. The (probably) first correlogram – a plot of the sample autocorrelation function against increasing time lag – can be found in a paper by G. U. Yule from 1926.

The understanding of the time series $(x_t)_{\mathcal{T}_0}$ as stemming from an underlying stochastic *process* $(X_t)_{\mathcal{T}}$, and the irregular component from a stationary one, evolved around that time too (Klein, 1997), enabling an inductive analysis of time series. Here, $(X_t)_{\mathcal{T}}$ is a sequence of random variables X_t, where $\mathcal{T}$ is a discrete and linearly ordered set with $\mathcal{T}_0 \subseteq \mathcal{T}$, while the observations $(x_t)_{\mathcal{T}_0}$ are part of the realization of the process $(X_t)_{\mathcal{T}}$. Major early steps towards the modeling of such stochastic processes are A. N. Kolmogorov's extension theorem

An Introduction to Discrete-Valued Time Series, First Edition. Christian H. Weiss.
© 2018 John Wiley & Sons Ltd. Published 2018 by John Wiley & Sons Ltd.
Companion website: www.wiley.com/go/weiss/discrete-valuedtimeseries

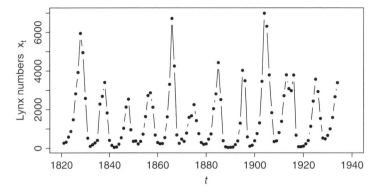

Figure 1.1 Annual number of lynx fur returns (1821–1934); see Elton & Nicholson (1942).

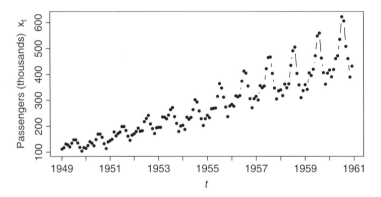

Figure 1.2 Monthly totals (in thousands) of international airline passengers (1949–1960); see Box & Jenkins (1970).

from 1933, the definitions of stationarity by A. Y. Khinchin and H. Wold in the 1930s, the development of the autoregressive (AR) model by G. U. Yule and G. T. Walker in the 1920s and 1930s, as well as of the moving-average (MA) model by G. U. Yule and E. E. Slutsky in the 1920s, their embedding into the class of linear processes by H. Wold in 1938, their combination to the full ARMA model by A. M. Walker in 1950, and, not to forget, the development of the concept of a Markov chain by A. Markov in 1906. All these approaches (see Appendix B for background information) are standard ingredients of modern courses on time series analysis, a fact which is largely due to G. E. P. Box and G. M. Jenkins and their pioneering textbook from 1970, in which they popularized the ARIMA models together with an iterative approach for fitting time series models, nowadays called the Box–Jenkins method. Further details on the history of time series analysis are provided in the books by Klein (1997) and Mills (2011), the history of ARMA models is sketched by Nie & Wu (2013), and more recent developments are covered by Tsay (2000) and Pevehouse & Brozek (2008).

From now on, let $(x_t)_{T_0}$ denote a time series stemming from the stochastic process $(X_t)_T$; to simplify notations, we shall later often use $T = \mathbb{Z} := \{\dots, -1, 0, 1, \dots\}$ (full set of integers) or $T = \mathbb{N}_0 := \{0, 1, \dots\}$ (set of non-negative integers). In the literature, we find several recent textbooks on time series analysis, for example the ones by Box et al. (2015), Brockwell & Davis (2016), Cryer & Chan (2008), Falk et al. (2012), Shumway & Stoffer (2011) amd Wei (2006). Typically, these textbooks assume that the random variables X_t are continuously distributed, with the possible outcomes of the process being real numbers (the X_t are assumed to have the range $\mathbb{R}$, where $\mathbb{R}$ is the set of real numbers). The models and methods presented there are designed to deal with such real-valued processes.

In many applications, however, it is clear from the real context that the assumption of a continuous-valued range is not appropriate. A typical example is the one where the X_t express a number of individuals or events at time t, such that the outcome is necessarily integer-valued and hence *discrete*. If the realization of a random variable X_t arises from counting, then we refer to it as a *count random variable*: a quantitative random variable having a range contained in the discrete set $\mathbb{N}_0$ of non-negative integers. Accordingly, we refer to such a discrete-valued process $(X_t)_T$ as a *count process*, and to $(x_t)_{T_0}$ as a *count time series*. These are discussed in Part I of this book. Note that also the two initial data examples in Figures 1.1 and 1.2 are discrete-valued, consisting of counts observed in time. Since the range covered by these time series is quite large, they are usually treated (to a good approximation) as being real-valued. But if this range were small, as in the case of "low counts", it would be misleading if ignoring the discreteness of the range.

An example of a low counts time series is shown in Figure 1.3, which gives the weekly number of active offshore drilling rigs in Alaska for the period 1990–1997; see Example 2.6.2 for further details. The time series consists of only a few different count values (between 0 and 6). It does not show an obvious trend or seasonal component, so the underlying process appears to be stationary. But it exhibits rather long runs of values that seem to be due to a strong degree of serial dependence. This is in contrast to the time series plotted in Figure 1.4, which concerns the weekly numbers of new infections with Legionnaires' disease in Germany for the period 2002–2008 (see Example 5.1.6). This has clear seasonal variations: a yearly pattern. Another example of a low counts time series with non-stationary behavior is provided by Figure 1.5, where the monthly number of "EA17" countries with stable prices (January 2000 to December 2006 in black, January 2007 to August 2012 in gray) is shown. As discussed in Example 3.3.4, there seems to be a structural change during 2007. If modeling such low counts time series, we need models that not only account for the discreteness of the range, but which are also able

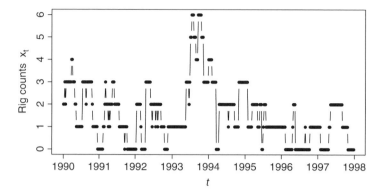

Figure 1.3 Weekly counts of active offshore drilling rigs in Alaska (1990–1997), see Example 2.6.2.

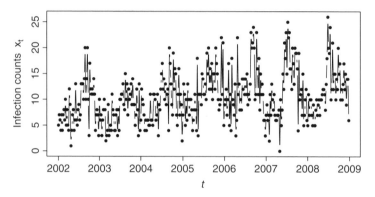

Figure 1.4 Weekly counts of new infections with Legionnaires' disease in Germany (2002–2008); see Example 5.1.6.

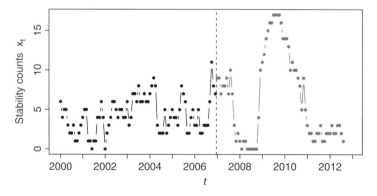

Figure 1.5 Monthly counts of "EA17" countries with stable prices from January 2000 to August 2012; see Example 3.3.4.

to deal with features of this kind. We shall address this topic in Part I of the present book.

All the data examples given above are count time series, which are the most common type of discrete-valued time series. But there is also another important subclass, namely *categorical time series*, as discussed in Part II of this book. For these, the outcomes stem from a qualitative range consisting of a finite number of categories. The particular case of only two categories is referred to as a *binary time series*. For the qualitative sleep status data shown in Figure 1.6, the six categories 'qt', ..., 'aw' exhibit at least a natural ordering, so we are concerned with an *ordinal* time series. In other applications, not even such an inherent ordering exists (*nominal* time series). Then a time series plot such as the one in Figure 1.6 is no longer possible, and giving a visualization becomes much more demanding. In fact, the analysis and modeling of categorical time series cannot be done with the common textbook approaches, but requires tailor-made solutions; see Part II.

For *real-valued* processes, *autoregressive moving-average* (ARMA) models are of central importance. With the (unobservable) *innovations*[1] $(\epsilon_t)_{\mathbb{Z}}$ being independent and identically distributed (i.i.d.) random variables (*white noise*; see Example B.1.2 in Appendix B), the observation at time t of such an ARMA process is defined as a weighted mean of past observations and innovations,

$$X_t = \alpha_1 \cdot X_{t-1} + \ldots + \alpha_p \cdot X_{t-p} + \epsilon_t - \beta_1 \cdot \epsilon_{t-1} - \ldots - \beta_q \cdot \epsilon_{t-q}. \tag{1.1}$$

In other words, it is explained by a part of its own past as well as by an interaction of selected noise variables. Further details about ARMA models are summarized in Appendix B.3. Although these models themselves can be applied only to particular types of processes (stationary, short memory, and

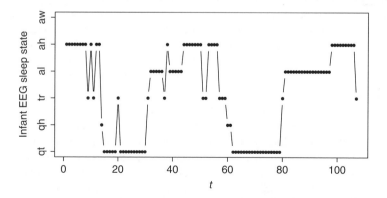

Figure 1.6 Successive EEG sleep states measured every minute; see Example 6.1.1.

---

1 For continuous-valued ARMA models, the innovations $(\epsilon_t)_{\mathbb{Z}}$ are commonly referred to as the error or noise process.

so on), they are at the core of several other models, such as those designed for non-stationary processes or processes with a long memory. In particular, the related *generalized autoregressive conditional heteroskedasticity* (GARCH) model, with its potential for application to financial time series, has become very popular in recent decades; see Appendix B.4.1 for further details. A comprehensive survey of models within the "ARMA alphabet soup" is provided by Holan et al. (2010). A brief summary and references to introductory textbooks in this field can be found in Appendix B.

In view of their important role in the modeling of real-valued time series, it is quite natural to adapt such ARMA approaches to the case of discrete-valued time series. This has been done both for the case of count data and for the categorical case, and such ARMA-like models serve as the starting point of our discussion in both Parts I and II. In fact, Part I starts with an integer-valued counterpart to the specific case of an AR(1) model, the so-called *INAR(1)* model, because this simple yet useful model allows us to introduce some general principles for fitting models to a count time series and for checking the model adequacy. Together with the discussion of forecasting count processes, also provided in Chapter 2, we are thus able to transfer the Box–Jenkins method to the count data case. In the context of introducing the INAR(1) model, the typical features of count data are also discussed, and it will become clear why integer-valued counterparts to the ARMA model are required; in other words, why we cannot just use the conventional ARMA recursion (1.1) for the modeling of time series of counts.

ARMA-like models using so-called "thinning operations", commonly referred to as *INARMA* models, are presented in Chapter 3. The INAR(1) model also belongs to this class, while Chapter 4 deals with a modification of the ARMA approach related to regression models; the latter are often termed *INGARCH* models, although this is a somewhat misleading name. More general regression models for count time series, and also hidden-Markov models, are discussed in Chapter 5. As this book is intended to be an introductory textbook on discrete-valued time series, its main focus is on simple models, which nonetheless are quite powerful in real applications. However, references to more elaborate models are also included for further reading.

In Part II of this book, we follow a similar path and first lay the foundations for analyzing categorical time series by introducing appropriate tools, for example for their visualization or the assessment of serial dependence; see Chapter 6. Then we consider diverse models for categorical time series in Chapter 7, namely types of Markov models, a kind of discrete ARMA model, and again regression and hidden-Markov models, but now tailored to categorical outcomes.

So for both count and categorical time series, a variety of models are prepared here to be used in practice. Once a model has been found to be adequate for the given time series data, it can be applied to forecasting future values.

The issue of forecasting is considered in several places throughout the book, as it constitutes the most obvious field of application of time series modeling. But in line with the seminal time series book by Box & Jenkins (1970), another application area is also covered here, namely the statistical monitoring of a process; see Part III. Chapter 8 addresses the monitoring of count processes, with the help of so-called *control charts*, while Chapter 9 presents diverse control charts for categorical processes. The aim of process monitoring (and particularly of control charts) is to detect changes in an (ongoing) process compared to a hypothetical "in-control" model. Initially used in the field of industrial statistics, approaches for process monitoring are nowadays used in areas as diverse as epidemiology and finance.

The book is completed with Appendix A, which is about some common count distributions, Appendix B, which summarizes some basics about stochastic processes and real-valued time series, and with Appendix C, which is on computational aspects (software implementation, datasets) related to this book.

Part I

Count Time Series

The first part of this book considers the most common type of discrete-valued time series, in which each observation arises from counting certain objects or events. Such count time series consist of time-dependent and quantitative observations from a range of non-negative integers (see also Appendix A). This topic has attracted a lot of research activity over the last three decades, and innumerable models and analytical tools have been proposed for such time series. To be quickly able to comprehensively discuss a first time series example, we start with a simple yet useful model for count time series in Chapter 2, namely the famous INAR(1) model, as proposed by McKenzie (1985). This model constitutes a counterpart to the continuous-valued AR(1) model, which cannot be applied to count time series because of the "multiplication problem" (a brief summary of conventional ARMA models is provided by Appendix B). To avoid this problem, the INAR(1) model uses the so-called binomial thinning operator as a substitute for the multiplication, thus being able to transfer the basic AR(1) recursion to the count data case. The INAR(1) model allows us, among other things, to introduce basic approaches for parameter estimation, model diagnostics, and statistical inference. These approaches are used in an analogous way in the more sophisticated models discussed in the later chapters of this book.

The thinning-based approach to count time series modeling is considered in more depth in Chapter 3, where higher-order ARMA-like models for counts are discussed, as are models with different types of thinning operation, and also thinning-based models for a finite range of counts and for multivariate counts. Chapter 4 then presents a completely different approach for stationary count processes: the so-called INGARCH models. Despite their (controversial) name, these models also focus on counts having an ARMA-like autocorrelation structure. But this time, the ARMA-like structure is not obtained by using

An Introduction to Discrete-Valued Time Series, First Edition. Christian H. Weiss.
© 2018 John Wiley & Sons Ltd. Published 2018 by John Wiley & Sons Ltd.
Companion website: www.wiley.com/go/weiss/discrete-valuedtimeseries

thinning operations, but by a construction related to regression models. Such regression models are covered in Chapter 5, which concludes Part I of this book by briefly presenting further popular models for count time series, including hidden-Markov models.

2

A First Approach for Modeling Time Series of Counts: The Thinning-based INAR(1) Model

As a first step towards the analysis and modeling of count time series, we consider an integer-valued counterpart to the conventional first-order autoregressive model, the INAR(1) model of McKenzie (1985). This constitutes a rather simple and easily interpretable Markov model for stationary count processes, but it is also quite powerful due to its flexibility and expandability. In particular, it allows us to introduce some basic approaches for parameter estimation, model diagnostics and statistical inference. These are used in an analogous way also for the more advanced models discussed in Chapters 3–5. The presented models and methods are illustrated with a data example in Section 2.5.

To prepare for our discussion about count time series, however, we start in Section 2.0 with a brief introduction to the notation used in this book, and with some remarks regarding characteristic features of count distributions in general (without a time aspect).

2.0 Preliminaries: Notation and Characteristics of Count Distributions

In contrast to the subsequent sections, here we remove any time aspects and look solely at separate random variables and their distributions. The first aim of this preliminary section is to acquaint the reader with the basic notation used in this book. The second one is to briefly highlight characteristic features of count distributions, which will be useful in identifying appropriate models for a given scenario or dataset. To avoid a lengthy and technical discussion, detailed definitions and surveys of specific distributions are avoided here but are provided in Appendix A instead.

Count data express the number of certain units or events in a specified context. The possible outcomes are contained in the set of non-negative integers, $\mathbb{N}_0 = \{0, 1, 2, \ldots\}$. These outcomes are not just used as labels; they arise from counting and are hence quantitative (ratio scale). Accordingly, we refer to a quantitative random variable X as a *count random variable* if its

An Introduction to Discrete-Valued Time Series, First Edition. Christian H. Weiss.
© 2018 John Wiley & Sons Ltd. Published 2018 by John Wiley & Sons Ltd.
Companion website: www.wiley.com/go/weiss/discrete-valuedtimeseries

range is contained in the set of non-negative integers, $\mathbb{N}_0 = \{0, 1, 2, \ldots\}$. Some examples of random count phenomena are:

- the number of emails one gets at a certain day (unlimited range $\mathbb{N}_0$)
- the number of occupied rooms in a hotel with n rooms (finite range $\{0, 1, \ldots, n\}$)
- the number of trials until a certain event happens (unlimited range $\mathbb{N} = \{1, 2, \ldots\}$).

A common way of expressing location and dispersion of a count random variable X is to use mean and variance, denoted as

$$\mu := E[X] = \sum_x x \cdot P(X = x), \qquad \sigma^2 := V[X] = E[(X - E[X])^2].$$

The definition and notation of more general types of moments are summarized in Table 2.1; note that $\mu = \mu_1$ is the mean of X, and $\sigma^2 = \overline{\mu}_2$ is the variance of X.

While such moments give insight into specific features of the distribution of X, the complete distribution is uniquely defined by providing its probability mass function (pmf), which we abbreviate as

$$p_k = P(X = k).$$

Similarly, $f_k = P(X \leq k)$ denotes the cumulative distribution function (cdf). An alternative way of completely characterizing a count distribution is to derive an appropriate type of *generating function*; the most common types are summarized in Table 2.2. The probability generating function (pgf), for instance, encodes the pmf of the distribution, but it also allows derivation of the factorial moments: the rth derivative satisfies $\mathrm{pgf}^{(r)}(1) = \mu_{(r)}$; in particular,

Table 2.1 Definition and notation of moments of a count random variable X.

For $n \in \mathbb{N}$, we refer to

$\mu_n := E[X^n]$ as the nth *moment* of X,

$\overline{\mu}_n := E[(X - \mu)^n]$ as the nth *central moment* of X,

$\mu_{(n)} := E[X_{(n)}] = E[X \cdots (X - n + 1)]$ as the nth *factorial moment* of X.

Table 2.2 Definition and notation of generating functions of a count r. v. X.

Generating functions of X:

Probability (pgf)	$\mathrm{pgf}(z) := E[z^X] = \sum_{k=0}^{\infty} p_k \cdot z^k$
Moment (mgf)	$\mathrm{mgf}(z) := \mathrm{pgf}(e^z) = 1 + \sum_{j=1}^{\infty} \mu_j/j! \cdot z^j$
Cumulant (cgf)	$\mathrm{cgf}(z) := \ln(\mathrm{pgf}(e^z)) =: \sum_{j=1}^{\infty} \kappa_j/j! \cdot z^j$
Factorial-cumulant (fcgf)	$\mathrm{fcgf}(z) := \ln(\mathrm{pgf}(1 + z)) =: \sum_{j=1}^{\infty} \kappa_{(j)}/j! \cdot z^j$

$\mathrm{pgf}'(1) = \mu$. The coefficients κ_j of $\mathrm{cgf}(z)$ are referred to as the *cumulants*. Particular cumulants are

$$\kappa_1 = \mu, \qquad \kappa_2 = \sigma^2, \qquad \kappa_3 = \overline{\mu}_3, \qquad \kappa_4 = \overline{\mu}_4 - 3\sigma^4;$$

that is, κ_3/σ^3 is the skewness and κ_4/σ^4 the excess of the distribution. The coefficients $\kappa_{(j)}$ of the factorial-cumulant generating function (fcgf) are referred to as the *factorial cumulants*.

A number of parametric models for count distributions are available in the literature. See Appendix A for a brief survey. There, the models are sorted according to the dimension of their ranges (univariate vs. multivariate), and according to size: in some applications, there exists a fixed upper bound $n \in \mathbb{N}$ that can never be exceeded, so the range is of finite size, taking the form $\{0, \dots, n\}$; otherwise, we have the unlimited range $\mathbb{N}_0$.

Distributions for the case of X being a univariate count random variable with the unlimited range $\mathbb{N}_0$ are presented in Appendix A.1. There, the Poisson distribution has an outstanding position (similar to the normal distribution in the continuous case) and often serves as the benchmark for the modeling of count data. One of its main characteristics is the *equidispersion* property, which means that its variance is always equal to its mean. If we define the (Poisson) *index of dispersion* as

$$I := I(\mu, \sigma^2) := \frac{\sigma^2}{\mu} \quad \in \quad (0; \infty) \tag{2.1}$$

for a random variable X with mean μ and variance σ^2, then the Poisson distribution always satisfies $I = 1$. Values for I deviating from 1, in turn, express a violation of the Poisson model: $I > 1$ indicates an *overdispersed* distribution, such as the negative binomial distribution from Example A.1.4 or Consul's generalized Poisson distribution from Example A.1.6. $I < 1$ expresses *underdispersion*, for example in the Good distribution from Example A.1.7 or the PL distribution from Example A.1.8.

Figure 2.1 illustrates the difference between the equidispersed Poisson distribution (black) and the overdispersed negative binomial distribution (NB; gray) or generalized Poisson distribution (GP; light gray), respectively. All distributions are calibrated to the same mean $\mu = 1.5$, but the plotted NB and GP models have dispersion indices $I = 2$ (that is, 100% overdispersion). It can be seen that both the NB and GP models have more probability mass for values ≥ 4, but also the zero probability is increased (Poi: ≈ 0.223, NB: ≈ 0.354, GP: ≈ 0.346); the latter phenomenon is discussed in more detail below.

Figure 2.2, in contrast, illustrates the effect of underdispersion, compared to the equidispersed Poisson distribution (black) with mean $\mu = 1.5$: the plotted Good distribution (gray; Example A.1.7) and the PL_1 distribution (light gray; Example A.1.8) both have mean ≈ 1.500 and dispersion index ≈ 0.500 (that is, 50% underdispersion). These underdispersed models concentrate most of

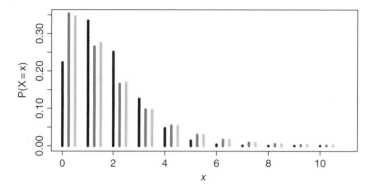

Figure 2.1 Count distributions with $\mu = 1.5$: Poisson in black; NB and GP distributions (both with 100% overdispersion) in gray.

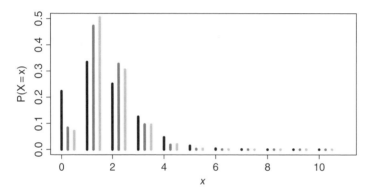

Figure 2.2 Count distributions with $\mu \approx 1.5$: Poisson in black; Good and PL_1 distributions (both with $\approx$ 50% underdispersion) in gray.

their probability mass on the values 1 and 2. In particular, the zero probability is much lower than in the Poisson case (Poi: ≈ 0.223, Good: ≈ 0.084, PL_1: ≈ 0.071).

When discussing Figures 2.1 and 2.2, it becomes clear that another characteristic property of the Poisson distribution is the probability of observing a zero, $p_0 := P(X = 0) = \exp(-E[X])$. Hence, the *zero index* (Puig & Valero, 2006)

$$I_{\text{zero}} := I_{\text{zero}}(\mu, p_0) := 1 + \frac{\ln p_0}{\mu} \quad \in \ (-\infty; 1), \tag{2.2}$$

as a function of mean μ and zero probability p_0, takes the value 0 for the Poisson distribution, but may differ otherwise. Values $I_{\text{zero}} > 0$ indicate *zero inflation* (excess of zeros with respect to a Poisson distribution), while $I_{\text{zero}} < 0$ refers to *zero deflation*. A useful approach for modifying a distribution's zero probability is described in Example A.1.9.

The previous discussion as well as the definition of the indices (2.1) and (2.2) are for distributions having an unlimited range $\mathbb{N}_0$. However, as mentioned before, sometimes the range is finite, $\{0, \ldots, n\}$ with fixed upper bound $n \in \mathbb{N}$. In such a case, the binomial distribution (Example A.2.1 in Appendix A.2) plays a central role. If we characterize its dispersion behavior in terms of the index of dispersion (2.1), the binomial distribution is underdispersed. However, since we are concerned with a different type of random phenomenon anyway – one with a finite range – it is more appropriate to evaluate the dispersion behavior in terms of the so-called *binomial index of dispersion*, defined by

$$I_{\mathrm{Bin}} := I_{\mathrm{Bin}}(n, \mu, \sigma^2) := \frac{\sigma^2}{\mu\,(1 - \mu/n)} \in (0; n) \tag{2.3}$$

for a random variable X with range $\{0, \ldots, n\}$, mean μ and variance σ^2. See also Hagmark (2009), and note that $I_{\mathrm{Bin}} \to I$ for $n \to \infty$. In view of this index, the binomial distribution always satisfies $I_{\mathrm{Bin}} = 1$, while a distribution with $I_{\mathrm{Bin}} > 1$ is said to exhibit *extra-binomial variation*. An example is the beta-binomial distribution from Example A.2.2. For illustration, Figure 2.3 shows a binomial and a beta-binomial distribution with range $\{0, \ldots, 15\}$ and the unique mean 6, but with the beta-binomial distribution exhibiting a strong degree of extra-binomial variation (420%).

Although this book, as an introductory course in discrete-valued time series, mainly focusses on the univariate case, in some places a brief account of possible multivariate generalizations is also provided. Therefore, Appendix A.3 presents multivariate extensions to some basic count models such as the Poisson or negative binomial. These extensions preserve the respective univariate distribution for their marginals, but they induce cross-correlation between the components of the multivariate count vector. An example is plotted in Figure 2.4. The bivariate Poisson (Example A.3.1) and negative

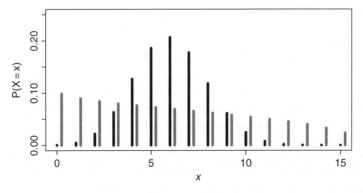

Figure 2.3 Binomial distribution Bin(15, 0.4) in black, and corresponding beta-binomial distribution with $\phi = 0.3$ ($I_{\mathrm{Bin}} = 5.2$) in gray.

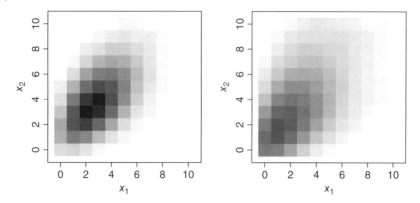

Figure 2.4 Bivariate Poisson (left) and negative binomial distribution (right) with mean $(3, 4)^\top$ and cross-correlation ≈ 0.535.

binomial distribution (Example A.3.2) shown there are adjusted to give the same mean $\mu = (3, 4)^\top$ and the same cross-correlation $\sqrt{2/7} \approx 0.535$, but the negative binomial model obviously shows more dispersion in both components: the dispersion indices are 2 and $2.\overline{3}$, respectively.

One of the multivariate count distributions, the multinomial distribution from Example A.3.3, will be of importance in Part II's consideration of *categorical* time series; see also the discussion in the appendix. The connection to compositional data (Remark A.3.4) is briefly mentioned in this context.

2.1 The INAR(1) Model for Time-dependent Counts

In 1985, in issues 4 and 5 of volume 21 of the *Water Resources Bulletin* (nowadays the *Journal of the American Water Resources Association*), a series of papers about time series analysis appeared, and were also published separately by the American Water Resources Association as the monograph *Time Series Analysis in Water Resources* (edited by K.W. Hipel). One of these papers, "Some simple models for discrete variate time series" by McKenzie (1985), introduced a number of AR(1)-like models for count time series. At this point, it is important to note that the conventional AR(1) recursion, $Y_t = \alpha \cdot Y_{t-1} + \varepsilon_t$, cannot be applied to count processes: even if the innovations ε_t are assumed to be integer-valued with range $\mathbb{N}_0$, the observations Y_t would still not be integer-valued, since the multiplication "$\alpha\cdot$" does not preserve the discrete range (the so-called *multiplication problem*). Therefore, the idea behind McKenzie's new models was to use different mechanisms for "reducing" Y_{t-1}. One such mechanism is the binomial thinning operator (Steutel & van Harn, 1979), which was used to define the *integer-valued AR(1) model*, or *INAR(1) model* for short. The binomial AR(1) model, discussed in Section 3.3, was also introduced in this context.

It seems that McKenzie's paper was overlooked in the beginning, possibly because the *Water Resources Bulletin* was not a typical outlet for time series papers: two years later, the INAR(1) model was proposed again by Al-Osh & Alzaid (1987), but now in the *Journal of Time Series Analysis*. Eventually, McKenzie's paper and the one by Al-Osh & Alzaid, turned out to be ground-breaking for the field of count time series, initiating innumerable research papers about thinning-based time series models (some of them are presented in Section 3) and attracting more and more attention to discrete-valued time series.

We shall now examine the important stochastic properties as well as relevant special cases of the INAR(1) model in great detail. This will allow for more compact presentations of many other models in the later chapters of this book.

2.1.1   Definition and Basic Properties

A way of avoiding the multiplication problem, as sketched above, is to use the probabilistic operation of *binomial thinning* (Steutel & van Harn, 1979), sometimes also referred to as *binomial subsampling* (Puig & Valero, 2007). If X is a discrete random variable with range $\mathbb{N}_0$ and if $\alpha \in (0;1)$, then the random variable $\alpha \circ X := \sum_{i=1}^{X} Z_i$ is said to arise from X by binomial thinning, and the Z_i are referred to as the *counting series*. They are i.i.d. binary random variables with $P(Z_i = 1) = \alpha$, which are also independent of X. So by construction, $\alpha \circ X$ can only lead to integer values between 0 and X. The boundary values $\alpha = 0$ and $\alpha = 1$ might be included in this definition by setting $0 \circ X := 0$ and $1 \circ X := X$. Since each Z_i satisfies $Z_i \sim \text{Bin}(1, \alpha)$ (see Example A.2.1), and since the binomial distribution is additive, $\alpha \circ X$ has a conditional binomial distribution given the value of X; that is, $\alpha \circ X | X \sim \text{Bin}(X, \alpha)$. In particular, using the law of total expectation, it follows that

$$E[\alpha \circ X] = E[\underbrace{E[\alpha \circ X \mid X]}_{\text{mean of binom. distr.}}] = E[\alpha \cdot X] \qquad (= \alpha\,\mu).$$

So the binomial thinning $\alpha \circ X$ and the multiplication $\alpha \cdot X$ have the same mean, which motivates us to use binomial thinning within a modified AR(1) recursion. However, they differ in many other properties; in particular, the multiplication is not a random operation. As an example, the law of total variance implies that

$$V[\alpha \circ X] = V[E[\alpha \circ X \mid X]] + E[V[\alpha \circ X \mid X]]$$
$$= V[\alpha \cdot X] + E[\alpha(1 - \alpha) \cdot X] \qquad (= \alpha^2\,\sigma^2 + \alpha(1 - \alpha)\,\mu),$$

so we have $V[\alpha \circ X] \neq V[\alpha \cdot X]$.

For the *interpretation* of the binomial thinning operation, consider a population of size X at a certain time t. If we observe the same population at a later time, $t + 1$, then the population may have shrunk, because some of the individuals had died between times t and $t + 1$. If the individuals survive independently of each other, and if the probability of surviving from t to $t + 1$ is equal to α for all individuals, then the number of survivors is given by $\alpha \circ X$.

Using the random operator "$\circ$", McKenzie (1985) and Al-Osh & Alzaid (1987) defined the INAR(1) process in the following way.

Definition 2.1.1.1 **(INAR(1) model)** Let the innovations $(\epsilon_t)_\mathbb{N}$ be an i.i.d. process with range $\mathbb{N}_0$, denote $E[\epsilon_t] = \mu_\epsilon$, $V[\epsilon_t] = \sigma_\epsilon^2$. Let $\alpha \in (0; 1)$. A process $(X_t)_{\mathbb{N}_0}$ of observations, which follows the recursion

$$X_t = \alpha \circ X_{t-1} + \epsilon_t,$$

is said to be an INAR(1) process if all thinning operations are performed independently of each other and of $(\epsilon_t)_\mathbb{N}$, and if the thinning operations at each time t as well as ϵ_t are independent of $(X_s)_{s<t}$.

Note that it would be more correct to write "$\circ_t$" in the above recursion to emphasize the fact that the thinning is realized at each time t anew. However, for the sake of readability, the time index is avoided.

The INAR(1) recursion of Definition 2.1.1.1 can be interpreted as follows (Al-Osh & Alzaid, 1987):

$$\underbrace{X_t}_{\text{Population at time } t} = \underbrace{\alpha \circ X_{t-1}}_{\text{Survivors from time } t-1} + \underbrace{\epsilon_t}_{\text{Immigration}} . \tag{2.4}$$

The INAR(1) process is a homogeneous Markov chain with the 1-step transition probabilities given by (McKenzie, 1985; Al-Osh & Alzaid, 1987):

$$p_{k|l} := P(X_t = k \mid X_{t-1} = l)$$

$$= \sum_{j=0}^{\min\{k,l\}} \binom{l}{j} \alpha^j (1 - \alpha)^{l-j} \cdot P(\epsilon_t = k - j). \tag{2.5}$$

For conditional mean and variance, we have (Alzaid & Al-Osh, 1988):

$$E[X_t \mid X_{t-1}] = \alpha \cdot X_{t-1} + \mu_\epsilon,$$
$$V[X_t \mid X_{t-1}] = \alpha(1 - \alpha) \cdot X_{t-1} + \sigma_\epsilon^2, \tag{2.6}$$

which are both linear functions of X_{t-1}. For the derivation of (2.5) and (2.6), note that $\alpha \circ X_{t-1}$ and ϵ_t are independent according to Definition 2.1.1.1. Since the conditional mean is linear in X_{t-1}, the INAR(1) model belongs to the class of *conditional linear AR(1)* models, or CLAR(1), as discussed by Grunwald et al. (2000). Note that the conditional variance differs from the AR(1) case as it varies with time (conditional heteroscedasticity; see the discussion before Definition B.4.1.1).

Let us now assume that the INAR(1) process is even stationary (Definition B.1.3). Conditions for guaranteeing a stationary solution of the INAR(1) recursion are discussed below. If we have given the innovations' distribution in terms of the pgf, then the observations' stationary marginal distribution is determined by the equation (Alzaid & Al-Osh, 1988):

$$\text{pgf}(z) = \text{pgf}(1 - \alpha + \alpha z) \cdot \text{pgf}_\epsilon(z). \tag{2.7}$$

See also the discussion in Section 2.1.3 below. Note that (2.7) is again obtained by applying the law of total expectation, as

$$E[z^{\alpha \circ X_{t-1} + \epsilon_t} \mid X_{t-1}] \overset{\text{indep.}}{=} E[z^{\alpha \circ X_{t-1}} \mid X_{t-1}] \cdot E[z^{\epsilon_t} \mid X_{t-1}]$$
$$\overset{\text{Ex. A.2.1}}{=} (1 - \alpha + \alpha z)^{X_{t-1}} \text{pgf}_\epsilon(z).$$

Equation 2.7 can be used to determine the marginal moments or cumulants of X_t; see Weiß (2013a). In particular, if $\mu_\epsilon, \sigma_\epsilon < \infty$, mean and variance are given by

$$\mu = \frac{\mu_\epsilon}{1 - \alpha} \quad \text{and} \quad \sigma^2 = \frac{\sigma_\epsilon^2 + \alpha\mu_\epsilon}{1 - \alpha^2}, \quad \text{that is,} \quad I = \frac{I_\epsilon + \alpha}{1 + \alpha}, \tag{2.8}$$

where I refers to the index of dispersion (2.1). It implies that X_t is over-/equi-/underdispersed iff ϵ_t is over-/equi-/underdispersed; that is, the dispersion behavior of the observations is determined by the one of the innovations.

The autocorrelation function (ACF; see Definition B.1.1) $\rho(k) :=$ $Corr[X_t, X_{t-k}]$ of a stationary INAR(1) process equals α^k (McKenzie, 1985; Al-Osh & Alzaid, 1987); that is, it is of AR(1) type. Expressions for higher-order joint moments in $(X_t)_{\mathbb{N}_0}$ are provided by Schweer & Weiß (2014).

Remark 2.1.1.2 (Branching process with immigration) A *branching process with immigration* (BPI) $(X_t)_{\mathbb{N}_0}$, also called a Galton–Watson process with immigration, is defined by the recursion (Venkataraman, 1982):

$$X_t = \underbrace{Z_{t;\, 1} + \ldots + Z_{t;\, X_{t-1}}}_{=0 \quad \text{if } X_{t-1}=0} + \epsilon_t,$$

where $X_0, Z_{t;\, r}, \epsilon_s$ are mutually independent count random variables. The *offspring* variables $Z_{t;\, r}$ are i.i.d. with pgf $A(z) = \sum_{k=0}^\infty a_k\, z^k$, and the *immigration* variables ϵ_s are i.i.d. with pgf $B(z) = \sum_{k=0}^\infty b_k\, z^k$. If the offspring mean satisfies $\mu_Z = A'(1) < 1$, then the BPI is said to be *subcritical*. The terminology "offspring" refers to the possible interpretation of $Z_{t;\, 1} + \ldots + Z_{t;\, X_{t-1}}$ as the reproduction generated by the generation $t - 1$. This interpretation is plausible if the $Z_{t;\, r}$ are allowed to also take values larger than 1.

If, however, $a_k = 0$ for all $k \geq 2$ – that is, if the $Z_{t;\, r}$ are Bernoulli-distributed according to $\text{Bin}(1, a_1)$ – then $Z_{t;\, 1} + \ldots + Z_{t;\, X_{t-1}}$ is nothing else than the binomial thinning $a_1 \circ X_{t-1}$. In this case, the interpretation of survivors (see Definition 2.1.1.1) is more appropriate. In particular, it becomes clear that the INAR(1) model according to Definition 2.1.1.1 can be understood as a special type of subcritical BPI; see Alzaid & Al-Osh (1988) and Kedem & Fokianos (2002, Section 5.1). As a consequence, results for subcritical BPIs can also be adapted to the INAR(1) process.

One such result is due to Heathcote (1966). Any BPI constitutes a homogeneous Markov chain; let $h(k) := \sum_{j=1}^{k} \frac{1}{j}$ denote the kth harmonic number. If the BPI is subcritical, if it is an irreducible and aperiodic Markov chain (Appendix B.2.2), and if $E[h(\epsilon_s)] < \infty$, then there exists a proper *stationary marginal distribution* for $(X_t)_{\mathbb{N}_0}$. Note that $E[h(\epsilon_s)] < \infty$ is automatically satisfied if ϵ_s has a finite mean. Another noteworthy result is the one by Pakes (1971) about the geometric ergodicity of subcritical BPIs, which can be used to derive mixing properties (see Definition B.1.5) for INAR(1) models; see also Example 2.1.3.3 below.

A further useful relationship of INAR(1) models is to certain queue length processes with an infinite number of servers. For instance, the Poisson INAR(1) model, which will be discussed in Section 2.1.2, corresponds to an $M/M/\infty$ queue observed at integer times (McKenzie, 2003).

2.1.2   The Poisson INAR(1) Model

The most popular instance of the INAR(1) family is the Poisson INAR(1) model, which was introduced by McKenzie (1985) and Al-Osh & Alzaid (1987). Here, it is assumed that the innovations $(\epsilon_t)_{\mathbb{N}}$ are i.i.d. according to the Poisson distribution Poi(λ), such that $\mu_\epsilon = \sigma_\epsilon^2 = \lambda$. Since all $P(\epsilon_t = j)$ are truly positive, this also holds for all transition probabilities $p_{k|l}$ from (2.5). Consequently, a Poisson INAR(1) process is an irreducible and aperiodic Markov chain (see Appendix B.2.2), such that Remark 2.1.1.2 implies a unique stationary marginal distribution for $(X_t)_{\mathbb{N}_0}$.

It is well known that this stationary marginal distribution is also a Poisson distribution, Poi(μ) with $\mu = \frac{\lambda}{1-\alpha}$. This follows from two important invariance properties of the Poisson distribution

- the invariance with respect to binomial thinning; that is, if $X \sim$ Poi(μ), then $\alpha \circ X \sim$ Poi($\alpha \mu$)
- the additivity; that is, if $Z \sim$ Poi($\alpha \mu$), $\epsilon \sim$ Poi($(1-\alpha)\mu$) and both are independent, then $Z + \epsilon \sim$ Poi($\alpha \mu + (1-\alpha)\mu$) = Poi($\mu$); see Example A.1.1.

Knowing both the conditional and the marginal distribution, we are able to easily simulate a stationary Poisson INAR(1) process – just initialize by Poi(μ) – and the full likelihood function is also directly available; see Remark B.2.1.2 and Example 2.2.2.1 below. Furthermore, the property of having both the observations and the innovations within the same distribution family is analogous to the case of a *Gaussian* AR(1) model. Another similarity between Poisson INAR(1) and Gaussian AR(1) processes, which distinguishes these special instances from other INAR(1) or AR(1) processes, respectively, is time reversibility, see Schweer (2015).

Example 2.1.2.1 (Sample paths) Figure 2.5 shows two sample paths for simulated Poisson INAR(1) processes. Both models were calibrated to give the same observational mean, $\mu = 3$, but the autocorrelation parameter α differs, and hence the innovations mean $\lambda = \mu (1 - \alpha)$. In Figure 2.5a, we have $\alpha = 0.5$ and $\lambda = 1.5$, and this moderate level of autocorrelation becomes visible through the short-term up and down movements. The situation in Figure 2.5b is much more extreme: $\lambda = 0.15$ implies that only rarely is a truly positive innovation (and hence an upward movement) generated. $\alpha = 0.95$ leads to $\alpha \circ X$ being equal to X most of the time, hence the constant segments, and

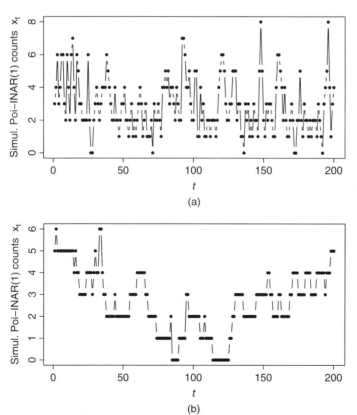

Figure 2.5 Simulated sample paths of Poisson INAR(1) processes with $\mu = 3$ and (a) $\alpha = 0.5$, (b) $\alpha = 0.95$, see Example 2.1.2.1.

otherwise to a slowly descending behavior (leisure extinction). The constant segments also go along with a very small and nearly constant conditional variance according to (2.6); note that the linear coefficient $\alpha(1 - \alpha)$ tends to 0 for either $\alpha \to 0$ or $\alpha \to 1$. This piecewise constant and slowly descending behavior is a characteristic feature of many binomial-thinning-based models (with large thinning probabilities). Other models, such as the INARCH(1) model, as discussed in Example 4.1.6, exhibit different behavior if highly correlated; see Remark 4.1.7.

2.1.3   INAR(1) Models with More General Innovations

The INAR(1) model becomes particularly simple if the innovations are chosen to be Poisson distributed; see Section 2.1.2. But much more flexibility in terms of marginal distributions is possible. One option is to select an appropriate model for the observations X_t and then compute the corresponding innovations' distribution from (2.7); see McKenzie (1985) and the details below. Another option is to choose the distribution of the innovations ϵ_t in order to obtain certain properties for the observations' distribution (Al-Osh & Alzaid, 1987; Alzaid & Al-Osh, 1988); this approach usually simplifies the computation of the transition probabilities (2.5). Generally, see (2.8), the dispersion behavior of the observations is easily controlled by that of the innovations. Also, the probability for observing a zero is influenced by the innovations: since the zero probability just equals pgf(0), (2.7) implies that

$$P(X = 0) = P(\epsilon = 0) \prod_{k=1}^{\infty} \mathrm{pgf}_\epsilon(1 - \alpha^k), \qquad (2.9)$$

see Jazi et al. (2012). So we are not only able to generate over- or underdispersion, but also zero inflation or deflation (see Equation 2.2).

Let us now look at special instances. The most natural extension beyond Poisson distributions is to consider the family of *discrete self-decomposable* (DSD) distributions for X_t (Steutel & van Harn, 1979), which includes, for example, the negative binomial (NB) distribution (see Example A.1.4) as well as the generalized Poisson (GP) distribution (Example A.1.6); see Zhu & Joe (2003) and Weiß (2008a) for more details. Here, a distribution is said to be DSD if its pgf satisfies:

$$\frac{\mathrm{pgf}(z)}{\mathrm{pgf}(1 - \alpha + \alpha\, z)} \quad \text{is itself a pgf for } \textit{all } \alpha \in (0; 1), \qquad (2.10)$$

that is, the coefficients of its power series expansion must be non-negative and add up to 1. In view of (2.7), (2.10) implies that DSD distributions are the marginal distributions of an INAR(1) process that can be preserved for *any* choice of α; the corresponding innovations' pgf is then given by (2.10). Note that any DSD distribution is also infinitely divisible (while the reverse statement does not hold). In other words, it is a particular type of compound Poisson (CP) distribution according to Example A.1.2. As a result, if it is not the Poisson distribution, a DSD distribution is overdispersed and zero-inflated.

Example 2.1.3.1 (Geometric INAR(1) process) The geometric distribution $\text{Geom}(\pi)$ (see Examples A.1.4 and A.1.5) is DSD, so it is a possible marginal distribution for an INAR(1) process $(X_t)_{\mathbb{N}_0}$; we refer to it as the *geometric INAR(1) process* (McKenzie, 1985). If $X \sim \text{Geom}(\pi)$, then $\alpha \circ X$ is also geometrically distributed with pgf

$$\text{pgf}_{\alpha \circ X}(z) = \text{pgf}(1 - \alpha + \alpha z) = \frac{\pi}{1 - (1 - \pi)(1 - \alpha + \alpha z)} = \frac{\frac{\pi}{\pi + \alpha (1-\pi)}}{1 - \frac{\alpha (1-\pi)}{\pi + \alpha (1-\pi)} z},$$

that is, $\alpha \circ X \sim \text{Geom}\left(\dfrac{\pi}{\pi + \alpha (1 - \pi)}\right)$. So (2.10) implies that the innovations $(\epsilon_t)_{\mathbb{N}}$ must have

$$\text{pgf}_\epsilon(z) = \frac{\text{pgf}(z)}{\text{pgf}(1 - \alpha + \alpha z)} = \alpha + (1 - \alpha) \frac{\pi}{1 - (1 - \pi) z},$$

which is the pgf of a *zero-inflated geometric distribution* (Example A.1.9), parent distribution $\text{Geom}(\pi)$ and inflation parameter α.

If we do not insist on having the same marginal distribution for all α, we can simply select a distribution for the innovations, thus controlling dispersion or zero behavior of the observations; see the above discussion. Following this strategy, the most straightforward extension is to choose ϵ_t to be CP-distributed, since then, as in the Poisson case, the observations are also CP-distributed, a characteristic which follows from the invariance properties described next (Schweer & Weiß, 2014).

Lemma 2.1.3.2 (Invariance properties) *Let the compounding order* $v \in \mathbb{N} \cup \{\infty\}$.

(i) *Additivity: If X_1, X_2 are independent with $X_i \sim CP_v(\lambda_i, H_i)$, then their sum $X_1 + X_2 \sim CP_v(\lambda, H)$ with*

$$\lambda \cdot H(z) = \sum_{x=1}^{v} (\lambda_1 h_{1;x} + \lambda_2 h_{2;x}) z^x,$$

including that $\lambda = \lambda_1 + \lambda_2$.

(ii) *Invariance with respect to binomial thinning:*
If $X \sim CP_v(\lambda, H)$, then $\alpha \circ X \sim CP_v(\mu, G)$, where

$$\mu \cdot G(z) = \lambda \sum_{j=1}^{v} \alpha^j \cdot \left(\sum_{i=j}^{v} h_i \binom{i}{j} (1 - \alpha)^{i-j} \right) \cdot z^j,$$

including that $\mu = \lambda (1 - H(1 - \alpha))$.

In fact, Puig & Valero (2007) showed that a count model being parametrized by its v first factorial cumulants $\kappa_{(1)}, \dots, \kappa_{(v)}$ is closed under addition and under binomial thinning *iff* it has a CP_v distribution. These invariance properties lead to the definition of the *compound Poisson INAR(1)* or CP-INAR(1) model.

Example 2.1.3.3 (CP-INAR(1) model) An INAR(1) process $(X_t)_{\mathbb{N}_0}$ according to Definition 2.1.1.1 is referred to as a *CP-INAR(1) process* if the innovations $(\epsilon_t)_{\mathbb{N}}$ are i.i.d. according to the $CP_v(\lambda, H)$ distribution from Example A.1.2 (possibly $v = \infty$). Since $I_\epsilon > 1$ for $v \geq 2$, it also follows from (2.8) that X_t is overdispersed.

In addition, the innovations $\epsilon_s \sim CP_v(\lambda, H)$ have a finite mean provided that $H'(1) < \infty$. Since a CP-INAR(1) process is also irreducible and aperiodic (Schweer & Weiß, 2014), we conclude that a CP-INAR(1) process with $H'(1) < \infty$ possesses a unique stationary marginal distribution. According to Lemma 2.1.3.2, this unique stationary marginal distribution is a compound Poisson one, having the same compounding order v as the innovations (Schweer & Weiß, 2014, Theorem 3.2.1). Hence the observations' distribution is indeed overdispersed but also zero-inflated; see Equation A.1. Formulae for the stationary marginal distribution and the h-step-ahead conditional distributions are provided by Schweer & Weiß (2014).

The relation to BPIs according to Remark 2.1.1.2 and, hence, the result by Pakes (1971) can be utilized to prove that a CP-INAR(1) process with $H'(1) < \infty$ is α-mixing (see Definition B.1.5) with geometrically decreasing weights (Schweer & Weiß, 2014, Theorem 3.4.1), a property that is useful for central limit theorems applied to CP-INAR(1) processes.

A widely used special instance of the CP-INAR(1) model is the *NB-INAR(1)* model, in which the innovations are negatively binomially distributed (Example A.1.4). Note that the marginal distribution of X_t is not an NB distribution, but just another type of CP_∞ distribution.

As mentioned above, a (non-Poisson) CP-INAR(1) model always has an overdispersed and zero-inflated marginal distribution. If, however, underdispersion or zero-deflation are required, then we have to choose the innovations from outside the CP family. Models with *underdispersed* innovations ϵ_t (following the Good distribution in Example A.1.7 or the PL distribution in Example A.1.8), and therefore with underdispersed observations X_t according to (2.8), are discussed by Weiß (2013a). Jazi et al. (2012) consider zero-modified innovations (Example A.1.9).

Remark 2.1.3.4 (MC approximation) Defining the INAR(1) model by specifying the innovations' distribution (as is commonly done in practice), one often does not obtain a closed-form expression for the observations' marginal distribution; a few exceptions are discussed in Schweer

& Weiß (2014). If one is only interested in the zero probability, and if the innovations' pgf is available (as for the distributions discussed in Appendix A.1), one can approximate this probability via (2.9); that is, by computing $P(X = 0) \approx P(\epsilon = 0) \prod_{k=1}^{M} \text{pgf}_{\epsilon}(1 - \alpha^k)$ with M sufficiently large.

If the complete marginal distribution is required – for example, to compute the full likelihood function – then one may utilize the Markov chain (MC) property. For M sufficiently large, define $\tilde{\mathbf{P}} := (p_{i|j})_{i,j=0,\ldots,M}$ with the transition probabilities (2.5). Then the marginal probabilities $(p_0, \ldots, p_M)^\top$ are approximated by the solution of the eigenvalue problem $\tilde{\mathbf{P}} \, \tilde{p} = \tilde{p}$; see the invariance equation (B.4). An alternative approach for approximation is described in Remark 2.6.3.

Example 2.1.3.5 (NB-INAR(1) model) Let us consider the NB-INAR(1) model to illustrate the approximations discussed in Remark 2.1.3.4; that is, where $\epsilon_t \sim \text{NB}(n, \pi)$ according to Example A.1.4. The innovations have dispersion index $I_\epsilon = 1/\pi$, so (2.8) implies for the observations: $I = (1/\pi + \alpha)/(1 + \alpha)$. Since $\text{pgf}_\epsilon(z) = \left(\frac{\pi}{1-(1-\pi)z} \right)^n$ and $P(\epsilon_t = 0) = \pi^n$, the zero probability is approximated by

$$p_0 \approx \pi^{(M+1)n} \Big/ \prod_{k=1}^{M} (1 - (1 - \pi)(1 - \alpha^k))^n,$$

see (2.9). The transition probabilities (2.5) are computed using

$$P(\epsilon_t = k) = \binom{n + k - 1}{k} \cdot (1 - \pi)^k \cdot \pi^n \qquad \text{for } k \in \mathbb{N}_0.$$

They can be used to apply the MC approximation for the pmf of the observations (with $M = 100$ in all the examples below).

For the NB-INAR(1) model with marginal mean $\mu = 1.5$ and $\alpha = 0.5$, one obtains $\mu_\epsilon = \mu(1 - \alpha) = 0.75$ according to (2.8), and this equals $n(1 - \pi)/\pi$ because of the NB assumption. For increasing n (note that $n \to \infty$ corresponds to the Poisson case), we compute

n	1	2	5	10	25	100	∞
π	0.571	0.727	0.870	0.930	0.971	0.993	1
I	1.500	1.250	1.100	1.050	1.020	1.005	1
p_0	0.292	0.261	0.239	0.231	0.226	0.224	0.223

All of these models have the same marginal mean and the same ACF, but (among others) dispersion index and zero probability differ. Figure 2.6 compares the pmf of the equidispersed Poisson INAR(1) model (gray) to the one of the NB-INAR(1) model with $(n, \pi) = (1.75, 0.7)$, so $I \approx 1.286$. The latter pmf has a higher zero probability as well as larger probabilities p_k for $k \geq 4$ (overdispersion).

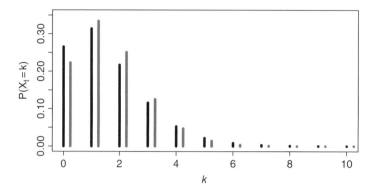

Figure 2.6 Marginal distribution ($\mu = 1.5$) of NB-INAR(1) model $(n, \pi, \alpha) = (1.75, 0.7, 0.5)$ in black, and of Poisson INAR(1) model $(\lambda, \alpha) = (0.75, 0.5)$ in gray.

2.2 Approaches for Parameter Estimation

The INAR(1) model is determined by the thinning parameter α on the one hand, and by further parameters characterizing the marginal distribution of the observations or innovations, respectively, on the other hand. Given the time series data $x_1, \ldots, x_T$, the task is to estimate the value of these parameters.

2.2.1 Method of Moments

Let $x_1, \ldots, x_T$ be a time series stemming from a stationary INAR(1) process. A quite pragmatic approach for parameter estimation is the *method of moments* (MM). Here, the idea is to select appropriate moment relations such that the true model parameters can be obtained by solving the resulting system of equations. For parameter estimation, the true moments are replaced by the corresponding sample moments (see Definition B.1.4), thus leading to the MM estimates.

For an INAR(1) model, one usually selects at least the marginal mean μ (to be estimated by the sample mean $\bar{x}$) as well as the first-order autocorrelation $\rho(1)$; the latter immediately leads to an MM estimator of α, defined as $\hat{\alpha}_{MM} := \hat{\rho}(1) := \hat{\gamma}(1)/\hat{\gamma}(0)$ with $\hat{\gamma}(k) = \frac{1}{T} \sum_{t=k+1}^{T} (X_t - \bar{X})(X_{t-k} - \bar{X})$ for $k \in \mathbb{N}_0$ (Definition B.1.4).

If we have to fit a *Poisson* INAR(1) model according to Section 2.1.2, then we only have one additional parameter besides α, which is either the observations' mean μ or the innovations' mean λ (depending on the chosen parametrization). So, applying (2.8), we define the required MM estimator by either $\hat{\mu}_{MM} := \bar{X}$ or $\hat{\lambda}_{MM} := \bar{X}(1 - \hat{\alpha}_{MM})$. If we have to fit an INAR(1) model with more general innovations, as in Section 2.1.3, then further moment relations are required. For instance, for the NB-INAR(1) model, we could consider the sample variance

$s^2 := \hat{\gamma}(0) = \frac{1}{T} \sum_{t=1}^{T} (x_t - \overline{x})^2$ (Definition B.1.4), because relation (2.8) offers a simple way to estimate the NB parameter π, since the innovations' index of dispersion just equals $I_\epsilon = 1/\pi$.

Example 2.2.1.1 **(Poisson INAR(1) model)** For a Poisson INAR(1) process, Freeland & McCabe (2005) showed that the above MM estimators are asymptotically normally distributed (with **0** denoting the zero vector):

$$\sqrt{T} \, (\hat{\alpha}_{\mathrm{MM}} - \alpha, \hat{\mu}_{\mathrm{MM}} - \mu)^\top \underset{\mathrm{a}}{\sim} \mathrm{N}\!\left(\mathbf{0}, \Sigma_{\alpha,\mu}\right), \qquad \text{where}$$

$$\Sigma_{\alpha,\mu} = \begin{pmatrix} 1 - \alpha^2 + \dfrac{\alpha}{\mu}\,(1-\alpha) & \alpha \\[2mm] \alpha & \mu\,\dfrac{1+\alpha}{1-\alpha} \end{pmatrix},$$

as well as

$$\sqrt{T} \, (\hat{\alpha}_{\mathrm{MM}} - \alpha, \hat{\lambda}_{\mathrm{MM}} - \lambda)^\top \underset{\mathrm{a}}{\sim} \mathrm{N}\!\left(\mathbf{0}, \Sigma_{\alpha,\lambda}\right), \qquad \text{where}$$

$$\Sigma_{\alpha,\lambda} = \begin{pmatrix} 1 - \alpha^2 + \dfrac{\alpha}{\lambda}\,(1-\alpha)^2 & -\lambda(1+\alpha) \\[2mm] -\lambda(1+\alpha) & \lambda\left(1 + \lambda\,\dfrac{1+\alpha}{1-\alpha}\right) \end{pmatrix}.$$

These relations can be utilized to compute approximate standard errors or confidence regions for the estimates, by plugging in the estimates instead of the true parameter values; for an asymptotic bias correction, see Weiß & Schweer (2016).

The estimators $\hat{\alpha}_{\mathrm{MM}} := \hat{\rho}(1)$ and $\hat{\lambda}_{\mathrm{MM}} := \overline{X}\,(1 - \hat{\alpha}_{\mathrm{MM}})$ are not only appropriate for the Poisson INAR(1) model, but more generally for any CLAR(1) model (Grunwald et al., 2000) that is parametrized with α, λ defined by $E[X_t \mid X_{t-1}] = \alpha \cdot X_{t-1} + \lambda$. So the MM estimators do not rely on the particular distribution of a Poisson INAR(1) process, as the ML estimators from Section 2.2.2 do, but only on this particular moment relation. Hence one may classify such an MM estimator as being semi-parametric, and one may expect it to be robust to mild violations of the model assumptions; see also Jung et al. (2005). Certainly, the (asymptotic) distribution of the MM estimators depends on the specific underlying model.

Remark 2.2.1.2 **(Conditional least squares estimation)** For the case of the Poisson INAR(1) model with parameters α and λ (innovations' mean), the *conditional least squares* (CLS) approach can also be used for parameter estimation (Al-Osh & Alzaid, 1987; Freeland & McCabe, 2005). Here, the idea is to accumulate the squared deviations between x_t and $E[X_t \mid x_{t-1}] \overset{(2.6)}{=} \alpha \cdot x_{t-1} + \lambda$

(with the latter being understood as the conditional mean forecast of x_t), and to choose α and λ such that this conditional sum of squares (*CSS*) is minimized:

$$(\hat{\alpha}_{\text{CLS}}, \hat{\lambda}_{\text{CLS}}) := \arg\min_{(\alpha, \lambda)} \quad CSS(\alpha, \lambda),$$

$$\text{where} \quad CSS(\alpha, \lambda) := \sum_{t=2}^{T} (x_t - \alpha \cdot x_{t-1} - \lambda)^2.$$

As shown by Klimko & Nelson (1978) and Al-Osh & Alzaid (1987), explicit expressions for the CLS estimators are given by

$$\hat{\alpha}_{\text{CLS}} = \frac{\sum_{t=2}^{T} X_t X_{t-1} - \frac{1}{T-1} \cdot \sum_{t=2}^{T} X_t \cdot \sum_{s=2}^{T} X_{s-1}}{\sum_{t=2}^{T} X_{t-1}^2 - \frac{1}{T-1} \cdot \left(\sum_{t=2}^{T} X_{t-1}\right)^2},$$

$$\hat{\lambda}_{\text{CLS}} = \frac{1}{T-1} \left(\sum_{t=2}^{T} X_t - \hat{\alpha}_{\text{CLS}} \cdot \sum_{t=2}^{T} X_{t-1} \right).$$

Their asymptotic distribution (assuming a Poisson INAR(1) process) was shown to be same as that of the MM estimators $(\hat{\alpha}_{\text{MM}}, \hat{\lambda}_{\text{MM}})$ given in Example 2.2.1.1 (Klimko & Nelson, 1978; Al-Osh & Alzaid, 1987; Freeland & McCabe, 2005). In contrast to the MM approach, however, it is more difficult to find CLS estimators for other types of INAR(1) processes, as some parameters might not be identifiable from the conditional mean.

The main advantage of the MM estimators is their simplicity (closed-form formulae) and robustness. But for the Poisson INAR(1) model, Al-Osh & Alzaid (1987), Jung et al. (2005) and Weiß & Schweer (2016) recommend using ML estimators instead, because they are less biased for small sample sizes.

2.2.2 Maximum Likelihood Estimation

Like the method of moments, also the *maximum likelihood* (ML) approach relies on a universal principle: one chooses the parameter values such that the observed sample becomes most "plausible". As shown in Remark B.2.1.2, the required (log-)likelihood function is easily computed for Markov processes. In the INAR(1) case with parameter vector θ – for example $\theta = (\alpha, \lambda)^\top$ in the Poisson case or $\theta = (\alpha, n, \pi)^\top$ in the NB case – the (full) *log-likelihood function* becomes

$$\ell(\theta) = \ln p_{x_1}(\theta) + \sum_{t=2}^{T} \ln p_{x_t | x_{t-1}}(\theta), \tag{2.11}$$

where the transition probabilities are computed according to (2.5). Sometimes, it is difficult to compute $p_{x_1}(\theta)$. While this is just a simple Poisson probability in the case of a Poisson INAR(1) model, a closed-form formula for, for example, an NB-INAR(1) model, is not available. Then, one may use the MC approximation from Remark 2.1.3.4 to obtain $p_{x_1}(\theta)$, or one may simply use the *conditional* log-likelihood function, which ignores the initial observation:

$$\ell(\theta \mid x_1) = \sum_{t=2}^{T} \ln p_{x_t \mid x_{t-1}}(\theta). \tag{2.12}$$

The (conditional) ML estimates are now computed as

$$\hat{\theta}_{\mathrm{ML}} = \arg\max_{\theta} \ell(\theta) \quad \text{or} \quad \hat{\theta}_{\mathrm{CML}} = \arg\max_{\theta} \ell(\theta \mid x_1),$$

respectively. In contrast to the CLS approach from Remark 2.2.1.2, it is difficult to find a closed-form solution to this optimization problem. Instead, a numerical optimization is typically applied, where, for example, the MM estimates described in Section 2.2.1 can be used as initial values for the optimization routine. If the optimization routine is able to compute the Hessian of ℓ at the maximum, standard errors can also be approximated; see Remark B.2.1.2 for details.

Example 2.2.2.1 **(Poisson INAR(1) model)** General results for the existence, consistency and asymptotic normality of the (C)ML estimators for discrete-valued Markov chains are available, for example in Part I of the book by Billingsley (1961); see Remark B.2.1.2 for further details. The particular case of a Poisson INAR(1) process was investigated in detail by Freeland & McCabe (2004a); see also Bu et al. (2008). In particular, asymptotic normality was established,

$$\sqrt{T-1}\,(\hat{\alpha}_{\mathrm{CML}} - \alpha, \hat{\lambda}_{\mathrm{CML}} - \lambda)^{\top} \underset{a}{\sim} \mathrm{N}(\mathbf{0}, \mathbf{I}^{-1}(\alpha, \lambda)),$$

where $\mathbf{I}$ denotes the expected Fisher information (Remark B.2.1.2).

A semi-parametric ML approach for INAR(1) processes, where the innovations' distribution is not further specified, was investigated by Drost et al. (2009).

2.3 Model Identification

Section 2.2 presented some standard approaches for fitting an INAR(1) model to a given count time series $x_1, \ldots, x_T$. The obtained estimates are meaningful only if the data indeed stem from an INAR(1) model. So an obvious question is how to identify an appropriate model class for the given data.

First, we look at the serial dependence structure. As for any CLAR(1) model, the ACF of the INAR(1) model is of AR(1) type, given by $\rho(k) = \alpha^k$ (Section 2.1.1). This, in turn, implies that the partial ACF (PACF) satisfies $\rho_{\mathrm{part}}(1) = \alpha$ and $\rho_{\mathrm{part}}(k) = 0$ for $k > 1$; see Theorem B.3.4. Hence to check if an INAR(1) model might be appropriate at all for the given time series data, we should compute the sample PACF (SPACF) to analyze if $\hat{\rho}_{\text{part}}(1)$ deviates significantly from 0, and if $\hat{\rho}_{\text{part}}(k)$ does not for any $k > 1$.

Remark 2.3.1 (Sample PACF) At this point, the (asymptotic) distribution of $(\hat{\rho}_{\text{part}}(1), \hat{\rho}_{\text{part}}(2), \ldots)$ becomes important. For stationary *linear processes* (Background B.3.1) with existing fourth-order moments, the asymptotic behavior of the sample ACF (SACF) is described by the well-known Bartlett's formula, and for an AR(1) process, it is known that the $\hat{\rho}_{\text{part}}(k)$ for $k \geq 2$ are asymptotically independent and normally distributed with mean 0 and variance $1/T$ (Brockwell & Davis, 1991). However, for non-linear processes, Bartlett's formula may be misleading, so one should use the more general result by Romano & Thombs (1996). For the case of a Poisson INAR(1) model, the asymptotic distribution of the $\hat{\rho}_{\text{part}}(k)$ was derived by Mills & Seneta (1991) (see Remark 2.1.1.2). Although the asymptotic variances are slightly larger than $1/T$, asymptotic independence still holds between the $\hat{\rho}_{\text{part}}(k)$ with $k \geq 2$. So the autocorrelation structure can be identified in a completely analogous way to the AR(1) case.

Further tests for serial dependence in count time series are discussed by Jung & Tremayne (2003).

Once we have identified the AR(1)-like autocorrelation structure, we should next analyze the marginal distribution. Here, an important question is if the simple Poisson model does well, or if the observed marginal distribution deviates significantly from a Poisson distribution. In the latter case, the type of deviation (overdispersion, zero-inflation, and so on) may help us to identify an appropriate model.

A rather general approach that allows us to detect diverse violations of the Poisson INAR(1) model are the pgf-based tests proposed by Meintanis & Karlis (2014). These tests compare the conjectured bivariate pgf – that is, $\text{pgf}_{X_t, X_{t-1}}(z_0, z_1) := E[z_0^{X_t} z_1^{X_{t-1}}]$ – with its sample counterpart. For the null of a Poisson INAR(1) model, (X_t, X_{t-1}) are bivariately Poisson distributed (Example A.3.1) with the bivariate pgf being given by (Alzaid & Al-Osh, 1988):

$$\text{pgf}_{X_t, X_{t-1}}(z_0, z_1) = \exp(\mu \, (z_0 + z_1 - 2 + \alpha \, (z_0 - 1)(z_1 - 1))), \qquad (2.13)$$

which is symmetric in z_0, z_1 in accordance with the time reversibility; note that (2.13) holds for any time lag $h \in \mathbb{N}$ if replacing α by α^h. Since the (asymptotic) distributions of the proposed test statistics are intractable, Meintanis & Karlis (2014) recommend a bootstrap implementation of the tests.

More simple diagnostic tests can be obtained by focussing on a particular type of violation of the Poisson model. Often, such violations go along with a violation of the equidispersion property,[1] and overdispersion in particular

---

1 A counterexample would be the Good distribution from Example A.1.7, the parameter values of which can be chosen such that it is non-Poisson but exhibits equidispersion.

is commonly observed in practice (Weiß, 2009c). An obvious test statistic for uncovering over- or underdispersion is the sample counterpart to the dispersion index (2.1); that is, $\hat{I} := S^2/\overline{X}$ (see Definition B.1.4). Under the null of a Poisson INAR(1) model, this test statistic is asymptotically normally distributed with

$$E[\hat{I}] \approx 1 - \frac{1}{T}\frac{1+\alpha}{1-\alpha}, \qquad V[\hat{I}] \approx \frac{2}{T}\frac{1+\alpha^2}{1-\alpha^2}, \tag{2.14}$$

see Schweer & Weiß (2014) and Weiß & Schweer (2015). Plugging in $\hat{\rho}(1)$ instead of α, the resulting normal approximation can be used for determining critical values or for computing P values.

Remark 2.3.2 (Sample variance) The negative bias expressed by (2.14) appears plausible in view of the general result by David (1985) about the bias of S^2:

$$E[S^2] = \sigma^2 - V[\overline{X}]. \tag{2.15}$$

Now consider the particular case of a Poisson INAR(1) process, where $\sigma^2 = \mu$ (equidispersion). Inserting the asymptotic variance of $\overline{X}$ from Example 2.2.1.1 into (2.15), and using (3.6) from Pickands & Stine (1997), it follows that S^2 is asymptotically normally distributed with

$$E[S^2] \approx \mu - \frac{1}{T}\frac{1+\alpha}{1-\alpha}\,\mu, \qquad V[S^2] \approx \frac{1}{T}\left(2\frac{1+\alpha^2}{1-\alpha^2}\,\mu^2 + \frac{1+\alpha}{1-\alpha}\,\mu\right).$$

If several candidate models have been identified as being relevant for the given data, a popular way to select a final model is to consider information criteria such as the AIC and BIC (see Remark B.2.1.1, Equation B.7, for the definitions), which are computed along with the ML estimates (Section 2.2.2). While the idea behind such information criteria is plausible, namely balancing goodness-of-fit against model size, they should be used with some caution in practice; see Emiliano et al. (2014). They may serve as guides for identifying a relevant model, but a decision to adopt a specific model should take into account further aspects; see Section 2.4. Other selection criteria include the conditional sum of squares, *CSS*, as computed during CLS estimation (Remark 2.2.1.2) or criteria related to forecasting (for example, realized coverage rates of prediction intervals); the topic of forecasting is discussed in Section 2.6. More generally, *scoring rules,* such as the ones discussed by Czado et al. (2009) and Jung & Tremayne (2011b) can be used for this purpose. Since some of these are closely related to tools for checking for model adequacy, we shall discuss them further; see Section 2.4 and Remark 2.4.1.

2.4   Checking for Model Adequacy

After having identified the best of the candidate models, it remains to check if they are really adequate for the analyzed data; that is, if the given time series constitutes a typical realization of the considered model. An obvious approach for checking the model adequacy is to compare some features of the fitted model with their sample counterparts, as computed from the available time series. Such a comparison should include the autocorrelation structure as well as marginal characteristics such as the mean, the dispersion ratio or the zero probability (see the corresponding formulae in Section 2.1). Besides merely comparing the respective numerical values, one may follow the idea of Tsay (1992) (see also Jung & Tremayne (2011a)) and compute *acceptance envelopes* for, for example, ACF or pmf, where the envelope is based on quantiles obtained from a parametric bootstrap for the fitted model.

More sophisticated tools relying on conditional distributions, which hence check the predictive performance, are presented in Jung & Tremayne (2011b) and Christou & Fokianos (2015). As a first approach, the *standardized Pearson residuals* (Harvey & Fernandes, 1989) should be analyzed; that is, the series

$$e_t := \frac{x_t - E[X_t \mid x_{t-1}]}{\sqrt{V[X_t \mid x_{t-1}]}} \qquad \text{for } t = 2, \dots, T, \tag{2.16}$$

where the conditional moments are given by (2.6). For models that are not Markov chains, the definition of e_t has to be adapted accordingly. For an adequate model, we expect these residuals to be uncorrelated, with a mean about 0 and a variance about 1. A variance larger/smaller than 1 indicates that the data show more/less dispersion than being considered by the model (Harvey & Fernandes, 1989). The variance of the Pearson residuals or their mean sum of squares ($\approx$ "normalized squared error score") are also sometimes used as a scoring rule for predictive model assessment (Czado et al., 2009). Instead of Pearson residuals, *forecast (mid-)pseudo-residuals* might also be used for checking the model adequacy (Zucchini & MacDonald, 2009, Section 6.2.3). But since these forecast pseudo-residuals are closely related to the PIT (described below), we shall not discuss this type of residual further here.

An approach that considers not only conditional moments, but the complete conditional distribution, is the (non-randomized) *probability integral transform* (PIT) (Czado et al., 2009; Jung & Tremayne, 2011b). Let $f_{\cdot|l} = (f_{k|l})_{k=0,1,\dots}$ with $f_{k|l} := P(X_t \leq k \mid X_{t-1} = l)$ denote the conditional cdf, conditioned on the last observation being $l \in \mathbb{N}_0$, where the $f_{k|l} = \sum_{j=0}^{k} p_{j|l}$ are computed from (2.5). Then the mean PIT is defined as (Czado et al., 2009; Jung &

Tremayne, 2011b):

$$\overline{F}(u) := \frac{1}{T-1} \sum_{t=2}^{T} F_t(u) \quad \text{for } u \in [0;1], \quad \text{where}$$

$$F_t(u) := \begin{cases} 0 & \text{if } u \leq f_{x_t-1|x_{t-1}}, \\ \dfrac{u - f_{x_t-1|x_{t-1}}}{f_{x_t|x_{t-1}} - f_{x_t-1|x_{t-1}}} & \text{if } f_{x_t-1|x_{t-1}} < u < f_{x_t|x_{t-1}}, \\ 1 & \text{if } u \geq f_{x_t|x_{t-1}}. \end{cases} \tag{2.17}$$

Here, we define $f_{-1|l} := 0$ for any $l \in \mathbb{N}_0$; note that the $f_{k|l}$ only needs to be computed for $k, l \leq \max\{x_1, \ldots, x_T\}$. The mean PIT now allows us to construct a histogram in the following way: dividing $[0;1]$ into the J subintervals $\left[\frac{j-1}{J}; \frac{j}{J}\right]$ for $j = 1, \ldots, J$ (say, $J = 10$), the jth rectangle is drawn with height $\overline{F}\left(\frac{j}{J}\right) - \overline{F}\left(\frac{j-1}{J}\right)$. If the fitted model is adequate, we expect the *PIT histogram* to look like that of a uniform distribution. Common deviations from uniformity are U-shaped histograms indicating that the fitted conditional distribution is underdispersed with respect to the data, while inverse-U shaped histograms indicate overdispersion (Czado et al., 2009), analogous to the variance of the Pearson residuals, as discussed above.

A related visual tool is the *marginal calibration diagram* (Czado et al., 2009), which compares the marginal frequencies of the time series, $\hat{p}_0, \hat{p}_1, \ldots$ where $\hat{p}_k := \frac{1}{T} \sum_{t=1}^{T} \mathbb{1}(x_t = k)$, with the aggregated conditional distributions, $\tilde{p}_k := \frac{1}{T} \sum_{t=1}^{T} p_{k|x_t}$, for example by plotting the differences $\tilde{p}_k - \hat{p}_k$ against $k \in \mathbb{N}_0$. Here, $\mathbb{1}(\cdot)$ denotes the indicator function. Analogously, one can compare the respective cumulative distributions with each other; that is, $\hat{f}_k := \frac{1}{T} \sum_{t=1}^{T} \mathbb{1}(x_t \leq k)$ and $\tilde{f}_k := \frac{1}{T} \sum_{t=1}^{T} f_{k|x_t}$.

Remark 2.4.1 **(Scoring rules)** As already mentioned, a number of scoring rules to assess the quality of predictive distributions have been proposed in the literature; see Czado et al. (2009) and Jung & Tremayne (2011b) for a detailed discussion. Typical scoring rules are of the form $s(p_{\cdot|x_{t-1}}, x_t)$, to compare the observation x_t realized at time t with the conditional distribution $p_{\cdot|x_{t-1}}$ based on the previous observation, where smaller score values express better agreement. The overall predictive performance of the model with respect to the time series $x_1, \ldots, x_T$ is evaluated by the mean score $\frac{1}{T-1} \sum_{t=2}^{T} s(p_{\cdot|x_{t-1}}, x_t)$.

A scoring rule that is closely related to the marginal calibration diagram is the *ranked probability score* (Czado et al., 2009; Jung & Tremayne, 2011b), which is defined as (the mean about) the squared deviations

$$s_{\text{rps}}(p_{\cdot|x_{t-1}}, x_t) := \sum_{k=0}^{\infty} (f_{k|x_{t-1}} - \mathbb{1}(x_t \leq k))^2 \tag{2.18}$$

between the conditional distribution and the actual observation. Other commonly used scoring rules are the *logarithmic score*

$$s_{ls}(p_{\cdot|x_{t-1}}, x_t) := -\ln p_{x_t|x_{t-1}}, \qquad (2.19)$$

which goes along with the conditional log-likelihood computation (2.12), and the *quadratic score*

$$s_{qs}(p_{\cdot|x_{t-1}}, x_t) := -2\, p_{x_t|x_{t-1}} + \sum_{k=0}^{\infty} p_{k|x_{t-1}}^2. \qquad (2.20)$$

Computing the mean score related to the fitted candidate models, a scoring rule might be used in the context of model selection; see the discussion in the end of Section 2.3.

2.5 A Real-data Example

To illustrate the models and methods discussed up until now, let us consider the dataset presented by Weiß (2008a). This is a time series expressing the daily number of downloads of a TEX editor for the period from June 2006 to February 2007 ($T = 267$). The plot in Figure 2.7 shows that these daily counts vary between 0 and 14, without any visible trend or seasonality. The up and down movements indicate a moderate autocorrelation level, which is confirmed by the SACF plot in Figure 2.8a. After further inspecting the SPACF, where only $\hat{\rho}_{part}(1)$ deviates significantly from 0, we conclude that an AR(1)-like model might be appropriate for describing the time series.

The observed marginal distribution is plotted in Figure 2.8b. The mean $\bar{x} \approx 2.401$ is clearly smaller than the variance $s^2 \approx 7.506$, so, at least empirically, we are concerned with a strong degree of overdispersion. This goes along with a high zero probability, $\hat{p}_0 \approx 0.277$, which is much larger than the

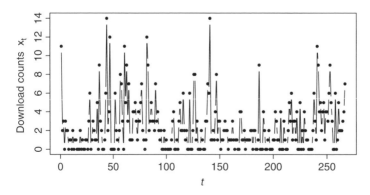

Figure 2.7 Plot of the download counts; see Section 2.5.

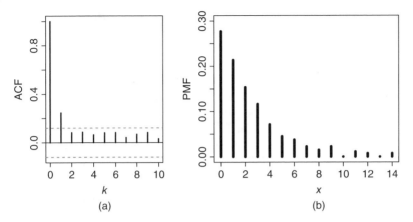

Figure 2.8 Sample autocorrelation (a) and marginal frequencies (b) of the download counts; see Section 2.5.

corresponding Poisson value $\exp(-\bar{x}) \approx 0.091$ (zero inflation). In summary, an INAR(1) model appears to be plausible for the data, possibly with an overdispersed (and zero-inflated) marginal distribution. As pointed out by Weiß (2008a), an INAR(1) model also seems plausible in view of interpretation (2.4): some downloads at day t might be initiated on the recommendation of users from the previous day $t - 1$ ("survivors"), the remaining downloads being due to users who became interested in the program on their own initiative ("immigrants").

To test for overdispersion within the INAR(1) model, we apply the dispersion test described in Section 2.3, plugging in $\hat{\alpha} = \hat{\rho}(1) \approx 0.245$ instead of α into Equation 2.14. Comparing the observed value $\hat{I} \approx 3.127$ with the approximate mean and standard deviation under the null of a Poisson INAR(1) model, given by about 0.994 and 0.092, respectively, it becomes clear that the overdispersion is indeed significant (P value ≈ 0). Therefore, we shall fit the NB-INAR(1) model to the data, but also the Poisson INAR(1) model and the corresponding i.i.d. models for illustration. The estimated mean and dispersion index of the innovations are given by $\hat{\mu}_\epsilon = \bar{x}\,(1 - \hat{\alpha}) \approx 1.813$ and $\hat{I}_\epsilon = \hat{I}\,(1 + \hat{\alpha}) - \hat{\alpha} \approx 3.647$, respectively.

Parameter estimation is done by a full likelihood approach, using the MC approximation for the initial probability in the case of the NB-INAR(1) model; see Example 2.1.3.5). As initial values for the numerical optimization routine, simple moment estimates are used (Section 2.2.2):

- $\hat{\alpha}_{\mathrm{MM}} \approx 0.245$ and $\hat{\lambda}_{\mathrm{MM}} \approx 1.813$ for the Poisson INAR(1)
- $\hat{\pi}_{\mathrm{MM}} = 1/\hat{I}_\epsilon \approx 0.274$, $\hat{n} = \hat{\mu}_\epsilon\,\hat{\pi}_{\mathrm{MM}}/(1 - \hat{\pi}_{\mathrm{MM}}) \approx 0.685$ for the NB-INAR(1).

The ML estimates $\hat{\theta}_{\mathrm{ML}}$ are now obtained by maximizing the respective full log-likelihood function, and the corresponding standard errors are

Table 2.3 Download counts: ML estimates and AIC and BIC values for different models.

Model	Parameter			AIC	BIC
	1	2	3		
i.i.d. Poisson	2.401			1323	1327
(μ)	(0.095)				
Poisson INAR(1)	1.991	0.174		1293	1300
(λ, α)	(0.110)	(0.033)			
i.i.d. NB	1.108	0.316		1103	1111
(n, π)	(0.158)	(0.034)			
NB-INAR(1)	0.835	0.291	0.154	1092	1103
(n, π, α)	(0.145)	(0.036)	(0.042)		

Figures in parentheses are standard errors. AIC and BIC values rounded.

approximated from the computed Hessian $\hat{\mathbf{J}} := \mathbf{H}_\ell(\hat{\theta}_{ML})$ as the square roots of the diagonal elements from the inverse $\hat{\mathbf{J}}^{-1}$ (Remark B.2.1.2). The obtained results are summarized in Table 2.3 together with the (rounded) values of the AIC and BIC from (B.7).

From the AIC and BIC values shown in Table 2.3, it becomes clear that the INAR(1) structure is always better than the respective i.i.d. model. In particular, the estimates for α are always significantly different from 0. Comparing the two INAR(1) models, the NB-INAR(1) model is clearly superior, as should be expected in view of the strong degree of overdispersion (and zero inflation). This decision is also supported by any of the scoring rules from Remark 2.4.1 ($\overline{s_{rps}}$: 1.399 vs. 1.309; $\overline{s_{ls}}$: 2.384 vs. 2.022; $\overline{s_{qs}}$: −0.121 vs. −0.179). Note that the parameter n of the fitted NB models is always close to 1; that is, these NB distributions are close to a geometric distribution (see Example A.1.5). While the NB-INAR(1) model is the best of the considered candidate models, it remains to check if it is also adequate for the data (Section 2.4).

For illustration, we also include the Poisson INAR(1) model in the remaining analyses. We start by computing the marginal properties of the fitted INAR(1) models. The means of both INAR(1) models – 2.411 (Poisson) and 2.407 (NB), according to (2.8) – are close to $\bar{x} \approx 2.401$. The observed index of dispersion $\hat{I} \approx 3.127$, however, is much better reproduced by the NB-INAR(1) model (3.111) than by the equidispersed Poisson INAR(1) model. The same applies to the zero probability, where $\hat{p}_0 \approx 0.277$ compared to 0.258 (NB model; see (2.9) and Example 2.1.3.5) and 0.090 (Poisson model). Also an analysis of the respective Pearson residuals (both series show no significant autocorrelation) supports use of the NB-INAR(1) model: the residuals variance for the NB, at 0.931, is close to 1, whereas for the Poisson, the residuals variance, at 2.871, is much too

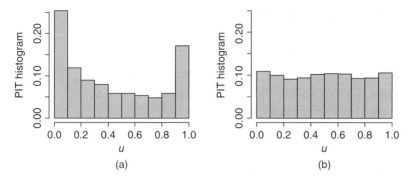

Figure 2.9 PIT histograms based on fitted Poisson and NB-INAR(1) model; see Section 2.5.

large, thus indicating that the data show more dispersion than described by the Poisson model.

Finally, let us have a look at the PIT histogram in Figure 2.9. The PIT histogram of the NB-INAR(1) model in (b) is close to uniformity, while the one of the Poisson INAR(1) model in (a) is strongly U-shaped (and also asymmetric). This U-shape indicates that the Poisson model does not show sufficient dispersion, confirming our previous analyses.

2.6 Forecasting of INAR(1) Processes

Given the model for the observed INAR(1) process, one of the main applications[2] of this model is to forecast future outcomes of the process. In other words, having observed $x_1, \dots, x_T$, we want to predict X_{T+h} for some $h \geq 1$. For real-valued processes, the most common type of *point forecast* is the conditional mean, as this is known to be optimal in the sense of the mean squared error. Applying the law of total expectation iteratively together with (2.6), it follows that the h-step-ahead conditional mean is given by

$$E[X_{T+h} \mid X_T] = \alpha^h \cdot X_T + \mu_\epsilon \frac{1 - \alpha^h}{1 - \alpha} = \alpha^h \cdot X_T + \mu \,(1 - \alpha^h). \qquad (2.21)$$

Note that this conditional mean only depends on X_T, but not on earlier observations, due to the Markov property (Appendix B.2.1). Conditional mean forecasting for INAR(1) processes was further investigated by Sutradhar (2008), and mean-based *forecast horizon aggregation* – that is, the forecasting of the sum $\sum_{j=1}^{h} X_{T+j}$ given X_T – was discussed by Mohammadipour & Boylan (2012),

---

2 Another important application area is *statistical process control*, where the process is monitored to detect possible changes in the process distribution; this kind of application is discussed later in Chapter 8.

including also other members of the INARMA family, the latter which are discussed in Section 3.1.

The main disadvantage of the mean forecast is that it will usually lead to a non-integer value, while X_{T+h} will certainly take an integer value from $\mathbb{N}_0$. Therefore, *coherent forecasting* techniques (that only produce forecasts in $\mathbb{N}_0$) are required for count processes (Freeland & McCabe, 2004b). For this purpose, the h-step-ahead conditional distribution of X_{T+h} given the past $X_T, \ldots, X_1$ needs to be computed for the INAR(1) model; that is, the h-step-ahead transition probabilities $p_{k|l}{}^{(h)} := P(X_{T+h} = k \mid X_T = l)$ (again only depending on X_T thanks to the Markov property). Once this distribution is available, the corresponding conditional median and mode can be used as a coherent point forecast. In fact, the conditional median also satisfies an optimality property, as it minimizes the mean absolute error.

So the essential question is how to compute the $p_{k|l}{}^{(h)}$. First note that Al-Osh & Alzaid (1987) have shown the following equality in distribution:

$$X_{T+h} \overset{d}{=} \alpha^h \circ X_T + \underbrace{\sum_{j=0}^{h-1} \alpha^j \circ \epsilon_{T+h-j}}_{=:\, \epsilon^{(h)}}.$$

So once the distribution of $\epsilon^{(h)}$ is available, the $p_{k|l}{}^{(h)}$ can be computed by adapting (2.5). Unfortunately, this distribution is generally not easily obtained. For the case of a CP-INAR(1) model, as introduced in Example 2.1.3.3 (the CP distribution is invariant with respect to binomial thinning according to Lemma 2.1.3.2), Schweer & Weiß (2014) showed that $\epsilon^{(h)}$ is CP-distributed, and they provided a closed-form expression for the pgf of $\epsilon^{(h)}$. After having done a numerical series expansion for $\mathrm{pgf}_{\epsilon^{(h)}}(z)$, the h-step-ahead transition probabilities $p_{k|l}{}^{(h)}$ are computed via (2.5) (replacing α by α^h).

Example 2.6.1 **(Forecasting Poisson INAR(1) processes)** In the particular case of a Poisson INAR(1) process, the results above further simplify, since now $\epsilon^{(h)} \sim \mathrm{Poi}(\mu(1-\alpha^h))$ (Freeland & McCabe, 2004b). So we explicitly obtain

$$p_{k|l}^{(h)} = \sum_{j=0}^{\min\{k,l\}} \binom{l}{j} \alpha^{h\,j}(1-\alpha^h)^{l-j} \cdot e^{-\mu(1-\alpha^h)}\frac{(\mu(1-\alpha^h))^{k-j}}{(k-j)!},$$

$$V[X_{T+h} \mid X_T] = \alpha^h(1-\alpha^h) \cdot X_T + \mu(1-\alpha^h),$$

also see (2.13). Note that with increasing h, this distribution just converges to the $\mathrm{Poi}(\mu)$ distribution; that is, to the stationary marginal distribution, as would be expected from the ergodicity of the process.

The h-step-ahead conditional distribution can certainly also be used to construct a *prediction interval* on level $1-\alpha$, based on the $\alpha/2$- and $(1-\alpha/2)$-quantile from this distribution in case of a two-sided interval, or based on the $(1 - \alpha)$-quantile for an upper-sided interval ("*worst-case* prediction").

Example 2.6.2   (Rig counts)   We analyze a time series of weekly counts of active rotary drilling rigs, where each count expresses the number of active off-shore drilling rigs in Alaska for the period 1990–1997 (length $T = 417$). The data are available from Baker Hughes.[3] These rig counts have been published for the USA and Canada since 1944, and international rig counts since 1975. They serve as an indicator of demand for products from the drilling industry.

A plot of the time series $x_1, \ldots, x_{417}$ is shown in Figure 2.10a. Obviously, we are concerned with low counts ($\overline{x} \approx 1.580$), and the long runs of values indicate a strong serial dependence. Indeed, looking at the SACF shown in Figure 2.10b,

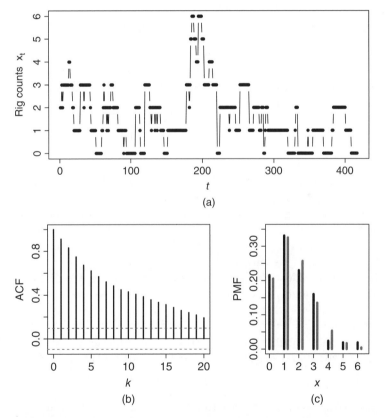

(a)

(b)

(c)

Figure 2.10 Plot of the rig counts in (a), their sample autocorrelation in (b), and marginal frequencies (black) together with a Poisson fit (gray) in (c); see Example 2.6.2.

---

3 phx.corporate-ir.net/phoenix.zhtml?c=79687&p=irol-rigcountsoverview.

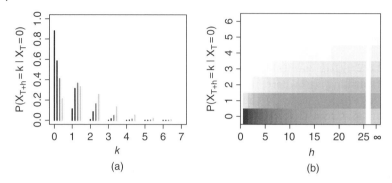

Figure 2.11 h-step-ahead forecasting distributions for horizons $h = 1, 5, 10, \infty$ (dark to light) in (a), and $h = 1, \ldots, 25, \infty$ in (b); see Example 2.6.2.

a high and slowly decreasing autocorrelation level becomes obvious ($\hat{\rho}(1) \approx 0.911$). An inspection of the SPACF reveals an approximate AR(1)-like structure such that an INAR(1) model appears to be reasonable for the data. Applying the dispersion test (2.14), it turns out that the observed (slight) degree of overdispersion ($\hat{I} \approx 1.110$) is not significant (P value ≈ 0.238). The histogram from Figure 2.10c, where the pmf of the Poi($\bar{x}$) distribution is shown in gray, confirms that a Poisson model might serve well for the data.

So we fit a Poisson INAR(1) model to the data, leading to the ML estimates $\hat{\lambda}_{\text{ML}} \approx 0.126$ (std. err. 0.018) and $\hat{\alpha}_{\text{ML}} \approx 0.918$ (std. err. 0.011). An analysis of the Pearson residuals and the PIT histogram confirms that the fitted Poisson INAR(1) model works reasonably well for the data. Using this fitted model, the h-step-ahead forecasting distributions, conditioned on the last observation $x_T = 0$, are easily computed using the results from Example 2.6.1. Due to the strong dependence structure, these distributions show little dispersion (see the term $1 - \alpha^h$ in the formula for $V[X_{T+h} \mid X_T]$), which is certainly attractive for forecasting, and they converge only slowly to the marginal Poisson distribution for increasing h. This is illustrated by Figure 2.11, where in (b), the distributions are represented by gray colors, with increasing darkness for increasing probability value, and with the gray colors in the last column referring to the marginal distribution ($h = \infty$). The median forecast for increasing h is equal to 0 for $h = 1, \ldots, 6$, and equal to 1 for $h \geq 7$. The 95%-quantile (as some kind of worst/best-case scenario) varies from 1 (lags 1–3) to 2 (lags 4–8) to 3 (lags 9–25) to 4 (lags $h \geq 26$).

Remark 2.6.3 (Approximate forecasting distribution) If closed-form expressions for $p_{k|l}^{(h)}$ are not available, one can make use of the Markov property. The MC approximation described in Remark 2.1.3.4 is easily modified for forecasting. If again M is sufficiently large, and if $\tilde{\mathbf{P}} := (p_{k|l})_{k,l=0,\ldots,M}$ with the transition probabilities (2.5), then the h-step-ahead transition probabilities

$(p_{k|l}{}^{(h)})_{k,l=0,...,M}$ are approximated by the matrix $\tilde{\mathbf{P}}^h$; see formula (B.3). Due to the ergodicity of the INAR(1) process (Remark 2.1.1.2), the columns of $\tilde{\mathbf{P}}^h$ converge to the approximate stationary marginal distribution $\tilde{p}$ from Remark 2.1.3.4, thus offering an alternative way of numerically computing $\tilde{p}$. Concerning the speed of convergence, see the Perron–Frobenius theorem, as described in Remark B.2.2.1.

Applied to the fitted NB-INAR(1) model from Section 2.5, where the last download count equals $x_T = 7$, the h-step-ahead forecasting distributions $p_{.|7}{}^{(h)}$ converge very quickly to the stationary marginal distribution as $h \to \infty$ (the quick convergence is not surprising in view of the weak autocorrelation level). This is illustrated by Figure 2.12, where the distributions for $h = 1$, $h = 2$ and $h = \infty$ (marginal distribution) are shown. The median forecast equals 2 for all forecasting horizons $h \geq 1$, while other quantiles may slightly change, say the lower quartile from 1 ($h = 1, 2$) to 0 ($h \geq 3$), the upper quartile from 4 ($h = 1, 2$) to 3 ($h \geq 3$), and the 95% quantile from 9 ($h = 1$) to 8 ($h \geq 2$). The latter could be used as the limit of an upper-sided 95% prediction interval. A two-sided interval is not possible since the zero probability is much larger than 2.5% for all h. The mode, another option for coherent point forecasting, equals 1 for $h = 1$, and 0 otherwise.

Up to now, we have assumed the INAR(1) model and its parameters, say θ, to be known. In practice, however, one has to estimate the parameters; that is, the forecasting distribution depends on the estimate $\hat{\theta}$. This causes uncertainty in the computed forecasting distribution. The case of a Poisson INAR(1) model, as in Example 2.6.2, is discussed by Freeland & McCabe (2004b). Here, the asymptotic distribution of, say, the ML estimator is known; see Section 2.2. It is an asymptotic normal distribution such that the asymptotic distribution

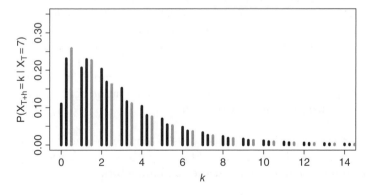

Figure 2.12 *h*-step-ahead forecasting distribution of fitted NB-INAR(1) model from Section 2.5, for forecasting horizon $h = 1$ in black, $h = 2$ in dark gray, and $h = \infty$ (marginal distribution) in light gray.

of $p_{k|l}^{(h)}(\hat{\boldsymbol{\theta}})$ can be determined by applying the Delta method. A closed-form expression for the asymptotic variance of $p_{k|l}^{(h)}(\hat{\boldsymbol{\theta}})$ was derived by Freeland & McCabe (2004b), and this can be used for computing a confidence interval for $p_{k|l}^{(h)}$. Jung & Tremayne (2006) extend this work to more general INAR models and investigate bootstrap-based methods for coherent forecasting under estimation uncertainty.

We conclude this chapter with a brief remark about how to simulate a stationary INAR(1) process.

Remark 2.6.4   (Simulation of INAR(1) process)   Since an INAR(1) process constitutes a Markov chain, the essential point for simulating a stationary INAR(1) process is its correct initialization. Because of the Markov property discussed in Appendix B.2.1, we have to ensure that the initial count stems from the stationary marginal distribution; if the remaining counts are then generated by using the one-step-ahead conditional distributions (2.5) – that is, by implementing the model recursion from Definition 2.1.1.1 – the whole process becomes stationary.

So how can we simulate the initial count according to the stationary marginal distribution? For the Poisson INAR(1) model, the stationary marginal distribution is explicitly known, being a simple Poisson distribution (Section 2.1.2). So we just use this Poisson distribution for the initial count, and the conditional distributions for the remaining counts.

If the stationary marginal distribution is not explicitly available, two approximate solutions are possible. First, one can compute the MC approximation $\tilde{\boldsymbol{p}} := (\tilde{p}_0, \ldots, \tilde{p}_M)^\top$ with sufficiently large M, as described in Remark 2.1.3.4; also see Remark 2.6.3 for an alternative approach. Then the initial count is generated according to the distribution $\tilde{\boldsymbol{p}}$ with finite support $\{0, \ldots, M\}$. Secondly, one may utilize the ergodicity of the INAR(1) process (Remark 2.1.1.2). The idea is to generate a prerun $X_{-r}, \ldots, X_0$, say by initializing $X_{-r} := \mathrm{round}(\mu)$ and by then generating $X_{-r+1}, \ldots, X_0$ from the one-step-ahead conditional distributions. Their corresponding marginal distributions then converge towards the stationary marginal distribution; see the discussion in Appendix B.2.2. If the length r of the prerun is sufficiently large – the values for r reported in the literature typically vary between 200 and 500 – then the distribution of X_0 is close to the required stationary marginal distribution.

The approach described in Remark 2.6.4 is easily adapted to other types of count processes, for example higher-order Markov processes, by considering the multivariate representation described after Definition B.1.7.

3

Further Thinning-based Models for Count Time Series

After having introduced important tasks and approaches for analyzing count time series, we shall now return to the question of how to model the underlying process. The main characteristic of the INAR(1) model in Section 2.1 is the use of the binomial thinning operator as a substitute for the multiplication, to be able to transfer the AR(1) recursion to the count data case. In Section 3.1, we shall see that this approach can also be used to define higher-order ARMA-like models. Furthermore, different types of thinning operation have been developed for such models; see Section 3.2. Finally, various thinning-based models to deal with count time series with a finite range (Section 3.3) and multivariate count time series (Section 3.4) are also available in the literature.

3.1 Higher-order INARMA Models

The INAR(1) model, as introduced in Section 2.1, was developed as an integer-valued counterpart to the conventional AR(1) model, mainly by replacing the multiplication in the AR(1) recursion with the binomial thinning operator. This idea is not limited to the first-order autoregressive case, but can also be used to mimic higher-order ARMA models; see Appendix B. The resulting models are then referred to as *INARMA models*.

Example 3.1.1 (INMA(1) model) An integer-valued counterpart to the MA(1) model $Y_t = \epsilon_t + \beta \cdot \epsilon_{t-1}$ was proposed by Al-Osh & Alzaid (1988), McKenzie (1988). Like the INAR(1) model discussed in Section 2.1, their *INMA(1) model* replaces the multiplication in the MA recursion by the binomial thinning operation (executed independently of any available random variables at time t), leading to the model recursion

$$X_t = \epsilon_t + \beta \circ \epsilon_{t-1} \qquad \text{with } \beta \in [0; 1], \tag{3.1}$$

where, as in Definition 2.1.1.1, the innovations $(\epsilon_t)_{\mathbb{Z}}$ are an i.i.d. process with range $\mathbb{N}_0$; again we denote $E[\epsilon_t] = \mu_\epsilon$ and $V[\epsilon_t] = \sigma_\epsilon^2$. We may interpret (3.1) by

An Introduction to Discrete-Valued Time Series, First Edition. Christian H. Weiss.
© 2018 John Wiley & Sons Ltd. Published 2018 by John Wiley & Sons Ltd.
Companion website: www.wiley.com/go/weiss/discrete-valuedtimeseries

analogy to (2.4): the population at time t consists of the immigrants at time t plus the survivors of the immigrants from the previous point in time $t - 1$; there are no survivors from former generations.

The stationary mean and variance of X_t follow immediately from the mean and variance of $\beta \circ \epsilon$ (Section 2.1.1) as

$$\mu = \mu_\epsilon (1 + \beta), \qquad \sigma^2 = (1 + \beta^2) \sigma_\epsilon^2 + \beta(1 - \beta) \mu_\epsilon.$$

The autocovariance function equals $\beta \sigma_\epsilon^2$ for lag 1, and 0 otherwise. These and further properties were derived by Al-Osh & Alzaid (1988).

The results simplify in the particular case of Poisson innovations $\epsilon_t \sim \text{Poi}(\lambda)$; see Al-Osh & Alzaid (1988), McKenzie (1988). Because of the Poisson's additivity and invariance with respect to binomial thinning (Section 2.1.2), it follows that the observations of such a Poisson INMA(1) process are also Poisson distributed, and because of the equidispersion, the autocorrelation function becomes $\rho(1) = \beta/(1 + \beta) \in [0; 0.5]$ and 0 otherwise. The Poisson INMA(1) process is also time-reversible with a linear 1-step-ahead conditional mean,

$$E[X_t \mid X_{t-1}] = \frac{\beta}{1 + \beta} X_{t-1} + \lambda,$$

see Al-Osh & Alzaid (1988); these properties do not hold in the non-Poisson case.

While both the INAR(1) and the INMA(1) recursion involve only one thinning operation, the higher-order models need more than one thinning operation at a time. As an example, the counterpart to the full MA(q) model, the *INMA(q) model*, is defined by a recursion of the form

$$X_t = \beta_0 \circ_t \epsilon_t + \beta_1 \circ_t \epsilon_{t-1} + \ldots + \beta_q \circ_t \epsilon_{t-q}, \qquad q \geq 1, \tag{3.2}$$

where the $q + 1$ thinnings at time t are performed independently of each other (in Example 3.1.1, we set $\beta_0 := 1$). Here, a time index t has been added below the operator " $\circ$ " to emphasize the fact that these are the thinnings being executed at time t. Now focussing on one particular innovation, say ϵ_t, it becomes clear that this innovation is altogether involved in $q + 1$ thinnings, namely $\beta_0 \circ_t \epsilon_t, \beta_1 \circ_{t+1} \epsilon_t, \ldots, \beta_q \circ_{t+q} \epsilon_t$. Since the thinning operations are *probabilistic* (in contrast to the multiplication used for an MA(q) model), the joint distribution of these thinnings has to be considered – that is, the conditional distribution of $(\beta_0 \circ_t \epsilon_t, \beta_1 \circ_{t+1} \epsilon_t, \ldots, \beta_q \circ_{t+q} \epsilon_t)$ given ϵ_t – thereby leading to different types of models of the same model order q.

Until now, a total of *four* different INMA(q) models have been proposed in the literature, each having slightly different interpretations and probabilistic properties; see Al-Osh & Alzaid (1988), McKenzie (1988), Brännäs & Hall (2001) and Weiß (2008b). To be more precise, the marginal properties are already fixed by definition (3.2), namely:

$$\text{pgf}(z) = \prod_{j=0}^{q} \text{pgf}_\epsilon(1 - \beta_j + \beta_j z),$$

$$\mu = \mu_\epsilon \, \beta_\bullet, \qquad \sigma^2 = \mu_\epsilon \, \beta_\bullet + (\sigma_\epsilon^2 - \mu_\epsilon) \sum_{j=0}^{q} \beta_j^2, \tag{3.3}$$

where $\beta_\bullet := \sum_{j=0}^{q} \beta_j$. The joint distributions, however, differ between the different types of INMA(q) processes, as we shall see below. But first, let us look at an example: as for the INAR(1) model (Section 2.1.2), the Poisson distribution plays an important role.

Example 3.1.2 **(Poisson INMA(q) model)** The marginal properties in (3.3) show that an equidispersed distribution for the innovations will lead to equidispersed observations. Furthermore, if the innovations are Poisson distributed, $\epsilon_t \sim \text{Poi}(\lambda)$, then the observations satisfy $X_t \sim \text{Poi}(\mu)$ with mean $\mu = \lambda \, \beta_\bullet$.

To be able to define different types of INMA(q) models, we follow the approach in Weiß (2008b) and look at the individual counting series. Let $(Z_{t;i}^{(j)})_{i=1,\dots,\epsilon_t}$ be the counting series of the thinning applied to ϵ_t at time $t+j$ for a $j = 0, \dots, q$, with $P(Z_{t;i}^{(j)} = 1) = \beta_j$. These $Z_{t;i}^{(j)}$ might be interpreted as indicators for the ϵ_t individuals being introduced to the considered system at time t ("generation t"). If $Z_{t;i}^{(j)} = 1$, then the ith individual of generation t is active at time $t+j$, where each individual has "lifetime" q.

In view of the general definition (3.2), the (q + 1)-dimensional vectors $Z_{t;i} :=$ $(Z_{t;i}^{(0)}, \dots, Z_{t;i}^{(q)})^\top$ (each corresponding to one specific individual) have to be i.i.d., but their components might be dependent, thus restricting the activation of an individual during its lifetime. The *independence model* by McKenzie (1988), for instance, assumes the components of $Z_{t;i}$ to be mutually independent (so there are no further restrictions concerning the activation of an individual), while the *sale model* by Brännäs & Hall (2001) requires $\beta_\bullet \le 1$ and defines the vector $(Z_{t;i}^{(0)}, \dots, Z_{t;i}^{(q)})^\top$ to be multinomially distributed according to $\text{MULT}^*(1; \beta_0, \dots, \beta_q)$; see Example A.3.3. So the sale model assumes that each individual becomes active at most once during its lifetime; an example would be if the "individuals" are perishable goods being produced at day t, and they become "active" when they are sold during their shelf-life. Weiß (2008b) analyzed the serial dependence structure of all these INMA(q) models; among other things, he showed that

$$\gamma(k) = (\sigma_\epsilon^2 - \mu_\epsilon) \sum_{j=k}^{q} \beta_j \beta_{j-k} + \mu_\epsilon \sum_{j=k}^{q} P(Z_{t-j,1}^{(j)} = Z_{t-j,1}^{(j-k)} = 1). \tag{3.4}$$

So the autocovariance function for different INMA(q) models differs according to the last term. For example, for the independence model by McKenzie (1988), we have $P(Z_{t-j,1}^{(j)} = Z_{t-j,1}^{(j-k)} = 1) = \beta_j \beta_{j-k}$, while $P(Z_{t-j,1}^{(j)} = Z_{t-j,1}^{(j-k)} = 1) = 0$ for the sale model because it is impossible that an individual becomes active twice.

Example 3.1.3 **(Poisson INMA(q) model)** Let us continue Example 3.1.2. Because of the equidispersion of the Poisson-distributed innovations, we have $\sigma_\epsilon^2 - \mu_\epsilon = 0$, so the formula (3.4) for the autocovariance function simplifies further. Furthermore, by analogy to (2.13) for the Poisson INAR(1) model, Weiß (2008b) showed that the pairs (X_t, X_{t-k}) with $k \in \mathbb{N}$ are bivariately Poisson distributed with pgf

$$\text{pgf}_{X_t, X_{t-k}}(z_0, z_1) = \exp\left(\mu\left(z_0 + z_1 - 2 + \rho(k)\left(z_0 - 1\right)\left(z_1 - 1\right)\right)\right).$$

So having specified the ACF, the whole bivariate distribution is fixed.

Two types of *INAR*(p) *models* have been proposed by Alzaid & Al-Osh (1990) and Du & Li (1991), both being based on the recursion

$$X_t = \alpha_1 \circ_t X_{t-1} + \ldots + \alpha_p \circ_t X_{t-p} + \epsilon_t, \tag{3.5}$$

where $\alpha_\bullet := \sum_{j=1}^{p} \alpha_j < 1$ is assumed. Obviously, the conditional distribution of $(\alpha_1 \circ_{t+1} X_t, \ldots, \alpha_p \circ_{t+p} X_t)$ given X_t now has to be specified. Du & Li (1991) assume conditional independence (by analogy to the *INMA*(q) *independence model*), while Alzaid & Al-Osh (1990) assume a conditional multinomial distribution (by analogy to the *INMA*(q) *sale model*):

$$(\alpha_1 \circ X_t, \ldots, \alpha_p \circ X_t)^\top \sim \text{MULT}^*(X_t; \alpha_1, \ldots, \alpha_p),$$

see Example A.3.3. Further specialized INAR(p) models can be defined by refining Equation 3.5 by analogy to the INMA(q) case; that is, by considering the counting series $(Z_{t,i}^{(j)})$ of the thinning applied to X_t ("generation t") at time $t + j$ and by specifying the distribution of $(Z_{t,i}^{(1)}, \ldots, Z_{t,i}^{(p)})$.

Remark 3.1.4 **(Interpretation of INAR(p) model)** For model order $p \geq 2$, a reasonable interpretation of the INAR(p) recursion in (3.5) is generally obtained by analogy to that of BPIs, see Remark 2.1.1.2; that is, by interpreting X_t as the total offspring at time t (caused by either previous generations or by immigration). This interpretation especially applies to the INAR(p) model by Du & Li (1991) (*DL-INAR*). For the INAR(p) model by Alzaid & Al-Osh (1990) (*AA-INAR*), the conditional multinomial distribution appears to be rather restrictive in view of reproduction. Here, one can think of some kind of renewal of the individuals from generation t: if such an individual is renewed at time $t + j$, then it becomes a member of generation $t + j$, but if it is not renewed within p time periods, then it is shut down.

Let us now look at properties of INAR(p) models. The stationary marginal mean is always given by $\mu = \mu_\epsilon/(1 - \alpha_\bullet)$. For the DL-INAR(p) model, the variance satisfies

$$\sigma^2 \cdot \left(1 - \sum_{i=1}^{p} \alpha_i \, \rho(i)\right) = \mu \sum_{j=1}^{p} \alpha_j(1 - \alpha_j) + \sigma_\epsilon^2; \qquad (3.6)$$

see Silva & Oliveira (2005) for higher-order joint moments. The ACF is obtained from the conventional AR(p) Yule–Walker equations (see (B.13)); that is,

$$\rho(k) = \sum_{i=1}^{p} \alpha_i \, \rho(|k - i|) \qquad \text{for } k \geq 1, \qquad (3.7)$$

as shown by Du & Li (1991). For the AA-INAR(p) model, in contrast, Alzaid & Al-Osh (1990) derived an ARMA(p, p − 1)-like autocorrelation structure, which seems to be the main reason why the DL-INAR(p) model is usually preferred in practice. But there are also reasons why the AA-INAR(p) model is attractive: if having Poisson innovations, its observations are also Poisson distributed and the whole process is time-reversible (Alzaid & Al-Osh, 1990; Schweer, 2015), which is analogous to the Gaussian AR(p) case. The DL-INAR(p) process, in contrast, is time-reversible only in trivial cases (Schweer, 2015), and equidispersed innovations generally do not imply equidispersed observations (Weiß, 2013a). Note that for $\sigma_\epsilon^2 = \mu_\epsilon$, (3.6) simplifies to

$$\sigma^2 \cdot \left(1 - \sum_{i=1}^{p} \alpha_i \, \rho(i)\right) = \mu \left(1 - \sum_{j=1}^{p} \alpha_j^2\right).$$

The conditional mean and variance of the DL-INAR(p) process are given by

$$E[X_t \mid X_{t-1}, \ldots] = \mu_\epsilon + \sum_{j=1}^{p} \alpha_j \, X_{t-j},$$

$$V[X_t \mid X_{t-1}, \ldots] = \sigma_\epsilon^2 + \sum_{j=1}^{p} \alpha_j(1 - \alpha_j) \, X_{t-j}. \qquad (3.8)$$

The transition probabilities of the DL-INAR(p) process, which is Markovian of order p, can be computed by utilizing the fact that the conditional distribution is a convolution between p binomial distributions and the innovations' distribution (Drost et al., 2009); see (3.12) below for illustration. Mixing and weak dependence properties for the DL-INAR(p) process are discussed by Doukhan et al. (2012, 2013) and a frequency-domain analysis is considered by Silva & Oliveira (2005).

Remark 3.1.5 (CINAR(p) model) An alternative AR(p)-like approach, the so-called *combined INAR(p)* (CINAR(p)) model, is due to Zhu & Joe (2006) and Weiß (2008c). A probabilistic mixing of lagged INAR(1) recursions is used:

$$X_t = D_{t,1} \cdot (\alpha \circ_t X_{t-1}) + \ldots + D_{t,p} \cdot (\alpha \circ_t X_{t-p}) + \epsilon_t, \tag{3.9}$$

where the $(D_{t,1}, \ldots, D_{t,p})$ are independent and multinomially distributed according to $\mathrm{MULT}(1; \phi_1, \ldots, \phi_p)$ with $\phi_1 + \ldots + \phi_p = 1$. By construction, the marginal distribution of such a process is that of the underlying INAR(1) process; that is, one with the same α and the same innovations' distribution. However, as for the INAR(p) model, different types of CINAR models are obtained for the same model order p if varying the joint distribution of the involved thinnings. In particular, if thinnings are independent, then the process is Markovian of order p, having the typical AR(p)-like autocorrelation structure with autoregressive parameters $\alpha \cdot \phi_i$. The conditional mean and transition probabilities are given by

$$E[X_t \mid X_{t-1}, \ldots] = \mu_\epsilon + \alpha \sum_{i=1}^{p} \phi_i X_{t-i},$$

$$P(X_t = x \mid X_{t-1} = x_{t-1}, \ldots) \tag{3.10}$$

$$= \sum_{y=0}^{x} P(\epsilon_t = y) \sum_{i=1}^{p} \phi_i \binom{x_{t-i}}{x-y} \alpha^{x-y} (1-\alpha)^{x_{t-i}-x+y}.$$

Higher-order (factorial) moments are easily computed by utilizing the binomial theorem (also see Example A.2.1), for example:

$$E\left[(X_t)_{(r)} \mid X_{t-1}, \ldots \right] = \sum_{i=1}^{p} \phi_i \sum_{j=0}^{r} \binom{r}{j} (X_{t-i})_{(j)} \, \alpha^j \, \mu_{\epsilon;\,(r-j)}. \tag{3.11}$$

While the marginal and autocorrelation properties of the *CINAR*(p) *independence model* are quite attractive, the data generating mechanism itself is somewhat artificial and difficult to interpret.

In a nutshell, while the INAR(1) model, with its intuitive interpretation and its simple stochastic properties, is very attractive for applications, higher-order extensions are not straightforward. Therefore, in Chapter 4, we discuss an alternative concept for ARMA-like count process models.

Example 3.1.6 (Gold particles) For illustration, we analyze a time series of counts of gold particles, which was originally published by Westgren (1916). These counts were measured in a fixed volume element of a colloidal solution over time, and the count values vary because of the Brownian motion of the particles. As in Jung & Tremayne (2006) and Weiß (2013a), we consider observations $501, \ldots, 880$ of series C (Westgren, 1916, pp. 12–13) and denote them

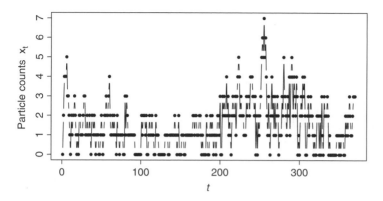

Figure 3.1 Plot of the particles counts; see Example 3.1.6.

as $x_1, \ldots, x_{380}$ for simplicity. The first $T = 370$ observations $x_1, \ldots, x_{370}$, which are plotted in Figure 3.1, are used for model fitting, while $x_{371}, \ldots, x_{380}$ serve as an exercise for forecasting.

The SACF shown in Figure 3.2a indicates an autoregressive autocorrelation structure. This is also confirmed by the SPACF in Figure 3.2b, where we have significant values at lags 1 and 2 but an abrupt drop towards zero with lag 3, so an autoregressive model of order p $\leq$ 2 appears plausible (see the discussion in Appendix B.3). Since we also have a significant SPACF value at lag 9, we shall later also try out some higher-order models. The empirical marginal pmf is shown in black in Figure 3.2c, together with the pmf of the Poi($\bar{x}$) distribution in gray, where $\bar{x} \approx 1.551$. The plot implies that a Poisson marginal distribution might be appropriate for the data, a suggestion which is confirmed by the index of dispersion $\hat{I} \approx 1.062$ (Poisson INAR(1) approximation according to (2.14): mean ≈ 0.990, std.err. ≈ 0.104).

As a result of the previous analyses, we shall fit the Poisson INAR(p) and CINAR(p) models of order p $\leq$ 2 to the data. To make the models more comparable, we shall use a slightly modified parametrization for the CINAR(p) model, with $\alpha_j := \alpha \, \phi_j$ for $j = 1, \ldots, p$ and, hence, $\alpha = \alpha_1 + \ldots + \alpha_p$ as well as $\phi_j = \alpha_j / \alpha$. After having computed moment estimates for initialization, the final estimates are obtained by maximizing the respective conditional log-likelihood function $\ell(\boldsymbol{\theta} \mid x_p, \ldots, x_1) = \sum_{t=p+1}^{T} \ln p(x_t \mid x_{t-1}, \ldots, x_{t-p})$ from (B.6). The required transition probabilities are given by (2.5) for INAR(1), by (3.10) for CINAR(2), and for the INAR(2) model by

$$P(X_t = k \mid X_{t-1} = l_1, X_{t-2} = l_2) = \sum_{j_1=0}^{\min\{k,l_1\}} \sum_{j_2=0}^{\min\{k-l_1,l_2\}}$$

$$\binom{l_1}{j_1} \alpha_1^{j_1} (1 - \alpha_1)^{l_1 - j_1} \cdot \binom{l_2}{j_2} \alpha_2^{j_2} (1 - \alpha_2)^{l_2 - j_2} \cdot P(\epsilon_t = k - j_1 - j_2).$$

$$\tag{3.12}$$

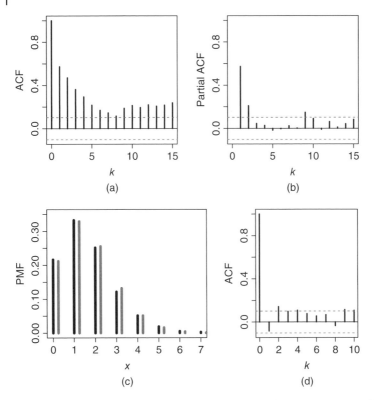

Figure 3.2 Particles counts; see Example 3.1.6: (a) SACF, (b) SPACF, (c) marginal frequencies (black) together with a Poisson fit (gray) (d) SACF of Pearson residuals based on fitted Poisson INAR(1) model.

For all models, we compute the CML estimates together with their approximate standard errors as well as the AIC and BIC, as described in Remark B.2.1.2. The results are summarized in Table 3.1. The first-order model not only performs worst in terms of AIC and BIC, but the corresponding Pearson residuals indicate that the model is not adequate for the data (Section 2.4): there are several significant SACF values in Figure 3.2d, which implies that a higher-order model should be used.

However, it is difficult to decide between the fitted Poisson INAR(2) and CINAR(2) model. While the latter has a Poisson marginal distribution (with mean ≈ 1.527 and dispersion index $= 1$), the former is slightly overdispersed (marginal mean ≈ 1.564, dispersion index ≈ 1.151); for the data, we observed $\bar{x} \approx 1.551$ and $\hat{I} \approx 1.062$. The Pearson residuals do not show significant SACF values for any of these models, while all the PIT histograms (see Figure 3.3) visibly deviate from uniformity (with a slight preference for the CINAR(2) model). Note that the conditional variance of the Poisson CINAR(2) model, as required

Table 3.1 Particles counts: CML estimates and AIC and BIC values for different models.

Model	Parameter			AIC	BIC
	1	2	3		
Poisson INAR(1)	0.734	0.531		1040	1047
(λ, α)	(0.063)	(0.036)			
Poisson INAR(2)	0.545	0.472	0.180	1027	1038
$(\lambda, \alpha_1, \alpha_2)$	(0.072)	(0.047)	(0.053)		
Poisson CINAR(2)	0.601	0.412	0.194	1027	1038
$(\lambda, \alpha_1, \alpha_2)$	(0.064)	(0.051)	(0.054)		

Figures in parentheses are standard errors. AIC and BIC values rounded.

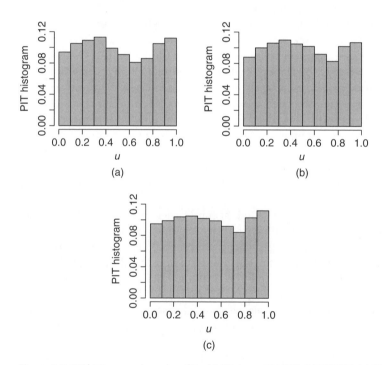

Figure 3.3 PIT histograms based on fitted (a) Poisson INAR(1), (b) INAR(2), (c) CINAR(2) models; see Example 3.1.6.

for the Pearson residuals, follows from (3.11) with $r = 2$ by using $\mu_{\epsilon;\ (r-j)} = \lambda^{r-j}$ (Example A.1.1). The ACF of the fitted INAR(2) model $(0.575, 0.451, 0.316, \ldots)$ is closer to the SACF $(0.574, 0.471, 0.364, \ldots)$ than the ACF of the CINAR(2) model $(0.511, 0.405, 0.266, \ldots)$; note that the estimate for α_1 is visibly smaller for the CINAR(2) model. So, in summary, none of the considered second-order models perfectly fits the data, but both seem to constitute reasonable approximations. If, for a practitioner, it is most important to imitate the observed autocorrelation structure, then the INAR(2) model should be preferred, while the CINAR(2) model, with its Poisson marginal distribution, might be more attractive in view of its equidispersion property.

At this point, let us also try out some higher-order models; we have seen in Figure 3.2b that there was also a significant SPACF value at lag 9. For simplicity, we concentrate on CINAR(p) models for this exercise. Starting with the CINAR(3) model, we estimate λ as 0.562 (standard error 0.069) and the autoregressive parameters $\alpha_1, \alpha_2, \alpha_3$ as 0.392 (0.054), 0.193 (0.055), 0.037 (0.044), respectively. It becomes clear that the estimate for α_3 is not significantly different from 0 at a 5% level. Also the BIC $\approx$ 1039 is larger than that of the second-order model, while the AIC further decreases to about 1024. Considering the fact that the AIC sometimes tends to overestimate the model order (see Katz (1981) and Remark B.2.1.2), the increased model order does not appear to be beneficial for the given data. This is even clearer for a fitted CINAR(4) model, where the estimate for α_4 is (numerically) identical to zero, and where now also the AIC is increased. All these analyses confirm that the second-order model seems to be superior for the data within the full CINAR(p) models. Only a special CINAR(9) model, where the autoregressive parameters $\alpha_3, \ldots, \alpha_8$ are set to zero (motivated by the significant SPACF value at lag 9), could be a reasonable alternative. This model has four parameters, $\lambda, \alpha_1, \alpha_2, \alpha_9$, and leads to a performance comparable to the CINAR(2) model, although certainly at the price of increased model complexity.

Let us now return to the fitted second-order models, and conclude this example with a forecasting exercise for the remaining observations $x_{371}, \ldots, x_{380}$. Based on the 1-step-ahead conditional distributions computed according to (3.12) and (3.10) for the fitted Poisson INAR(2) and CINAR(2) model, respectively, the 5%- and the 95%-quantiles were computed; see Table 3.2. It can be seen that the CINAR(2) model produces slightly narrower quantile bands, which is plausible in view of also it having less marginal dispersion (see above). Neither observation violates any of the quantile bands.

Remark 3.1.7 (**Further extensions**)   The use of thinning operations is not limited to development of stationary and ARMA-like count models. As an example, by defining the thinning parameter or further model parameters to depend on time, say through a *time-dependent mean* (McKenzie, 1985; Moriña et al., 2011), through *periodically varying parameters* (Monteiro et al., 2010) or through incorporating *covariate information* (Azzalini, 1994; Freeland & McCabe, 2004a; Sutradhar, 2008), one obtains a non-stationary extension of

Table 3.2 Particles counts: 5% and 95% quantiles of 1-step-ahead forecasting distributions for $x_{371}, \ldots, x_{380}$ based on fitted Poisson INAR(2) and CINAR(2) models.

	t	371	372	373	374	375	376	377	378	379	380
95%	INAR(2)	5	5	4	4	4	3	4	4	3	4
	CINAR(2)	4	4	4	4	4	3	3	4	3	3
	x_t	3	2	3	2	1	2	2	1	2	1
5%	INAR(2)	1	1	0	0	0	0	0	0	0	0
	CINAR(2)	1	1	0	1	0	0	0	0	0	0

the INAR(1) model. A related approach is to use *state-dependent parameters*; see Weiß (2015a) and the references therein. In particular, the idea of a *self-exciting threshold* (SET) autoregression (Turkman et al., 2014) can be adapted to the integer-valued case, too; see Monteiro et al. (2012) for such a model.

As another example, thinning operations can also be used to transfer bilinear models (Turkman et al., 2014) to the integer-valued case, leading to the *INBL models*; see, for example Doukhan et al. (2006), who discuss the INBL$(1, 0, 1, 1)$ model in great detail. It is defined by the recursion [1]

$$X_t = \alpha \circ X_{t-1} + \beta \circ (X_{t-1}\, \epsilon_{t-1}) + \epsilon_t, \tag{3.13}$$

where thinnings are performed independently of each other. A (strictly) stationary solution exists if $\alpha + \beta\, \mu_\epsilon < 1$, having the mean $\mu = (\mu_\epsilon + \beta\, \sigma_\epsilon^2)/(1 - \alpha - \beta\, \mu_\epsilon)$. The variance exists (and hence weak stationarity holds) if $(\alpha + \beta\, \mu_\epsilon)^2 + \beta^2\, \sigma_\epsilon^2 < 1$ and $\mu_{\epsilon,4} < \infty$ holds. For further results, including the ACF in the case of Poisson-distributed innovations, see Doukhan et al. (2006).

To illustrate the potential of some of the extensions mentioned in Remark 3.1.7, let us look at simulated sample paths. The first two parts of Figure 3.4 show paths of non-stationary Poisson INAR(1) processes, where the thinning parameter α is kept constant in time, but the innovations' mean varies according to $\lambda_t = \exp(\gamma^\top z_t)$, with z_t representing the covariate information at time t (Freeland & McCabe, 2004a). Such trend and seasonality are often observed in practice; corresponding real-data examples are presented in Examples 5.1.6 and 5.1.7. Figure 3.4c shows a sample path with a piecewise pattern, which was generated by a simple SET extension of the Poisson INAR(1) model with one threshold and delay 1; see Monteiro et al. (2012) for further background.

---

1 In fact, Doukhan et al. (2006) define this model by using the concept of generalized thinning as described in Section 3.2.

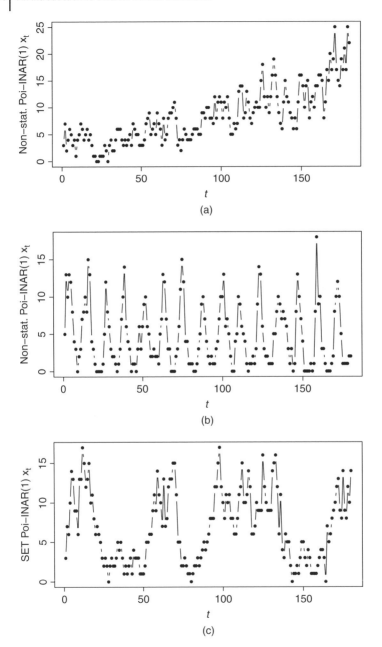

Figure 3.4 Simulated Poisson INAR(1) process ($\alpha = 0.5$) with innovations $\epsilon_t \sim \text{Poi}(\lambda_t)$:
(a) having trend $\lambda_t = \exp(\gamma_0 + \gamma_1\, t)$ with $\gamma_0 = \ln 1.5$, $\gamma_1 = 0.01$; (b) having seasonality
$\lambda_t = \exp(\gamma_0 + \gamma_2\, \sin(2\pi\, t/12))$ with $\gamma_0 = \ln 1.5$, $\gamma_2 = 1.5$; (c) having SET mechanism $\lambda_t = 1.5$
if $X_{t-1} \leq 5$ and $\lambda_t = 5$ otherwise.

3.2 Alternative Thinning Concepts

The idea behind a thinning operation (and the related time series models) can be modified in diverse ways; see the surveys by Weiß (2008a) and Scotto et al. (2015). Such modified thinning concepts allow for different stochastic properties and alternative interpretation schemes. We shall pick out two of these alternative thinning concepts here, but many further approaches are described in Weiß (2008a) and Scotto et al. (2015).

In view of the BPIs according to Remark 2.1.1.2, the *generalized thinning* operation, as proposed by Latour (1998), appears reasonable:

$$\alpha \bullet_\beta X := \sum_{j=1}^{X} Z_j \qquad \text{with } \alpha \in (0; 1) \text{ and } \beta > 0, \tag{3.14}$$

where the random variables Z_j (counting series) are allowed to have the full range $\mathbb{N}_0$ instead of only $\{0, 1\}$. Here, the Z_j are required to have mean α and variance β. Since now the Z_j may become larger than 1, the interpretation (2.4) as "survival indicators" is no longer appropriate, but they can be understood as describing a reproduction mechanism: as in Remark 2.1.1.2, Z_j might be the number of children being generated by the jth individual of the population behind X.

Using their generalized thinning operation (which includes binomial thinning as a special case), Latour (1998) extended the INAR(p) model (3.5) by Du & Li (1991) to the *GINAR(p) model*, which is defined as

$$X_t = \alpha_1 \bullet_{\beta_1} X_{t-1} + \ldots + \alpha_p \bullet_{\beta_p} X_{t-p} + \epsilon_t, \tag{3.15}$$

where again $\alpha_\bullet := \sum_{j=1}^{p} \alpha_j < 1$ is assumed. In (3.15), a time index for the thinning operations is omitted for the sake of readability, but it is understood that all thinnings are performed independently, as in the model by Du & Li (1991).

Although (3.15) has been generalized with respect to (3.5), $\mu = \mu_\epsilon/(1 - \alpha_\bullet)$ as well as the Yule–Walker equations (3.7) still hold, while formula (3.6) for the variance has to be modified (Latour, 1998):

$$\sigma^2 \cdot \left(1 - \sum_{i=1}^{p} \alpha_i \, \rho(i)\right) = \mu \sum_{j=1}^{p} \beta_j + \sigma_\epsilon^2. \tag{3.16}$$

Setting $\beta_j := \alpha_j(1 - \alpha_j)$, we would obtain (3.6) again.

A particular instance of generalized thinning, which has received considerable interest in the literature, is the *negative binomial thinning operator* "$\alpha *$" of Ristić et al. (2009), for which the Z_j are geometrically distributed with parameter $\pi := 1/(1 + \alpha)$ (hence $E[Z_j] = \alpha$; see Example A.1.5). Therefore, $\alpha * X$ is conditionally $NB(X, 1/(1 + \alpha))$-distributed due to the additivity of the NB distribution (Example A.1.4). Ristić et al. (2009) use this operation to construct a first-order process having a geometric marginal distribution, and refer to it as the *new geometric INAR(1) process*, abbreviated as NGINAR(1). Also the Poisson INARCH models, discussed later in Section 4.1, might be understood as particular GINAR models using *Poisson thinning* (Rydberg & Shephard, 2000; Weiß, 2015a).

Another family of thinning operations assumes the counting series to be Bernoulli distributed, but now the thinning probability α is allowed to be random itself. The resulting thinning operation is then called *random coefficient* (RC) thinning, and it has been applied in the context of count time series modeling by a number of authors, including Joe (1996) and Zheng et al. (2007). As a specific instance (Joe, 1996), we consider the case where α_ϕ follows the BETA $\left(\frac{1-\phi}{\phi} \alpha, \frac{1-\phi}{\phi} (1 - \alpha) \right)$ distribution with $\alpha, \phi \in (0; 1)$; that is, where $E[\alpha_\phi] = \alpha$ and $\sigma_\alpha^2 := V[\alpha_\phi] = \phi \, \alpha(1 - \alpha)$. Then the conditional distribution of $\alpha_\phi \circ X$ given X is a beta-binomial distribution (see Example A.2.2), and the thinning operation is referred to as *beta-binomial thinning* accordingly.

A counterpart to the INAR(1) model from Definition 2.1.1.1 using a general random coefficient thinning operation "$\alpha_t \circ$" (that is, where the distribution of α_t on $[0; 1)$ is only required to have mean α and a certain variance σ_α^2, but is not further specified) was investigated by Zheng et al. (2007). Their *RCINAR(1) model* is defined by

$$X_t = \alpha_t \circ_t X_{t-1} + \epsilon_t. \tag{3.17}$$

While the (conditional) mean and ACF remain as in the INAR(1) case (Section 2.1.1), the effect of the additional uncertainty manifests itself in the (conditional) variance (Zheng et al., 2007):

$$V[X_t \mid X_{t-1}] = \sigma_\alpha^2 \, (X_{t-1}^2 - X_{t-1}) + \alpha(1 - \alpha) \, X_{t-1} + \sigma_\epsilon^2,$$

$$\sigma^2 = V[X_t] = \mu \, \frac{1 - \alpha^2 + (\mu - 1) \, \sigma_\alpha^2 + (1 - \alpha) \left(\frac{\sigma_\epsilon^2}{\mu_\epsilon} - 1 \right)}{1 - \alpha^2 - \sigma_\alpha^2}. \tag{3.18}$$

Note that $V[X_t \mid X_{t-1}]$ is a quadratic function of X_{t-1}, and the observations are overdispersed even if the innovations are equidispersed.

Joe (1996) and Sutradhar (2008) considered that instance of the RCINAR(1) model where beta-binomial thinning "$\alpha_\phi \circ$" is used. If $\epsilon_t \sim \mathrm{NB}(n(1 - \alpha), \pi)$ and if $\phi := 1/(n + 1)$, then the stationary marginal distribution of

$$X_t = \alpha_{1/(n+1)} \circ X_{t-1} + \epsilon_t, \tag{3.19}$$

omitting the time index at the thinning operation, is $X_t \sim \mathrm{NB}(n, \pi)$. So innovations and observations are from the same family of distributions, $\mathrm{NB}(\cdot, \pi)$, and have a unique index of dispersion, $I = 1/\pi$, both being completely analogous to the case of a Poisson INAR(1) model (Section 2.1.2).

Example 3.2.1 (**Download counts**) Let us consider again the time series of download counts as discussed in Section 2.5. When analyzing the data, Weiß

(2008a) recommended to use the NB-RCINAR(1) model (3.19) for these data. Therefore, we now also fit this model to the data by a full likelihood approach, where the transition probabilities are computed by analogy to (2.5) by replacing the pmf of the Bin(l, α) distribution by that of the BB($l; \alpha, \frac{1}{n+1}$) distribution (see Example A.2.2). We obtain the estimates (std.err. in parentheses)

$$\hat{n}_{\text{ML}} \approx 1.134 \ (0.174), \quad \hat{\pi}_{\text{ML}} \approx 0.315 \ (0.038), \quad \hat{\alpha}_{\text{ML}} \approx 0.274 \ (0.058).$$

In particular, the NB-RCINAR(1) model indeed leads to the lowest values of AIC (≈ 1085) and BIC (≈ 1096). The marginal distribution of the fitted model is an NB distribution with mean ≈ 2.463, dispersion index ≈ 3.172, and zero probability ≈ 0.270. The Pearson residuals do not contradict the fitted model, and also the PIT histogram is reasonable close to uniformity (although it looks a bit worse than the one of the NB-INAR(1) model in Figure 2.9b).

Zheng et al. (2006) extended the RCINAR(1) model to the pth-order RCINAR(p) model, analogous to the INAR(p) model of Du & Li (1991); that is, with thinnings being performed independently (Section 3.1).

Remark 3.2.2 **(Convolution-closed models)** Another way of generalizing the binomial-thinning-based Poisson INAR(1) model to different marginal distributions and to higher-order dependence structures was proposed by Joe (1996). The idea behind this approach can be explained by looking back to the construction of the Poisson INAR(1) model, as described at the beginning of Section 2.1.2. There, we made use of the additivity of the Poisson distribution; that is, the property that the sum of two independent Poisson random variables is again Poisson distributed. Generally, the pmf of the sum of two independent random variables is obtained by *convoluting* their individual pmfs. Using this terminology, the additivity of the Poisson distribution is equivalently stated by saying that the Poisson distribution is *convolution-closed*. However, there are further convolution-closed distributions in Appendix A, such as the negative binomial (Example A.1.4) or the generalized Poisson (Example A.1.6). In fact, the whole compound Poisson family (Example A.1.2) shares some kind of additivity; see Lemma 2.1.3.2. Therefore, the approach by Joe (1996) mainly concentrates on convolution-closed and infinitely divisible marginal distributions.

So let us look back to Section 2.1.2. The stationary Poisson INAR(1) model generates the count $X_t \sim \text{Poi}(\mu)$ at time t as the sum of the independent random variables $\alpha \circ X_{t-1} \sim \text{Poi}(\alpha \mu)$ and $\epsilon_t \sim \text{Poi}((1-\alpha)\mu)$. As a result – see (2.13) – we know that the successive observations (X_t, X_{t-1}) are bivariately Poisson distributed according to $\text{MPoi}(\alpha \mu; (1-\alpha)\mu, (1-\alpha)\mu)$ (Example A.3.1). So we have the equality in distribution

$$(X_t, \ X_{t-1}) \stackrel{\text{d}}{=} (Y_2 + Y_{12}, \ Y_1 + Y_{12}), \tag{3.20}$$

where Y_1, Y_2, Y_{12} are independent with $Y_1, Y_2 \sim \text{Poi}((1 - \alpha)\mu)$ and $Y_{12} \sim \text{Poi}(\alpha \mu)$. In this construction, one may interpret Y_{12} as a common latent component of X_t and X_{t-1} inducing the serial dependence, while Y_1, Y_2 reflect the innovation part (Joe, 1997; Jung & Tremayne, 2011a). To obtain first-order models with non-Poisson marginals, Joe (1996) used the same construction as in (3.20), but with Y_1, Y_2, Y_{12} stemming from another type of convolution-closed and infinitely divisible distribution:

- the negative binomial, leading to the NB-RCINAR(1) model (3.19), or
- the generalized Poisson, leading to the model by Alzaid & Al-Osh (1993), which can be expressed by using the quasi-binomial thinning operation.

For the latter model, also see Jung & Tremayne (2011a), who refer to it as the *GPJ(1) model*. Generally, as shown by Joe (1996), all these first-order models can be understood as some kind of thinning-based model, by using an appropriate type of generalized thinning, and they are all Markov chains with CLAR(1) structure (Grunwald et al., 2000).

The construction in (3.20) can also be extended to higher-order autoregressions, but then "the notation is a bit cumbersome" (Joe, 1996, p. 671). So the reader is referred to the works by Joe (1996) and Jung & Tremayne (2011a) for more details; it is just noted here that such higher-order autoregressions do not possess a CLAR structure anymore. Finally, it is pointed out that Markov count models can also be defined using copulas (Joe, 1997, 2016), background information on which is provided by Joe (1997) and Genest & Nešlehová (2007). A Markov chain with the X_t having cdf F, for instance, is constructed by specifying the bivariate distribution of (X_t, X_{t-1}) in the form $C(F(x_0), F(x_1))$, where C is an appropriate bivariate copula. A potential benefit of such copula-based models is the possibility for negative autocorrelations.

Remark 3.2.3 ($\mathbb{Z}$-**valued time series**)   Although they have received less attention in the literature than count models, it should be emphasized that models for time series having the *full* set of integers $\mathbb{Z}$ as their range (hereafter referred to as $\mathbb{Z}$-*valued time series*) have also been discussed by some authors. Probably the first contribution concerning $\mathbb{Z}$-valued time series was the one by Kim & Park (2008), who introduced the *signed binomial thinning operator*, defined as $\alpha \odot X := \text{sgn}(\alpha) \, \text{sgn}(X) \, (|\alpha| \circ |X|)$, where $\alpha \in (-1; 1)$, and where $\text{sgn}(z) = 1$ for $z \geq 0$ and -1 otherwise. Using this operator, Kim & Park (2008) defined signed counterparts to both the AA- and the DL-INAR(p) models (Section 3.1). These are referred to as *INARS(p) models*. The DL-INARS(p) model, where the thinnings in

$$X_t = \alpha_1 \odot X_{t-1} + \ldots + \alpha_p \odot X_{t-p} + \epsilon_t$$

are assumed to be all independent, has a stationary solution if the roots of $\alpha(z) := 1 - \alpha_1 z - \ldots - \alpha_p z^p$ are outside the unit circle (see also Appendix B.3);

then it exhibits the typical AR(p)-autocorrelation structure. Andersson & Karlis (2014) investigated a special case of the INARS(1) model, where the innovations are assumed to stem from a Skellam distribution (see Example A.1.10). References to further models for $\mathbb{Z}$-valued time series can be found in Section 3.4 of Scotto et al. (2015).

3.3 The Binomial AR Model

In many applications, it is known that the observed count data cannot become arbitrarily large; their range has a natural upper bound $n \in \mathbb{N}$ that can never be exceeded. The models discussed up to now – the INAR(1) model from Definition 2.1.1.1 and all its extensions – can only be applied to count processes having an infinite range $\mathbb{N}_0$. As an example, if we want to guarantee that $X_t = \alpha \circ X_{t-1} + \epsilon_t$ does not become larger than n, then the range of ϵ_t at time t would have to be restricted to $\{0, \ldots, n - X_{t-1}\}$, which would contradict the innovations' i.i.d. assumption. So different solutions are required for "finite-valued" counts.

One such solution for time series of counts supported on $\{0, \ldots, n\}$, with a fixed upper limit $n \in \mathbb{N}$, is the *binomial AR(1) model*, which was proposed by McKenzie (1985) together with the INAR(1) model. It replaces the INAR(1)'s innovation term by an additional thinning, $\beta \circ (n - X_{t-1})$, such that this term cannot become larger than $n - X_{t-1}$.

Definition 3.3.1 **(Binomial AR(1) model)** Let $\rho \in \left(\max\left\{ \frac{-\pi}{1-\pi}, \frac{1-\pi}{-\pi} \right\}; 1 \right)$ and $\pi \in (0; 1)$. Define $\beta := \pi (1 - \rho)$ and $\alpha := \beta + \rho$. Fix $n \in \mathbb{N}$.
The process $(X_t)_{\mathbb{Z}}$, defined by the recursion

$$X_t = \alpha \circ X_{t-1} + \beta \circ (n - X_{t-1}),$$

where all thinnings are performed independently of each other, and where the thinnings at time t are independent of $(X_s)_{s<t}$, is referred to as a *binomial AR(1) process* .

The condition on ρ guarantees that the derived parameters α, β satisfy $\alpha, \beta \in (0; 1)$; that is, these parameters can indeed serve as thinning probabilities.

The binomial AR(1) model of Definition 3.3.1 is easy to interpret (Weiß, 2009a). Suppose that we have a system of n mutually independent units, each being either in state "1" or state "0". Let X_{t-1} be the number of units being in state "1" at time $t - 1$. Then $\alpha \circ X_{t-1}$ is the number of units still being in state "1" at time t, with individual transition probability α ("survival probability"). $\beta \circ (n - X_{t-1})$ is the number of units, which moved from state "0" to state "1" at time t, with individual transition probability β ("revival probability").

It is known that $(X_t)_{\mathbb{Z}}$ is a stationary, ergodic and ϕ-mixing finite Markov chain (also see Appendix B.2.2). Its marginal distribution is $\text{Bin}(n, \pi)$, and the (truly positive) 1-step-ahead transition probabilities are given by

$$p_{k|l} = \sum_{m=\max\{0, k+l-n\}}^{\min\{k,l\}} \binom{l}{m} \binom{n-l}{k-m} \alpha^m (1-\alpha)^{l-m} \beta^{k-m} (1-\beta)^{n-l+m-k},$$

(3.21)

see McKenzie (1985) and Weiß (2009a). Conditional mean and variance are both linear in X_{t-1}, and given by

$$\begin{aligned}
E[X_t \mid X_{t-1}] &= \rho \cdot X_{t-1} + n\beta, \\
V[X_t \mid X_{t-1}] &= \rho(1-\rho)(1-2\pi) \cdot X_{t-1} + n\beta(1-\beta).
\end{aligned}$$

(3.22)

Closed-form expressions for the corresponding h-step-ahead regression properties are obtained by replacing ρ by ρ^h (Weiß & Pollett, 2012). Also note that the eigenvalues of the transition matrix $\mathbf{P}$ according to (3.21) are just given by $1, \rho, \ldots, \rho^n$ (Weiß, 2009a). So the speed of convergence of $\mathbf{P}^h \to \boldsymbol{p}\mathbf{1}^\top$ is determined by the value of ρ according to the Perron–Frobenius theorem (Remark B.2.2.1). The ACF of the binomial AR(1) model is of AR(1)-type, given by $\rho(k) = \rho^k$ for $k \geq 0$ (note that ρ might become negative). Closed-form expressions for higher-order joint moments in such a process are provided by Weiß & Kim (2013); the binomial AR(1) process is also time-reversible (McKenzie, 1985).

Remark 3.3.2 (**Renewal-based models**) If $n = 1$, the binomial AR(1) model according to Definition 3.3.1 reduces to the binary Markov chain, which will be discussed in Example 7.1.3. But for $n > 1$, there is also a close relationship between both models due to the additive structure of the involved thinning operations. Each of the n independent "units" might be associated its own binary Markov chain, indicating if this unit is in state "0" or "1". The whole binomial AR(1) process might then be thought of as a sum of these n independent binary Markov chains. This idea was picked up by Cui & Lund (2009, 2010), who defined count processes by superpositioning stationary renewal processes. Such models can be designed to give, for example, binomially or Poisson-distributed marginals, and they include the binomial AR(1) model as a special case.

A pth-order autoregressive version, which uses a probabilistic mixing approach analogous to the CINAR(p) model from Remark 3.1.5 and thus preserves the binomial marginal distribution, was proposed by Weiß (2009b). For the resulting *binomial AR(p) model*, defined by

$$X_t = \sum_{i=1}^{\text{p}} D_{t,i} \left(\alpha \circ_t X_{t-i} + \beta \circ_t (n - X_{t-i}) \right)$$

(3.23)

completely analogous to (3.9), the special case of independent thinnings is again relevant. It leads to a pth-order Markov process, the ACF of which satisfies the Yule–Walker equations

$$\rho(k) = \rho \sum_{i=1}^{p} \phi_i \cdot \rho(|k - i|). \tag{3.24}$$

The conditional mean and transition probabilities are given by

$$E[X_t \mid X_{t-1}, \ldots] = \mu\,(1 - \rho) + \rho \sum_{i=1}^{p} \phi_i\, X_{t-i},$$

$$P(X_t = x \mid X_{t-1} = x_{t-1}, \ldots) = \sum_{i=1}^{p} \phi_i \cdot$$

$$\sum_{y=0}^{x} \binom{x_{t-i}}{y}\, \alpha^y\,(1 - \alpha)^{x_{t-i}-y} \binom{n - x_{t-i}}{x - y}\, \beta^{x-y}\,(1 - \beta)^{n-x_{t-i}-x+y}. \tag{3.25}$$

Higher-order (factorial) moments follow in an analogous way to (3.11),

$$E[(X_t)_{(r)} \mid X_{t-1}, \ldots]$$
$$= \sum_{i=1}^{p} \phi_i \sum_{j=0}^{r} \binom{r}{j}\, (X_{t-i})_{(j)} \alpha^j\,(n - X_{t-i})_{(r-j)} \beta^{r-j}. \tag{3.26}$$

The binomial AR(1) model and its pth-order extension have a binomial marginal distribution; in particular, their binomial index of dispersion according to (2.3) satisfies $I_{\mathrm{Bin}} = 1$. To allow for time-dependent and finite-range counts with $I_{\mathrm{Bin}} > 1$ (*extra-binomial variation*), Weiß & Kim (2014) proposed replacing the binomial thinning operations in Definition 3.3.1 (or in (3.23), respectively) by beta-binomial ones (see Section 3.2). The resulting *beta-binomial AR(1) model* is characterized by the recursion

$$X_t = \alpha_\phi \circ X_{t-1} + \beta_\phi \circ (n - X_{t-1}), \tag{3.27}$$

where both thinnings use a unique dispersion parameter $\phi \in (0; 1)$. Many stochastic properties are analogous to those of the binomial AR(1) model, but the beta-binomial AR(1) model exhibits extra-binomial variation (Weiß & Kim, 2014):

$$I_{\mathrm{Bin}} = 1 + \frac{(n - 1) \cdot (1 - 2\pi(1 - \pi)(1 - \rho))}{\left(\frac{1}{\phi} - 1\right)(1 + \rho) + (1 - 2\pi(1 - \pi)(1 - \rho))}. \tag{3.28}$$

The transition probabilities are obtained from (3.21) by replacing the involved binomial distributions by beta-binomial ones (Example A.2.2). They can be used to compute the stationary marginal distribution as well as any *h*-step-ahead forecasting distribution *exactly* as described in Appendix B.2.1, since the beta-binomial AR(1) process is a *finite* Markov chain. The conditional mean is the same as in (3.22) (thus we also have the same ACF), while the conditional variance is now a quadratic function of X_{t-1} (Weiß & Kim, 2014):

$$V[X_t \mid X_{t-1}] = (\rho(1-\rho)(1-2\pi)(1-\phi) - 2n\beta(1-\beta)\phi) X_{t-1}$$
$$+ \phi (\alpha(1-\alpha) + \beta(1-\beta)) X_{t-1}^2 + n\beta(1-\beta)(1+\phi(n-1)). \tag{3.29}$$

Extensions of the binomial AR(1) model using state-dependent parameters (also see Remark 3.1.7) have been proposed by Weiß & Pollett (2014), for example to describe binomial underdispersion, and by Möller (2016) for SET-type models.

Remark 3.3.3   (Model fitting)   The approaches for parameter estimation, model identification and adequacy checking described in Sections 2.2–2.4 are easily adapted to the binomial AR case. As the only exception, the dispersion test (2.14) in Section 3.3 is not reasonable in the context of a finite range. Therefore, Weiß & Kim (2014) considered a sample version of the *binomial index of dispersion* (2.3), defined as $\hat{I}_{\text{Bin}} := n\,S^2/(\overline{X}\,(n-\overline{X}))$. Under the null of a binomial AR(1) model, this test statistic is asymptotically normally distributed with

$$E\left[\hat{I}_{\text{Bin}}\right] \approx 1 - \frac{1}{T}\left(1 - \frac{1}{n}\right)\frac{1+\rho}{1-\rho}, \qquad V\left[\hat{I}_{\text{Bin}}\right] \approx \frac{2}{T}\left(1 - \frac{1}{n}\right)\frac{1+\rho^2}{1-\rho^2}, \tag{3.30}$$

see Weiß & Kim (2014) and Weiß & Schweer (2015).

Example 3.3.4   (Price Stability)   Weiß & Kim (2014) analyzed a time series for the EA17, a group of countries in the Euro area, consisting of $n = 17$ member states. Starting in January 2000, it was determined for each month, how many of these countries showed stable prices (that is, an inflation rate below 2%), leading to monthly counts with range $\{0, \ldots, 17\}$. As in Weiß & Kim (2014), we first restrict to the time series $x_1, \ldots, x_T$ corresponding to the period from January 2000 to December 2006 (length $T = 84$). The data collected for January 2007 to August 2012 (shown in gray in Figure 3.5a) is later used for forecasting.

The SACF in Figure 3.5b shows significant values for lags 1 and 2, and the SPACF indicates an AR(1)-like autocorrelation structure. The estimated pmf is shown in black in Figure 3.5c. The marginal mean equals $\overline{x} \approx 4.274$, so the overall proportion of countries having stable prizes is (only) $\hat{\pi}_{\text{MM}} \approx 0.251$. The corresponding binomial distribution $\text{Bin}(n, \hat{\pi}_{\text{MM}})$ is shown in gray in Figure 3.5c, and it looks less dispersed than the sample pmf. This is also confirmed by the binomial index of dispersion, given by $\hat{I}_{\text{Bin}} \approx 1.521$. Compared to

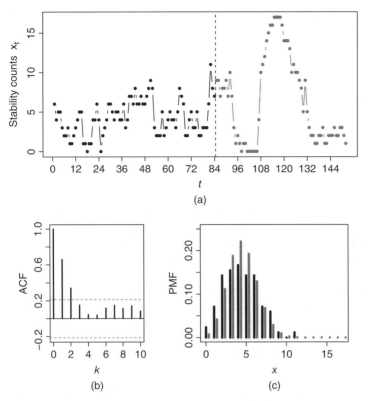

Figure 3.5 EA17 prices: (a) price stability counts (2000–2006 in black), (b) their sample autocorrelation, (c) marginal frequencies (black) and a binomial fit (gray). See Example 3.3.4.

a binomial AR(1) model with parameter values $\hat{\pi}_{\text{MM}}$ and $\hat{\rho}_{\text{MM}} := \hat{\rho}(1) \approx 0.658$, where approximate mean and standard deviation for $\hat{I}_{\text{Bin}}$ would be given by about 0.946 and 0.238, respectively, according to (3.30), we realize a significant degree of extra-binomial variation (P value ≈ 0.008). So the beta-binomial AR(1) model might be more appropriate for the data.

Both models are fitted to the data by the (full) ML approach; see Table 3.3 for a summary of the results. In terms of the AIC, the beta-binomial AR(1) model is superior, while the BIC prefers the binomial AR(1) model. Furthermore, the estimate for ϕ is not significantly different from 0, but this might just be due to the small sample size. On the other hand, the binomial index of dispersion within the fitted beta-binomial AR(1) model (≈ 1.315) better meets the observed value $\hat{I}_{\text{Bin}} \approx 1.521$ (the same holds for $\rho(1)$). Also the Pearson residuals are slightly better for the beta-binomial model, with the variance ≈ 1.040 (vs. ≈ 1.275 for the binomial model) being pretty close to 1. The PIT histograms only roughly agree with uniformity in both cases.

Table 3.3 Price stability counts: ML estimates and AIC and BIC values for different models.

Model	Parameter			AIC	BIC
	1	2	3		
Binomial AR(1)	0.256	0.578		327.3	332.2
(π, ρ)	(0.022)	(0.058)			
Beta-binomial AR(1)	0.260	0.622	0.037	326.7	334.0
(π, ρ, ϕ)	(0.028)	(0.064)	(0.028)		

Figures in parentheses are standard errors.

In view of the significant degree of extra-binomial variation for $x_1, \ldots, x_{84}$, let us discuss the fitted beta-binomial model in more detail, while being aware that this model is far from being perfect for the data. The estimate $\hat{\pi}_{\mathrm{ML}}$ implies that the overall probability for an EA17 country satisfying the stability criterion is only about 26%. If such a country has stable prices in a month $t - 1$, it will also have stable prices in month t with about 72% probability, since the survival probability is estimated as $\hat{\alpha}_{\mathrm{ML}} \approx 0.721$. In contrast, if there are no stable prices in month $t - 1$, prices will become stable one month later with a probability of only about 10% (revival probability $\hat{\beta}_{\mathrm{ML}} \approx 0.098$).

Next, we use the fitted beta-binomial model for forecasting "future" values; that is, the values after $x_{84} = 7$. Caused by the rather strong degree of dependence, the h-step-ahead forecasting distributions (given $x_{84} = 7$) converge relatively slowly to the marginal distribution. For instance, the 5% quantile equals 3 for $h = 1$, then 2 for $h = 2, 3, 4$, and 1 for $h \geq 5$. Similarly, the median changes from 6 ($h = 1$) to 5 ($h = 2, \ldots, 5$) to 4 ($h \geq 6$), and the 95% quantile from 9 ($h = 1, 2, 3$) to 8 ($h \geq 4$). The speed of convergence is characterized by the Perron–Frobenius theorem (Remark B.2.2.1); here, the second largest eigenvalue of the fitted model's transition matrix takes the value $\hat{\rho}_{\mathrm{ML}} \approx 0.622$.

The result of continued 1-step-ahead forecasting is shown in Figure 3.6. After a few observations, the data appear to be misplaced with respect to the quantile bands. As argued by Weiß & Kim (2014), there is indeed external evidence for a structural change after $t = 84$ (that is, for January 2007 and later), for example increased inflation due to strongly surging oil prices in 2007, and the beginning of the financial crisis in 2008 and the bankruptcy of Lehman Brothers. Tools for detecting such changes in a count process are presented in Chapter 8.

3.4 Multivariate INARMA Models

In many applications, we do not observe a single feature over time, but instead a number of related features simultaneously, thus leading to a multivariate

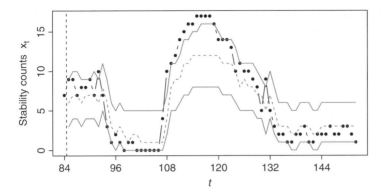

Figure 3.6 Plot of the price stability counts for $t \geq 84$, together with 1-step-ahead median (dashed) plus 5%- and 95%-quantiles (solid); see Example 3.3.4.

time series. If these multivariate observations are truly real-valued, then, for example, vector autoregressive moving-average (VARMA) models might be appropriate for describing the data; see Appendix B.4.2 for a brief summary. Here, we consider the count data case again; that is, a d-variate time series $x_1, \ldots, x_T$ with the range being $\mathbb{N}_0^d$ (or a subset of it). For such multivariate count time series, a number of thinning-based models (designed as counterparts to the VARMA model) have been discussed in the literature; see Section 4 in Scotto et al. (2015) for a survey. These approaches mainly differ in the way of defining a multivariate type of thinning operation. The most widely discussed approach, which is defined in a close analogy to conventional matrix multiplication, is *matrix-binomial thinning*, which was introduced by Franke & Subba Rao (1993).

Let X be a d-dimensional count random vector; that is, having the range $\mathbb{N}_0^d$. Let $\mathbf{A} \in [0; 1]^{d \times d}$ be a matrix of thinning probabilities. Then Franke & Subba Rao (1993) define the ith component of the d-dimensional count random vector $\mathbf{A} \circ X$ by

$$(\mathbf{A} \circ X)_i = \sum_{j=1}^{d} a_{ij} \circ X_j \qquad \text{for } i = 1, \ldots, d, \tag{3.31}$$

where the d^2 univariate binomial thinning operations (Section 2.1.1) are performed independently of each other. A generalized version of this operator, using univariate *generalized* thinnings (3.14) instead of binomial ones, was proposed by Latour (1997). We denote this operation by "$\mathbf{A} \bullet_{\mathbf{B}}$", with the (i, j)th thinning "$a_{ij} \bullet_{b_{ij}}$" satisfying $a_{ij} \in [0; 1]$ and $b_{ij} > 0$ (for binomial thinning, we have $b_{ij} = a_{ij}(1 - a_{ij})$). It satisfies $E[\mathbf{A} \bullet_{\mathbf{B}} X] = \mathbf{A} \, E[X]$ and

$$E[(\mathbf{A} \bullet_{\mathbf{B}} X)(\mathbf{A} \bullet_{\mathbf{B}} X)^{\top}] = \mathbf{A} \, E[X \, X^{\top}] \, \mathbf{A}^{\top} + \text{diag}(\mathbf{B} \, E[X]).$$

In an analogy to Definition B.4.2.1, a multivariate extension of the INAR(1) model (*MINAR(1) model*) according to Definition 2.1.1.1 is given by the recursion

$$X_t = \mathbf{A} \circ X_{t-1} + \boldsymbol{\epsilon}_t \qquad \text{(or, analogously, by using } \mathbf{A} \bullet_{\mathbf{B}}), \tag{3.32}$$

where $(\boldsymbol{\epsilon}_t)_{\mathbb{Z}}$ is an i.i.d. d-dimensional count process with finite mean $\boldsymbol{\mu}_{\boldsymbol{\epsilon}}$ and covariance matrix $\boldsymbol{\Sigma}_{\boldsymbol{\epsilon}}$; see Franke & Subba Rao (1993), Latour (1997) and Pedeli & Karlis (2013a). Latour (1997) showed that (Condition (B.25) in Appendix B.4.2):

$$\det(\mathbf{I} - \mathbf{A}\, z) \neq 0 \qquad \text{for } |z| \leq 1,$$

where $\mathbf{I}$ denotes the identity matrix, still guarantees the existence of a unique stationary solution. The marginal mean is given by $\boldsymbol{\mu} = (\mathbf{I} - \mathbf{A})^{-1}\, \boldsymbol{\mu}_{\boldsymbol{\epsilon}}$, and the autocovariance function is obtained from a slightly modified version of (B.30),

$$\boldsymbol{\Gamma}(k) = \mathbf{A}^k\, \boldsymbol{\Gamma}(0) =: \mathbf{A}^k\, \boldsymbol{\Sigma}, \quad \boldsymbol{\Sigma} = \mathbf{A}\, \boldsymbol{\Sigma}\, \mathbf{A}^{\top} + \mathrm{diag}\,(\mathbf{B}\, \boldsymbol{\mu}) + \boldsymbol{\Sigma}_{\boldsymbol{\epsilon}}. \tag{3.33}$$

If the thinning matrix $\mathbf{A}$ of the MINAR(1) model (3.32) is not a diagonal matrix, then the component processes $(X_{t,i})_{\mathbb{Z}}$ are generally not univariate INAR(1) processes. Therefore, Pedeli & Karlis (2011, 2013b) concentrate on the diagonal case $\mathbf{A} = \mathrm{diag}(a_1, \ldots, a_d)$ of matrix-binomial thinning (thus also reducing the number of model parameters) such that the cross-correlation between $X_{t,i}$ and $X_{t,j}$ is solely caused by the innovations. Note that (3.33) simplifies in this case, because $\mathbf{A}^k$ becomes $\mathrm{diag}(a_1^k, \ldots, a_d^k)$:

$$\gamma_{ij}(k) = a_i^k\, \sigma_{ij}, \quad (1 - a_i a_j)\, \sigma_{ij} = \delta_{i,j}\, a_i(1 - a_i)\, \mu_i + \sigma_{\boldsymbol{\epsilon};ij}, \tag{3.34}$$

where $\delta_{i,j}$ denotes the Kronecker delta. As particular instances of the MINAR(1) model with *diagonal-matrix-binomial thinning*, they consider the cases with $\boldsymbol{\epsilon}_t$ stemming from a multivariate Poisson or negative binomial distribution (Examples A.3.1 and A.3.2). The corresponding component processes $(X_{t,i})_{\mathbb{Z}}$ then follow the univariate Poisson INAR(1) (Section 2.1.2) or NB-INAR(1) model (Example 2.1.3.3), respectively. Note that Pedeli & Karlis (2011, 2013b) use a slightly different parametrization for the MNB distribution from Example A.3.2, with $\beta, \lambda_1, \ldots, \lambda_d$ obtained from the relations $n = \beta^{-1}$ and $\pi_j = \beta\, \lambda_j \big/ \big(1 + \beta\, \sum_{i=1}^d \lambda_i\big)$ for $j = 1, \ldots, d$.

Remark 3.4.1 (**MINAR(p) model**) Latour (1997) also considered the d-dimensional extension of the DL-INAR(p) model (3.5) by Du & Li (1991) and showed that it can be embedded into the MINAR(1) model by using the thinning matrix $\mathbf{D}_{11}$ defined as in (B.28) in Appendix B.4.2. In particular, the univariate DL-INAR(p) model (3.5) by Du & Li (1991) can be represented as a

MINAR(1) model with $X_t := (X_t, \ldots, X_{t-p+1})^\top$, with $\boldsymbol{\epsilon}_t := (\epsilon_t, 0, \ldots, 0)^\top$, and with thinning matrix

$$
A := \begin{pmatrix}
\alpha_1 & \alpha_2 & \cdots & \alpha_{p-1} & \alpha_p \\
1 & 0 & \cdots & 0 & 0 \\
0 & 1 & \ddots & \vdots & \vdots \\
\vdots & \ddots & \ddots & 0 & \vdots \\
0 & \cdots & 0 & 1 & 0
\end{pmatrix}.
$$

Example 3.4.2 (BINAR(1) model)   The particular instance of the *bivariate INAR(1)* (BINAR(1)) model with diagonal-matrix-binomial thinning was discussed by Pedeli & Karlis (2011) in great detail. For the innovations $(\boldsymbol{\epsilon}_t)_{\mathbb{Z}}$, they considered either the bivariate Poisson or negative binomial distribution; see Examples A.3.1 and A.3.2 (a copula-based model was developed by Karlis & Pedeli (2013)). Combining the moment properties for MPoi($\lambda_0; \lambda_1, \lambda_2$) and MNB($n, \pi_1, \pi_2$), as given in Examples A.3.1 and A.3.2, respectively, with formula (3.34), the moment properties of the resulting BINAR(1) models are immediately obtained. The transition probabilities $p_{k|l} := P(X_t = k \mid X_{t-1} = l)$ follow by extending (2.5) to (Pedeli & Karlis, 2011):

$$
p_{k|l} = \sum_{j_1=0}^{\min\{k_1, l_1\}} \sum_{j_2=0}^{\min\{k_2, l_2\}} \binom{l_1}{j_1} a_1^{j_1} (1-a_1)^{l_1-j_1}
$$
$$
\cdot \binom{l_2}{j_2} a_2^{j_2} (1-a_2)^{l_2-j_2} \cdot P(\boldsymbol{\epsilon}_t = k - j).
$$

In the Poisson case, the components $(X_{t,i})_{\mathbb{Z}}$ follow the Poisson INAR(1) model with $\epsilon_{t,i} \sim \text{Poi}(\lambda_i + \lambda_0)$; in the NB case, they follow the NB-INAR(1) model with $\epsilon_{t,i} \sim \text{NB}(n, \pi_0/(\pi_0 + \pi_i))$.

For illustration, we briefly discuss the data example presented by Pedeli & Karlis (2011). Each bivariate count x_t expresses the number of daytime $(x_{t,1})$ and nighttime $(x_{t,2})$ road accidents in the Schiphol area (Netherlands) for the days $t = 1, \ldots, 365$ in 2001; see the plot in Figure 3.7. Both components $(x_{t,i})$ show significant values for the SACF at lag 1 ($\hat{\rho}_{11}(1) \approx 0.125$ and $\hat{\rho}_{22}(1) \approx 0.134$), and the sample cross-correlation is computed as $\hat{\rho}_{12}(0) \approx 0.142$.

The respective marginal frequencies are shown in Figure 3.8, with the zero frequencies being 0.014 and 0.258, respectively. The estimated means are $\bar{x}_1 \approx 7.277$ and $\bar{x}_2 \approx 1.504$. We not only have more daytime accidents in the mean, but also a much stronger daytime dispersion, with $\hat{I}_1 \approx 2.869$ and $\hat{I}_2 \approx 1.245$ (approximate upper-sided critical values at the 5% level according to (2.14) would be given by 1.120 for both components). Therefore, we shall consider both the Poisson and the negative binomial BINAR(1) models as candidate models. The results of a CML estimation are summarized in

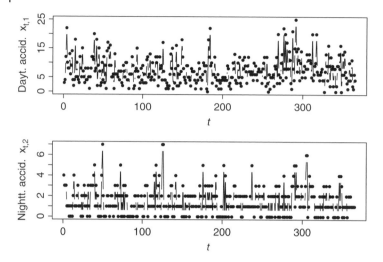

Figure 3.7 Plot of the daily number of daytime and nighttime road accidents; see Example 3.4.2.

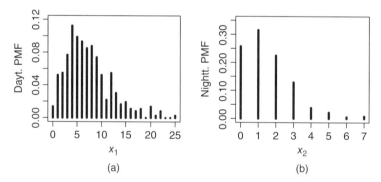

Figure 3.8 Plot of the marginal frequencies for the daily number of (a) daytime and (b) nighttime road accidents; see Example 3.4.2.

Table 3.4. Since both models have the same number of parameters, only the maximal log-likelihood is shown, which prefers the NB model.

If computing the components' Pearson residuals based on the fitted Poisson-BINAR(1) model from Table 3.4, the variances of 2.857 and 1.215 indicate that the Poisson model is not able to capture the dispersion within the data, especially with respect to the first component (daytime accidents). Here, the fitted NB-BINAR(1) model (with residual variances of 1.139 and 0.963) performs much better than the Poisson one. Generally, the observations' marginal properties are matched quite well by the fitted NB-BINAR(1) model,

Table 3.4 Accidents counts: CML estimates and maximal log-likelihoods for different models.

Model	Parameter					ℓ_{max}
	1	2	3	4	5	
Poisson BINAR(1)	6.460	1.099	0.268	0.077	0.087	−1745
$(\lambda_1, \lambda_2, \lambda_0, a_1, a_2)$	(0.241)	(0.125)	(0.105)	(0.025)	(0.043)	
Negative binomial BINAR(1)	4.449	0.547	0.104	0.043	0.114	−1619
$(n, \pi_1, \pi_2, a_1, a_2)$	(0.588)	(0.025)	(0.008)	(0.038)	(0.044)	

Figures in parentheses are standard errors.

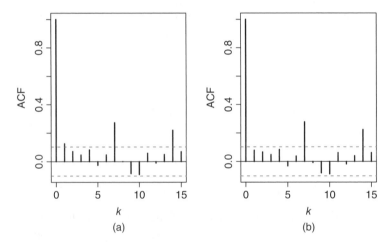

(a) (b)

Figure 3.9 Plot of SACF for daytime accidents (a) and for corresponding Pearson residuals (b) from NB-BINAR(1) model, see Example 3.4.2.

with means of 7.289 and 1.496, dispersion indices of 2.504 and 1.267, and zero probabilities of 0.011 and 0.265, respectively.

Despite these nice marginal features, the NB-BINAR(1) model is also not perfectly adequate for the data, which becomes clear by inspecting the SACF of the Pearson residuals in Figure 3.9b. There, we find significant SACF values for lags 7, 14, …, which could also have been observed by inspecting the SACF for the original daytime counts in Figure 3.9a; see also Liu (2012). So there seems to be a "day-of-week effect" causing this seasonal pattern. For the nighttime counts, neither the original counts nor their residuals show such seasonality. Computing the means of $x_{1+j\cdot 7,1}$, of $x_{2+j\cdot 7,1}$, and so on, we obtain the values

$$7.827, \quad 8.038, \quad 8.750, \quad 8.827, \quad 4.250, \quad 5.192, \quad 8.115,$$

showing an increased mean number of daytime accidents on weekdays. So a refined model able to deal with the apparent seasonality is required for the data. For instance, using a separate set of parameters for the NB-BINAR(1) model's innovations on weekdays and on the weekend leads to a visible improvement concerning the seasonality ($\ell_{\max} \approx -1578$).

Finally, let us turn to the case of multivariate counts having a finite range; that is, a range of the form $\{0, \ldots, n\} := \{0, \ldots, n_1\} \times \ldots \times \{0, \ldots, n_d\}$. At first glance, it appears to be possible to combine Definition 3.3.1 for a binomial AR(1) model with the concept of matrix-binomial thinning (3.31), and to define a model according to the recursive scheme $X_t = \mathbf{A} \circ X_{t-1} + \mathbf{B} \circ (n - X_{t-1})$. However, if the matrices $\mathbf{A}, \mathbf{B}$ are non-diagonal, it may happen that the upper limit n is violated, since, for example, $a_{11} \circ X_{t-1,1} + a_{12} \circ X_{t-1,2} + \ldots$ might become larger than n_1. So we would have to restrict ourselves to diagonal thinning matrices to ensure the upper limit n. But then, due to the absence of an innovations term, the component series would be independent of each other. So, in summary, a non-trivial multivariate binomial AR(1) model using matrix-binomial thinning is not possible.

For this reason, restricting to the bivariate case, Scotto et al. (2014) introduced another type of thinning operation, based on the *bivariate binomial distribution of Type II* (Example A.3.5). Let X be a bivariate count random variable with range $\{0, \ldots, n\} := \{0, \ldots, n_1\} \times \{0, \ldots, n_2\}$. The *bivariate binomial thinning* operation "$\otimes$" is defined by requiring the conditional distribution

$$\boldsymbol{\alpha} \otimes X | X \sim \mathrm{BVB}_{\mathrm{II}}(X_1, X_2, \min\{X_1, X_2\}; \alpha_1, \alpha_2, \phi_\alpha), \tag{3.35}$$

where the thinning parameter $\boldsymbol{\alpha} = (\alpha_1, \alpha_2, \phi_\alpha)^\top$ has to satisfy the restrictions given in Example A.3.5. Note that in contrast to matrix-binomial thinning, bivariate binomial thinning (3.35) generates cross-correlation (if $\phi_\alpha \neq 0$) and preserves the upper bound X at the same time. Furthermore, the marginals of $\boldsymbol{\alpha} \otimes X$ just correspond to univariate binomial thinnings: $(\boldsymbol{\alpha} \otimes X)_i | X \sim \mathrm{Bin}(X_i, \alpha_i)$.

Using this operation, Scotto et al. (2014) defined the *bivariate binomial AR(1) model* with upper limit n by extending Definition 3.3.1 as follows: let π_i, ρ_i for $i = 1, 2$ be separate sets of parameters according to Definition 3.3.1, derive α_i, β_i accordingly. Choose ϕ_α, ϕ_β such that $\boldsymbol{\alpha} = (\alpha_1, \alpha_2, \phi_\alpha)^\top$ and $\boldsymbol{\beta} = (\beta_1, \beta_2, \phi_\beta)^\top$ become valid BVB-parameters. Then the $\mathrm{BVB}_{\mathrm{II}}$-AR(1) process $(X_t)_{\mathbb{Z}}$ is defined by the recursion

$$X_t = \boldsymbol{\alpha} \otimes X_{t-1} + \boldsymbol{\beta} \otimes (n - X_{t-1}), \tag{3.36}$$

where the thinnings are again performed independently of each other. By construction, the components of this process are just univariate binomial AR(1) processes with parameters n_1, π_1, ρ_1 and n_2, π_2, ρ_2, respectively, and these components are cross-correlated to each other (Scotto et al., 2014):

$$
Cov[X_{t,1}, X_{t,2}] = \phi_\alpha \frac{\sqrt{\alpha_1 \alpha_2 (1 - \alpha_1)(1 - \alpha_2)}}{1 - \rho_1 \rho_2} E\left[\min\{X_{t,1}, X_{t,2}\}\right]
$$

$$
+ \phi_\beta \frac{\sqrt{\beta_1 \beta_2 (1 - \beta_1)(1 - \beta_2)}}{1 - \rho_1 \rho_2} E\left[\min\{n_1 - X_{t,1}, n_2 - X_{t,2}\}\right].
$$

$$(3.37)$$

To numerically compute these and further properties of the model, it should be noted that $(X_t)_{\mathbb{Z}}$ constitutes a finite Markov chain (so (B.4) holds) with transition probabilities

$$
p_{k|l} = \sum_{a_1=0}^{\min\{k_1, l_1\}} \sum_{a_2=0}^{\min\{k_2, l_2\}} p_{(l;\alpha)}(a) \cdot p_{(n-l;\beta)}(k - a), \tag{3.38}
$$

where $p_{(n;\alpha)}(x)$ abbreviates the pmf of the $\text{BVB}_{\text{II}}(n_1, n_2, \min\{n_1, n_2\}; \alpha_1, \alpha_2, \phi_\alpha)$ distribution as given in Example A.3.5. Note that $(X_t)_{\mathbb{Z}}$ is even a primitive Markov chain and hence ϕ-mixing with a uniquely determined stationary marginal distribution. Further properties and a data example are provided by Scotto et al. (2014).

4

INGARCH Models for Count Time Series

The models discussed in Chapter 3 used types of thinning operations to transfer the ARMA model to the count data case. Another popular approach for modeling such stationary processes of counts are the INGARCH models, the definition of which is related to linear regression models (also see Section 5.1). Despite their controversial name, these models are particularly attractive for overdispersed counts with an ARMA-like autocorrelation structure. Results concerning the basic model with a conditional Poisson distribution are presented, but generalizations with, for example, a binomial or negative binomial conditional distribution are also considered.

4.1 Poisson Autoregression

Due to the multiplication problem discussed in Section 2.1, the ARMA models of Definition B.3.5 are not applicable to the count data case. The models presented in Chapters 2 and 3 circumvented this problem by replacing the multiplications with a type of thinning operation, that ensures that these modified model recursions always produce integer values. The INGARCH models to be presented in this chapter use another solution to the multiplication problem: a linear regression of the conditional means $M_t := E[X_t | X_{t-1}, \ldots]$. To construct an AR(1)-like model, for instance, the AR(1) recursion $Y_t = \alpha_1\, Y_{t-1} + \varepsilon_t$ is transferred to the level of conditional means as $M_t = \alpha_1\, X_{t-1} + \beta_0$. Then the count at time t is generated by using, say, a conditional Poisson distribution – that is, $X_t \sim \text{Poi}(M_t)$ – thus guaranteeing that the outcomes are always integer values. This approach is not only related to linear regression; it also shares analogies with the definition of an ARCH(1) model (Definition B.4.1.1), where the autoregression is defined at the level of conditional variances: $\sigma_t^2 = \beta_0 + \alpha_1\, Y_{t-1}^2$. Such ARCH models are extended beyond pure autoregression by also including past conditional variances in the model recursion; see the GARCH equation (B.20). For example, the GARCH(1, 1) model is determined by $\sigma_t^2 = \beta_0 + \alpha_1\, Y_{t-1}^2 + \beta_1\, \sigma_{t-1}^2$. Picking up this idea, an

An Introduction to Discrete-Valued Time Series, First Edition. Christian H. Weiss.
© 2018 John Wiley & Sons Ltd. Published 2018 by John Wiley & Sons Ltd.
Companion website: www.wiley.com/go/weiss/discrete-valuedtimeseries

INGARCH(1, 1) model is defined by also including a feedback term, now with respect to the previous conditional mean: $M_t = \beta_0 + \alpha_1 X_{t-1} + \beta_1 M_{t-1}$.

The full INGARCH model was introduced by Rydberg & Shephard (2000), Heinen (2003) and Ferland et al. (2006). The name indicates that, as mentioned before, this model can be understood as an integer-valued counterpart to the conventional GARCH model, but also see the discussion in Remark 4.1.2 below. Conditioned on the past observations, the INGARCH model assumes an ARMA-like recursion for the conditional mean. Depending on the choice of the conditional distribution family, different INGARCH models are obtained. The basic INGARCH model, which is discussed in this section, assumes a conditional Poisson distribution.

Definition 4.1.1 (INGARCH model) Let $(X_t)_\mathbb{Z}$ be a process with range $\mathbb{N}_0$. The process $(X_t)_\mathbb{Z}$ follows the (Poisson) INGARCH(p, q) model with p $\geq$ 1 and q $\geq$ 0 if

(i) X_t, conditioned on $X_{t-1}, X_{t-2}, \ldots$, is Poisson distributed according to Poi(M_t), where
(ii) the conditional mean $M_t := E[X_t \mid X_{t-1}, \ldots]$ satisfies

$$M_t = \beta_0 + \sum_{i=1}^{p} \alpha_i X_{t-i} + \sum_{j=1}^{q} \beta_j M_{t-j}$$

with $\beta_0 > 0$ and $\alpha_1, \ldots, \alpha_p, \beta_1, \ldots, \beta_q \geq 0$.

If q = 0, the model of Definition 4.1.1 is referred to as the *INARCH(p) model*.

Remark 4.1.2 (Terminology) There is a lot of confusion in the literature about how to refer to the models given through Definition 4.1.1. In Rydberg & Shephard (2000), they are referred to as *BIN models*, in Heinen (2003) as *ACP models* (autoregressive conditional Poisson), and in Fokianos et al. (2009) as *(linear) Poisson autoregressive models*. The name INGARCH seems to be due to Ferland et al. (2006), and it is motivated by the analogy between condition (ii) in Definition 4.1.1 and the GARCH equation (B.20); see the initial discussion. Certainly, condition (ii) refers to conditional means, while (B.20) refers to conditional variances, which seems to be the main reason why some authors refuse to use the INGARCH terminology. On the other hand, for the particular *Poisson* INGARCH model, condition (ii) also applies to the conditional variances in view of the equidispersion property of the conditional Poisson distribution. Although the analogy between GARCH and INGARCH models is far from being perfect, we shall use the name *INGARCH models* in the sequel, as this name seems to be more often used in the literature than any of its competitors.

Although having the equidispersed Poisson distribution as a conditional distribution, the INGARCH model is well suited for overdispersed counts, since it satisfies

$$\mu_t := E[X_t] = E[M_t], \qquad V[X_t] = \mu_t + V[M_t] > \mu_t,$$

provided that these moments exist. In fact, Ferland et al. (2006) showed that for $\alpha_\bullet + \beta_\bullet := \sum_{i=1}^p \alpha_i + \sum_{j=1}^q \beta_j < 1$, the INGARCH process exists and is strictly stationary, with finite first- and second-order moments. In this case, the mean equals

$$\mu = \frac{\beta_0}{1 - \alpha_\bullet - \beta_\bullet}, \tag{4.1}$$

and variance and autocovariances are determined by a set of Yule–Walker-like equations (Weiß, 2009c):

$$\gamma(k) = \sum_{i=1}^p \alpha_i \, \gamma(|k-i|) + \sum_{j=1}^{\min\{k-1,q\}} \beta_j \, \gamma(k-j)$$

$$\qquad + \sum_{j=k}^q \beta_j \, \gamma_M(j-k) \qquad \text{for } k \geq 1,$$

$$\gamma_M(k) = \sum_{i=1}^{\min\{k,p\}} \alpha_i \, \gamma_M(|k-i|) + \sum_{i=k+1}^p \alpha_i \, \gamma(i-k) \tag{4.2}$$

$$\qquad + \sum_{j=1}^q \beta_j \, \gamma_M(|k-j|) \qquad \text{for } k \geq 0,$$

where $\gamma(k) := Cov[X_t, X_{t-k}]$ and $\gamma_M(k) := Cov[M_t, M_{t-k}]$. Note the analogy to equations (B.18) for the conventional ARMA model (Appendix B.3), and note the difference to the ARCH case (Appendix B.4.1), where the non-squared observations are uncorrelated.

Remark 4.1.3   (INGARCH vs. INARMA)   Equation 4.2 shows that, despite their name, the INGARCH models also form an integer-valued counterpart to the ARMA models, just like the INARMA models discussed in Section 3.1. So one may certainly ask about the possible advantages and disadvantages of these two "competitors". While the INARMA approach allows for counterparts to pure AR and MA models (note that a pure MA-like model is *not* included in the INGARCH family), a reasonable definition of a full ARMA-like model is not that obvious but easily created within the INGARCH framework; also see Example 4.1.4. Even for the purely autoregressive case, where both approaches can be used, the generated sample paths will often differ from each other in their structure; see Remark 4.1.7 for illustration. As we shall also later see, analytical expressions for marginal properties are more difficult in the INGARCH than in the INARMA case, while the INGARCH model (by definition) has simple conditional distributions. The latter is also useful in the context of conditional ML estimation, because the likelihood function factorizes to

$$L(\theta) = P(X_T = x_T \,|\, x_{T-1}, \ldots) \cdot P(X_{T-1} = x_{T-1} \,|\, x_{T-2}, \ldots) \cdots ,$$

with the conditional probabilities $P(X_t = x_t \,|\, x_{t-1}, \ldots)$ stemming from the Poisson distribution $\text{Poi}(M_t)$. The computation of such conditional probabilities is more demanding for INARMA models; see (3.12) for the INAR(2) model as an example. On the plus side of INARMA models, in contrast, their interpretability has to be noted. This may allow for a deeper understanding of the data-generating process in some applications. Further pros and cons could be listed here, but the present consideration already makes clear that a general recommendation of one or other of these approaches cannot be given.

Further results concerning likelihood estimation, especially on the asymptotic properties of the resulting ML estimators, are provided by Ferland et al. (2006), Fokianos et al. (2009) and Cui & Wu (2016).

Example 4.1.4 (INGARCH(1, 1) model) The particular case of the stationary INGARCH(1, 1) model, where part (ii) of Definition 4.1.1 simplifies to $M_t = \beta_0 + \alpha_1 X_{t-1} + \beta_1 M_{t-1}$, was further investigated by Heinen (2003), Ferland et al. (2006) and Fokianos et al. (2009), among others. It was shown that all moments exist, where the variance equals

$$\sigma^2 = \frac{1 - (\alpha_1 + \beta_1)^2 + \alpha_1^2}{1 - (\alpha_1 + \beta_1)^2} \cdot \mu,$$

and the ACF is given by

$$\rho(k) = (\alpha_1 + \beta_1)^{k-1} \frac{\alpha_1\,(1 - \beta_1(\alpha_1 + \beta_1))}{1 - (\alpha_1 + \beta_1)^2 + \alpha_1^2} \qquad \text{for } k \geq 1,$$

also see (4.2). Furthermore, Theorem 3.1 in Neumann (2011) implies that such a process is α-mixing with geometrically decreasing weights (see Definition B.1.5).

Fokianos (2011) emphasized that the INGARCH(1, 1) process, although defined using the hidden conditional means $(M_t)_{\mathbb{Z}}$ (feedback mechanism), is *observation-driven* in the sense of Cox (1981); that is, the serial dependence can be explained by past observations (for a *parameter-driven* process, in contrast, serial dependence is caused by a latent process; see also Remark 5.2.1 below). In particular, it follows that

$$M_t = \alpha_1 \sum_{k=1}^{t} \beta_1^{k-1} X_{t-k} + \beta_1^t M_0 + \beta_0 \frac{1 - \beta_1^t}{1 - \beta_1},$$

that is, the current observation is influenced by all past observations, but with a weight decreasing exponentially with increasing time lag k. So the INGARCH(1, 1) model offers a parsimoniously parametrized way of accounting for a long memory.

Example 4.1.5   (Transactions counts)   We analyze a part of a dataset that was originally published by Brännäs & Quoreshi (2010). For the working days between 2 and 22 July 2002, it provides the number of transactions of the Ericsson B stock per minute between 9:35 and 17:14. As we shall see below, the data are characterized by a slowly decaying SACF (long memory). In Brännäs & Quoreshi (2010), the data are modeled by INMA(q) models (Section 3.1) with a high model order q. But parts of the data have also been modeled by the more parsimoniously parametrized INGARCH(1, 1) model; see Fokianos et al. (2009), Zhu (2012c), Christou & Fokianos (2014) and Davis & Liu (2016), among others. We shall follow this latter approach and restrict our analyses to the counts observed on 2 July 2002, which constitute a time series of length $T = 460$. A plot of these data is shown in Figure 4.1a.

The SACF in Figure 4.1b takes moderate but slowly decaying values, and the SPACF, showing several significant values, indicates that a purely

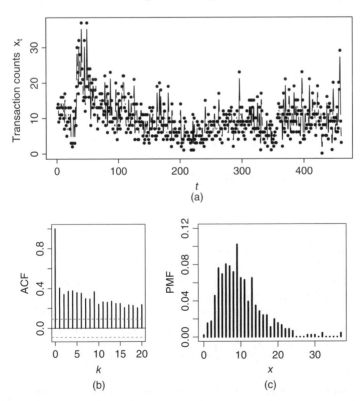

Figure 4.1 Ericsson stocks: (a) transactions counts, (b) sample autocorrelation, (c) marginal frequencies. See Example 4.1.5.

Table 4.1 Transactions counts: CML estimates for Poisson INGARCH(1,1) model, together with ℓ_{max}.

$\hat{\beta}_{0,CML}$	$\hat{\beta}_{1,CML}$	$\hat{\alpha}_{1,CML}$	$\hat{m}_{1,CML}$	ℓ_{max}
0.292	0.832	0.139	12.058	−1430
(0.100)	(0.023)	(0.018)		

autoregressive model will not be appropriate for the data. Therefore, see the discussion in Example 4.1.4, it is indeed reasonable to try to fit the INGARCH(1, 1) model to the data. The marginal distribution of the data, (Figure 4.1c), has mean $\bar{x} \approx 9.909$ and shows a strong degree of overdispersion, $\hat{I} \approx 3.307$.

As the initial step of parameter estimation, we compute moment estimates for the parameters $\beta_0, \beta_1, \alpha_1$, in the following way: the quotient $\hat{\rho}(2)/\hat{\rho}(1)$ estimates $\alpha_1 + \beta_1$ ($\rho(\hat{1}) \approx 0.405$, $\hat{\rho}(2) \approx 0.340$), which is used to compute $\hat{\beta}_{0,MM} \approx 1.596$ from (4.1). The formula for $\rho(1)$ leads to a quadratic equation in α_1 (given the value for $\alpha_1 + \beta_1$), which finally results in $\hat{\alpha}_{1,MM} \approx 0.286$ and $\hat{\beta}_{1,MM} \approx 0.553$. These estimates can now be used as initial values for CML estimation; see also Remark 4.1.3. This, however, is not a trivial task, because the conditioning requires not only $x_1 = 13$, but also the value m_1 of the initial conditional mean M_1:

$$L(\theta \,|\, x_1, m_1) = \prod_{t=2}^{T} p(x_t \,|\, x_{t-1}, m_{t-1}),$$

where we update $m_t = \beta_0 + \alpha_1 \, x_{t-1} + \beta_1 \, m_{t-1}$.

One solution is to specify m_1 as, say, $m_1 := \bar{x}$ or $m_1 := 0$ (Fokianos et al., 2009), but this choice turns out to have a significant effect on the resulting CML estimates:

$$(\hat{\beta}_{0,CML}, \hat{\beta}_{1,CML}, \hat{\alpha}_{1,CML}) \approx (0.581, 0.745, 0.199) \text{ with } \ell_{max} \approx -1449$$

if using $m_1 := 0$, while $m_1 := \bar{x}$ leads to

$$(\hat{\beta}_{0,CML}, \hat{\beta}_{1,CML}, \hat{\alpha}_{1,CML}) \approx (0.291, 0.831, 0.140) \text{ with } \ell_{max} \approx -1431.$$

Therefore, we shall follow the suggestion of Ferland et al. (2006) here and treat m_1 as a further parameter during estimation. We obtain the figures shown in Table 4.1 (standard errors in parentheses), which are quite close to the estimates obtained by initializing with $m_1 := \bar{x}$.

The mean of the fitted model (≈ 9.858) is close to the observed one, and also its ACF is slowly decaying ($0.344, 0.334, 0.324, 0.315, \ldots$). Consequently, the SACF of the Pearson residuals (computed using $\hat{m}_{1,CML}$ as the initial conditional mean) indicates an adequate model. However, the dispersion index of the fitted model (≈ 1.329) is much too low, which goes along with the Pearson residuals having a variance (≈ 2.330) clearly larger than 1, and with the strongly U-shaped PIT histogram shown in Figure 4.2a. So while the Poisson

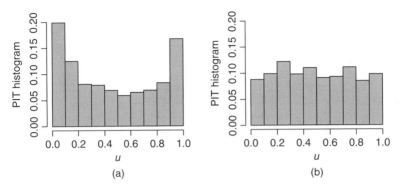

Figure 4.2 PIT histograms based on (a) fitted Poisson INGARCH(1, 1) and (b) GP-INGARCH(1, 1) model. See Examples 4.1.5 and 4.2.4.

INGARCH(1, 1) model is able to describe the observed autocorrelation structure, it cannot explain the strong volatility of the data. Therefore, we shall pick up the suggestion by Zhu (2012c) and use modified types of INGARCH models with a conditional non-Poisson distribution; see Example 4.2.4 below.

It should be mentioned that the slow decay of the SACF of the transactions counts from Example 4.1.5 is not necessarily caused by a long memory, but might also be explained by, for example, change points in the process (Kirch & Kamgaing, 2016, 10.5.2.1). Generally, it is well known that transactions data often exhibit an intraday pattern because of higher trading activity at the beginning and at the end of a trading day; see, for example, Wood et al. (1985). But here, for illustration, we continue with the INGARCH(1, 1) modeling.

An important subfamily of the INGARCH models from Definition 4.1.1 is the class of purely autoregressive (Poisson) *INARCH(p) models*, where q = 0 and

$$M_t = \beta_0 + \sum_{i=1}^{p} \alpha_i X_{t-i}, \quad \alpha_\bullet < 1. \tag{4.3}$$

Such INARCH models were discussed by Rydberg & Shephard (2000) and Weiß (2009c) in some detail. The INARCH(p) model constitutes a pth-order Markov model, thus being a competitor to the DL-INAR(p) model (3.5). The Poisson INARCH(p) model can also be understood as a particular GINAR(p) model, see the discussion in Section 3.2. It has simple Poisson transition probabilities,

$$P(X_t = x \mid x_{t-1}, \ldots) = \exp\left(-\beta_0 - \sum_{i=1}^{p} \alpha_i x_{t-i}\right) \frac{\left(\beta_0 + \sum_{i=1}^{p} \alpha_i x_{t-i}\right)^x}{x!}, \tag{4.4}$$

which is attractive for CML estimation according to (B.6). It also has a linear conditional mean and variance, both given by $\beta_0 + \sum_{i=1}^{p} \alpha_i x_{t-i}$ (also see (3.8) for the DL-INAR(p) model). In particular, equations (4.2) simplify to

$$\gamma(k) = \sum_{i=1}^{p} \alpha_i \, \gamma(|k - i|) + \delta_{k,0} \, \mu \qquad \text{for } k \geq 0, \tag{4.5}$$

that is, we have the typical AR(p) autocorrelation structure (see (B.13)). As a consequence, the model order p can be identified by inspecting the (S)PACF.

Comparing with the discussion in Section 3.1, it becomes obvious that the INGARCH approach is easier to use when handling a higher-order ARMA-like autocorrelation structure than the INARMA approach; see also Remark 4.1.3. On the other hand, closed-form expressions for the stationary marginal distribution or for h-step ahead forecasting distributions are difficult to find, even in the simplest case of an INARCH(1) model.

Example 4.1.6 (INARCH(1) model)   The INARCH(1) model constitutes a counterpart to the INAR(1) model discussed in Section 2.1, and it is a boundary case of the INGARCH(1, 1) model discussed in Example 4.1.4. Denoting its model parameters by $\beta := \beta_0 > 0$ and $\alpha := \alpha_1 \in (0; 1)$, the INARCH(1) model requires X_t to be conditionally Poisson-distributed in the following way:

$$X_t \mid X_{t-1}, X_{t-2}, \ldots \; \sim \; \text{Poi}(\beta + \alpha \, X_{t-1}).$$

Hence, the transition probabilities, as required for likelihood computation (see (B.6)), are simply given by

$$p_{k|l} = e^{-\beta - \alpha \, l} \, (\beta + \alpha \, l)^k \, / k!. \tag{4.6}$$

The conditional variance and mean coincide, and they are both linear in the previous observation, given by $\beta + \alpha \, x_{t-1}$. The latter implies that the INARCH(1) model belongs to the class of CLAR(1) models (Grunwald et al., 2000).

An INARCH(1) process is a stationary, ergodic and α-mixing Markov chain (Neumann, 2011). All moments of an INARCH(1) process exist (Ferland et al., 2006). The marginal cumulants can be determined according to the recursive scheme provided by Weiß (2009b):

$$\kappa_1 = \frac{\beta}{1 - \alpha}, \qquad \kappa_n = -(1 - \alpha^n)^{-1} \sum_{j=1}^{n-1} s_{n,j} \, \kappa_j \quad \text{for } n \geq 2,$$

where the coefficients $s_{n,j}$ are the Stirling numbers of the first kind, given by

$$s_{n,0} = 0, \qquad s_{n,n} = 1, \qquad s_{n+1,j} = s_{n,j-1} - n \, s_{n,j}$$

for $j = 1, \ldots, n$ and $n \geq 1$. In particular, the marginal mean and variance of an INARCH(1) process are given by

$$\mu = \frac{\beta}{1 - \alpha} \quad \text{and} \quad \sigma^2 = \frac{\beta}{(1 - \alpha)(1 - \alpha^2)}, \qquad \text{that is,} \quad I = \frac{1}{1 - \alpha^2} > 1.$$

The autocorrelation function equals $\rho(k) = \alpha^k$ as in the standard AR(1) case, and closed-form expressions for the joint (central) moments and cumulants up to order 4 are provided by Weiß (2010a).

While the 1-step-ahead conditional properties of the INARCH(1) model are very simple, there is no closed-form formula for the stationary marginal distribution, or for the h-step-ahead conditional properties with $h \geq 2$. To obtain these, at least numerically, the MC approximation of Remarks 2.1.3.4 and 2.6.3 has to be adopted.

Remark 4.1.7 (Comparison to INAR(1) model) At first glance, the Poisson INAR(1) and INARCH(1) model are very similar; choosing $\beta = \lambda$ and a unique value of α, they have the same marginal mean and the same ACF (see Section 2.1.2). But while the Poisson INAR(1) model is unconditionally equidispersed, the INARCH(1) model shows increasing overdispersion with increasing α. Also the whole sample paths generated by these models differ more and more from each other with increasing α. This can be seen by comparing Figure 2.5b (Example 2.1.2.1) with Figure 4.3, where both sample paths refer to the marginal mean $\mu = 3$ and the strong autocorrelation level $\alpha = 0.95$. In contrast to Figure 2.5b, the INARCH(1) process leads to long runs only for the value 0, while we have vivid fluctuations otherwise (note that the linear term $\alpha\, x_{t-1}$ of the conditional variance is much larger than 0 except for $x_{t-1} = 0$). Also more extreme counts are observed in the INARCH(1) case, which can be explained to some extent by the strong level of overdispersion, $I \approx 10.256$.

Figure 4.4 highlights the difference between the conditional variances of the Poisson INAR(1) and INARCH(1) models, given by $\alpha(1 - \alpha) \cdot x_{t-1} + \mu\,(1 - \alpha)$ (see (2.6)) and $\alpha \cdot x_{t-1} + \mu\,(1 - \alpha)$ (see Example 4.1.6), respectively. This

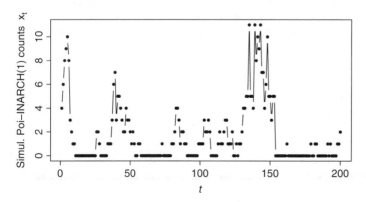

Figure 4.3 Simulated sample path of Poisson INARCH(1) process with $\mu = 3$ and $\alpha = 0.95$; see Remark 4.1.7.

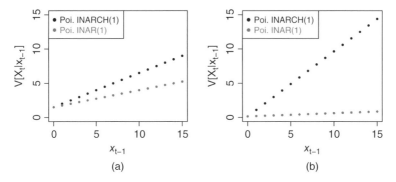

Figure 4.4 Conditional variances of Poisson INAR(1) and INARCH(1) process with $\mu = 3$ and (a) $\alpha = 0.50$, (b) $\alpha = 0.95$; see Remark 4.1.7.

difference is much larger for $\alpha = 0.95$ than for $\alpha = 0.50$. For the Poisson INAR(1) model, the conditional variance shows nearly constant and low values for $\alpha = 0.95$, explaining the overall tendency to produce long runs (see Example 2.1.2.1). In contrast, it quickly tends to large values for the Poisson INARCH(1) model, so runs are observed mainly for zero, with vivid fluctuations otherwise. Further information about the relation between Poisson INAR(1) and INARCH(1) models can be found in Weiß (2015a).

Example 4.1.8 (Strike counts) We analyze the monthly number of work stoppages (strikes and lock-outs) of 1000 or more workers, as published by the US Bureau of Labor Statistics.[1] We restrict ourselves to the period 1994–2002, leading to a time series of length $T = 108$, as was analyzed by Jung et al. (2005) and Weiß (2010b), among others. The plot in Figure 4.5a shows similar fluctuations as in the plotted INARCH(1) sample path in Figure 4.3. An analysis of the SPACF (the SACF is shown in Figure 4.5b) indicates an AR(1)-like autocorrelation structure, with $\hat{\rho}(1) \approx 0.573$. The marginal distribution has mean $\bar{x} \approx 4.944$ and is significantly overdispersed ($\hat{I} \approx 1.587$) according to the test (2.14); also see the plot in Figure 4.5c. So altogether, the INARCH(1) model appears to be a reasonable candidate for the data, but we also consider the Poisson and NB-INAR(1) model as well as the NB-RCINAR(1) model as further candidates (Section 2.5 and Example 3.2.1).

The models' parameters are estimated by a full maximum likelihood approach; see Table 4.2 for a summary of the results. The (equidispersed) Poisson INAR(1) model not only performs worst in terms of AIC and BIC, but also in respect of its Pearson residuals (variance ≈ 1.354) and its

---

1 http://www.bls.gov/wsp/.

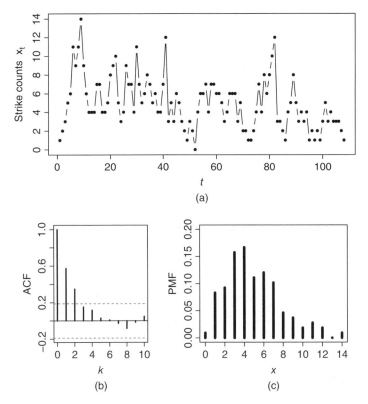

Figure 4.5 Strikes: (a) counts, (b) sample autocorrelation, (c) marginal frequencies; see Example 4.1.8.

U-shaped PIT histogram (see Figure 4.6a; only $J = 5$ intervals are used since the time series is rather short). So models that are able to reproduce the overdispersion are required. The remaining INARCH(1), NB-INAR(1) and NB-RCINAR(1) model are more adequate in these respects, with marginal dispersion indices of 1.408, 1.522 and 1.704, respectively, with the Pearson residuals having variances of 1.026, 0.998 and 1.000, respectively, and with the PIT histograms being close to uniformity (see the INARCH(1) PIT histogram in Figure 4.6b for illustration). The AIC and BIC point to the parsimoniously parametrized INARCH(1) model as being preferred among the candidate models.

The h-step-ahead conditional distributions (conditioned on $x_T = 1$) and the stationary marginal distribution (corresponding to $h = \infty$ due to the ergodicity) for the fitted INARCH(1) model are computed by the MC approximation of Remarks 2.1.3.4 and 2.6.3. The convergence of the conditional distributions is illustrated by Figure 4.7 (increasing darkness for increasing

Table 4.2 Strike counts: ML estimates, AIC and BIC for different models.

Model	Parameter			AIC	BIC
	1	2	3		
Poisson INARCH(1)	1.723	0.643		470	475
(β, α)	(0.382)	(0.080)			
Poisson INAR(1)	2.423	0.503		480	485
(λ, α)	(0.297)	(0.056)			
Negative-binomial INAR(1)	3.473	0.613	0.548	475	484
(n, π, α)	(2.065)	(0.129)	(0.057)		
Negative-binomial RCINAR(1)	9.331	0.657	0.592	473	481
(n, π, α)	(4.259)	(0.105)	(0.062)		

Figures in parentheses are standard errors.

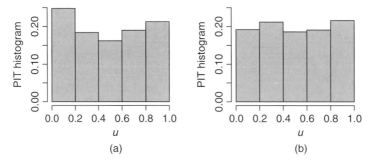

(a)   (b)

Figure 4.6 PIT histograms based on fitted (a) Poisson INAR(1) and (b) INARCH(1) model; see Example 4.1.8.

probability value), where the gray colors in the last column refer to the marginal distribution.

Remark 4.1.9 (Further extensions) As for the thinning-based models (see Remark 3.1.7), the basic INGARCH approach can be extended in several ways. Extensions of the INARCH(1) model to account for trend and seasonality are discussed by Held et al. (2005, 2006). Conditional linear models as before, but with non-Poisson conditional distributions, are presented in Section 4.2, but the *estimating functions approach* described by Thavaneswaran & Ravishanker (2016) should also be mentioned in this context. Models where the linear recursion in Definition 4.1.1(ii) is replaced by a *log-linear* one are discussed

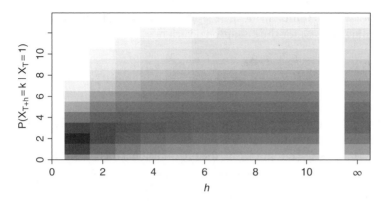

Figure 4.7 h-step-ahead conditional distributions (conditioned on $x_T = 1$) and stationary marginal distribution; see Example 4.1.8.

by Fokianos & Tjøstheim (2011), while Fokianos & Tjøstheim (2012) consider more general *non-linear* autoregressions. Some of these models are also briefly discussed in Section 5.1 in the context of *regression models*.

Keeping the conditionally linear structure but allowing for the inclusion of covariate information, Agosto et al. (2016) proposed extending the INGARCH recursion in part (ii) of Definition 4.1.1 to

$$M_t = \beta_0 + \sum_{i=1}^{p} \alpha_i X_{t-i} + \sum_{j=1}^{q} \beta_j M_{t-j} + h(\boldsymbol{\gamma}, \boldsymbol{Z}_t),$$

where the response function h takes only non-negative real values (also see Section 5.1). For the case where the covariates $\boldsymbol{Z}_t$ are not deterministic but stem from a Markov chain, Agosto et al. (2016) derive conditions for the existence of a stationary solution and analyze the asymptotic properties of the ML estimator.

Finally, a *self-exciting threshold* (SET) extension of the Poisson INGARCH model has been proposed by Wang et al. (2014).

4.2   Further Types of INGARCH Models

The standard INGARCH model with its conditional Poisson distribution exhibits unconditional overdispersion, but the degree of overdispersion is determined by the actual autocorrelation structure (say, $I = 1/(1 - \alpha^2)$ for the Poisson INARCH(1) model from Example 4.1.6). As a consequence, this model was not able to describe the strong volatility of the transactions counts in Example 4.1.5. To overcome this limitation, Xu et al. (2012) proposed the family of *dispersed INARCH models* (DINARCH), which again assume a linear relationship for the conditional mean (see (4.3)), but include an additional (constant) scaling factor η for the conditional variance:

$$E[X_t \mid X_{t-1}, \ldots] = \beta_0 + \alpha_1 X_{t-1} + \ldots + \alpha_p X_{t-p},$$
$$V[X_t \mid X_{t-1}, \ldots] = \eta \cdot E[X_t \mid X_{t-1}, \ldots]. \tag{4.7}$$

So the characteristic feature is a *time-invariant* conditional dispersion index, being equal to η. Obviously, the Poisson INARCH model is an instance of the DINARCH model with $\eta = 1$.

For the case $p = 1$ (see Example 4.1.6), the unconditional mean and variance are given by

$$\mu = \frac{\beta}{1 - \alpha} \quad \text{and} \quad \sigma^2 = \frac{\eta}{1 - \alpha^2} \cdot \mu, \tag{4.8}$$

that is, η allows control of the (unconditional) degree of dispersion independently of α (Xu et al., 2012).

Example 4.2.1 (NB-INGARCH models) As a particular instance of the DINARCH family, Xu et al. (2012) proposed a conditional negative binomial model (see Example A.1.4), where X_t given $X_{t-1}, \ldots$ follows the $NB(n_t, \pi)$ distribution with $n_t = M_t \frac{\pi}{1-\pi}$ and with the conditional mean M_t satisfying (4.3). So the conditional dispersion parameter η is given by $1/\pi$ and fixed over time, while the NB parameter n varies according to the observed past.

A different type of *NB-INARCH model* (even a full NB-INGARCH model) is proposed by Zhu (2011), who assumes the parameter n of the conditional NB distribution to be fixed while π varies with time: X_t given $X_{t-1}, \ldots$ follows the $NB(n, \pi_t)$ distribution with $\pi_t = 1/(1 + M_t/n)$, that is,

$$n \frac{1 - \pi_t}{\pi_t} = M_t := \beta_0 + \sum_{i=1}^{p} \alpha_i X_{t-i} + \sum_{j=1}^{q} \beta_j M_{t-j}.$$

Note that in the original parametrization of Zhu (2011), the parameters $\beta_0, \ldots, \alpha_p$ do not directly refer to the conditional mean but to the odds $(1 - \pi_t)/\pi_t$. But to keep it consistent with Definition 4.1.1, the above parametrization is preferred here.

For Zhu's NB-INGARCH model, the conditional dispersion index varies with time according to $1/\pi_t = 1 + M_t/n$, so this type of NB-INGARCH model differs from the DINARCH approach (4.7).

Comparing Zhu's NB-INGARCH(1, 1) model with the standard INGARCH (1, 1) model from Example 4.1.4, we have the same unconditional mean and ACF, but the unconditional variance now equals (Zhu, 2011):

$$\sigma^2 = \frac{1 - (\alpha_1 + \beta_1)^2 + \alpha_1^2}{1 - (\alpha_1 + \beta_1)^2 - \alpha_1^2/n} \cdot \mu \left(1 + \frac{\mu}{n}\right).$$

Table 4.3 Specific INGARCH models, where the conditional mean $M_t := E[X_t \mid X_{t-1}, \ldots]$ satisfies $M_t = \beta_0 + \sum_{i=1}^{p} \alpha_i X_{t-i} + \sum_{j=1}^{q} \beta_j M_{t-j}$.

Model	Conditional distribution	Conditional dispersion index
Poi. INGARCH	$\mathrm{Poi}(M_t)$	1
NB^{Xu}-INGARCH	$\mathrm{NB}(n_t, \pi)$ with $n_t = M_t \frac{\pi}{1-\pi}$	$1/\pi$
NB^{Zhu}-INGARCH	$\mathrm{NB}(n, \pi_t)$ with $\pi_t = 1/(1 + M_t/n)$	$1 + M_t/n$
GP-INGARCH	$\mathrm{GP}(\lambda_t, \theta)$ with $\lambda_t = (1 - \theta) M_t$	$1/(1 - \theta)^2$
ZIP-INGARCH	$\mathrm{ZIP}(\lambda_t, \omega)$ with $\lambda_t = \frac{1}{1-\omega} M_t$	$1 + \frac{\omega}{1-\omega} M_t$

The variance corresponding to the INARCH(1) case (Example 4.1.6) follows by setting $\beta_1 = 0$.

A brief overview of the different INGARCH models is provided in Table 4.3. Both types of NB-INGARCH model (as well as the GP-INGARCH model to be discussed in the next example) are instances of the *CP-INGARCH model* (see Example A.1.2 about the compound Poisson distribution) introduced by Gonçalves et al. (2015). It is given by

$$\mathrm{pgf}_{X_t \mid X_{t-1}, \ldots}(z) = \exp\left(\frac{M_t}{H_t'(1)} \left(H_t(z) - 1\right) \right), \tag{4.9}$$

where $H_t(z)$ denotes the pgf of the compounding distribution (assumed to be normalized to $H_t(0) = 0$ for uniqueness), which is generally allowed to depend on time t through past observations. If $H_t(z) = H(z)$ is constant in time, then the above condition $\alpha_\bullet + \beta_\bullet < 1$ still guarantees the existence of a strictly stationary and ergodic solution for the CP-INGARCH model, having finite first- and second-order moments (Gonçalves et al., 2015). Further restricting to the case $q = 0$, the resulting CP-INARCH model becomes an instance of the DINARCH model, where $\eta = 1 + H''(1)/H'(1)$.

Example 4.2.2 (GP-INGARCH model) Zhu (2012a) proposed an INGARCH model with a conditional generalized Poisson distribution (Example A.1.6). Here, X_t given the past observations $X_{t-1}, \ldots$ is assumed to follow the $\mathrm{GP}(\lambda_t, \theta)$ distribution with $\lambda_t = (1 - \theta) M_t$, such that the conditional dispersion index, given by $1/(1 - \theta)^2$, is constant in time. Therefore, the *GP-INARCH model* belongs to the DINARCH family (4.7) (with $\eta = 1/(1 - \theta)^2$) such that (4.8) holds for the GP-INARCH(1) case. For the GP-INGARCH(1, 1) model, we have (Zhu, 2012a)

$$\sigma^2 = \frac{1 - (\alpha_1 + \beta_1)^2 + \alpha_1^2}{1 - (\alpha_1 + \beta_1)^2} \cdot \frac{\mu}{(1 - \theta)^2},$$

that is, compared to Example 4.1.4, the unconditional variance is modified by a factor $1/(1-\theta)^2$.

Note that Zhu (2012a) also allows θ to become negative (conditional under-dispersion), but this case has to be considered with caution in view of the problems discussed below (Example A.1.6).

The INGARCH approach also allows generation of zero inflation.

Example 4.2.3 (ZIP-INGARCH model) Zhu (2012b) proposed to use a zero-modified distribution (Example A.1.9) as the conditional distribution for the INGARCH approach. Among others, he considered the *ZIP-INGARCH model* defined by the conditional distribution ZIP(λ_t, ω) with $\lambda_t = \frac{1}{1-\omega} M_t$ (as in Example 4.2.1, we use a slightly different parametrization than in the original proposal by Zhu (2012b)). As a result, the conditional variance is given by

$$V[X_t \mid X_{t-1}, \ldots] = \left(1 + \frac{\omega}{1-\omega} M_t\right) M_t,$$

that is, the ZIP-INARCH model does not belong to the DINARCH class (4.7). Picking up Example 4.1.4 again, the unconditional variance of the ZIP-INGARCH(1, 1) model is obtained as (Zhu, 2012b)

$$\sigma^2 = \frac{1 - (\alpha_1 + \beta_1)^2 + \alpha_1^2}{1 - (\alpha_1 + \beta_1)^2 - \frac{\omega}{1-\omega} \alpha_1^2} \cdot \mu \left(1 + \frac{\omega}{1-\omega} \mu\right).$$

The variance corresponding to the INARCH(1) case (Example 4.1.6) follows by setting $\beta_1 = 0$.

Example 4.2.4 (Transactions counts) Let us continue Example 4.1.5 about the transactions counts. Since zeros are observed quite seldom, the ZIP-INGARCH model is not plausible for the data, but both types of NB- and the GP-INGARCH(1, 1) model (Examples 4.2.1 and 4.2.2) are reasonable candidate models (also see Zhu (2012c)). CML estimation is done by analogy to Example 4.1.5, and the results are summarized in Table 4.4.

Note that the estimates for $\beta_0, \beta_1, \alpha_1$ are very similar for all models. Furthermore, any of the INGARCH(1, 1) models with additional dispersion leads to a considerable improvement compared to the Poisson INGARCH(1, 1) model from Example 4.1.5. For instance, the dispersion indices of these models are 2.931, 3.029 and 2.961, respectively, and the variances of their Pearson residuals are 1.035, 1.049 and 1.022, respectively. Furthermore, all PIT histograms are reasonably close to uniformity; see Figure 4.2b as an example. A decision on one of these models is difficult; the maximized log-likelihood suggests the GP-INGARCH(1, 1) model.

Finally, let us have a look at the case of counts having the finite range $\{0, \ldots, n\}$ with some fixed upper limit $n \in \mathbb{N}$ (see also the discussion in

Table 4.4 Transactions counts: CML estimates for different models, together with ℓ_{max}.

Model	Parameter				$\hat{m}_1$	ℓ_{max}
	1	2	3	4		
Poi. INGARCH(1, 1)	0.292	0.832	0.139		12.058	−1430
$(\beta_0, \beta_1, \alpha_1)$	(0.100)	(0.023)	(0.018)			
NB^{Xu}-INGARCH(1, 1)	0.295	0.836	0.134	0.444	12.939	−1328
$(\beta_0, \beta_1, \alpha_1, \pi)$	(0.145)	(0.034)	(0.027)	(0.030)		
NB^{Zhu}-INGARCH(1, 1)	0.270	0.845	0.127	7.861	12.038	−1329
$(\beta_0, \beta_1, \alpha_1, n)$	(0.142)	(0.034)	(0.026)	(0.959)		
GP-INGARCH(1, 1)	0.293	0.838	0.132	0.338	13.099	−1327
$(\beta_0, \beta_1, \alpha_1, \theta)$	(0.144)	(0.034)	(0.026)	(0.023)		

Figures in parentheses are standard errors.

Section 3.3). None of the above models can be used in such a situation, since the respective conditional distribution has an unbounded range.

Example 4.2.5 (Binomial INARCH(1) model) A version of the INARCH(1) model suitable for finite-valued counts was proposed by Weiß & Pollett (2014). For their *binomial INARCH(1) model*, they assume

$$X_t \mid X_{t-1}, X_{t-2}, \ldots \sim \text{Bin}\left(n, \beta + \alpha \frac{X_{t-1}}{n}\right), \tag{4.10}$$

where $\beta, \beta + \alpha \in (0; 1)$ has to be satisfied. Analogous to the binomial AR(1) model from Section 3.3, this gives a stationary, ergodic and ϕ-mixing Markov chain, but now with simple binomial 1-step-ahead transition probabilities:

$$p_{k|l} = \binom{n}{k} \left(\beta + \alpha \frac{l}{n}\right)^k \left(1 - \beta - \alpha \frac{l}{n}\right)^{n-k}. \tag{4.11}$$

The conditional mean and variance are obtained from the conditional binomial distribution as

$$E[X_t \mid X_{t-1}] = \alpha \cdot X_{t-1} + n\beta,$$
$$V[X_t \mid X_{t-1}] = -\frac{\alpha^2}{n} \cdot X_{t-1}^2 + \alpha(1 - 2\beta) \cdot X_{t-1} + n\beta(1 - \beta), \tag{4.12}$$

that is, in contrast to (3.22) for the binomial AR(1) model, the conditional variance is now a quadratic function in X_{t-1}. Unconditional mean and variance are given by (Weiß & Pollett, 2014):

$$\mu = \frac{n\beta}{1-\alpha}, \quad \sigma^2 = \frac{\mu\left(1 - \frac{\mu}{n}\right)}{1 - \left(1 - \frac{1}{n}\right)\alpha^2}, \quad \text{that is,} \quad I_{\text{Bin}} = \frac{1}{1 - \left(1 - \frac{1}{n}\right)\alpha^2}.$$

(4.13)

Note that the binomial index of dispersion I_{Bin} (see the definition in (2.3)) can only take values in $[1; n)$. So, analogous to the case of the Poisson INARCH(1) model from Example 4.1.6, but in contrast to the binomial AR(1) model, we observe extra-binomial variation, the degree of which is determined through the autocorrelation parameter α. The autocorrelation function is given by $\rho(k) = \alpha^k$. Note that α and hence $\rho(k)$ might also take negative values, which is in contrast to the case of the INARCH(1) models. Another difference to the Poisson INARCH(1) model from Example 4.1.6 is the fact that the conditional variance in (4.13) is not a linear but a quadratic function in X_{t-1}.

As for the other INARCH(1) models, there are no closed-form expressions available for the stationary marginal distribution or the h-step-ahead conditional distributions with $h \geq 2$. But due to the finite range, and in complete analogy to the case of the beta-binomial AR(1) model (see the discussion in Section 3.3), these can be exactly computed numerically by utilizing the Markov property; see Appendix B.2.1 for details.

Example 4.2.6 (Hantavirus infections) The Robert-Koch-Institut (2016) collects data about cases of notifiable diseases in Germany. With *SurvStat@RKI 2.0*, Robert-Koch-Institut (2016) offers a web interface that allows retrieval of data from their disease database. Here, we shall follow an application presented by Weiß & Pollett (2014) and analyze some data about infections by the hantavirus, which is mainly carried by rodents.[2] According to Heyman et al. (2009), hemorrhagic fever with renal syndrome, caused by the hantavirus and with a mortality rate of up to 12%, affects tens of thousands of individuals each year in Europe, and numbers of human cases are rising, perhaps because of mild winters. As an indicator of the regional spread of hantavirus infections, we consider the weekly number x_t of territorial units (out of $n = 38$ territorial units according to the "NUTS Level 2") with at least one new case of a hantavirus infection. As in Weiß & Pollett (2014), we restrict ourselves to the 2011 data ($T = 52$ weeks). Note, however, that we consider updated data (data status at 7 January 2016: two of the counts have been increased by 1 in the meantime); that is, the later results are slightly different from the ones reported by Weiß & Pollett (2014).

---

2 A factsheet about the virus by the European Centre for Disease Prevention and Control can be found at: http://ecdc.europa.eu/en/healthtopics/hantavirus/.

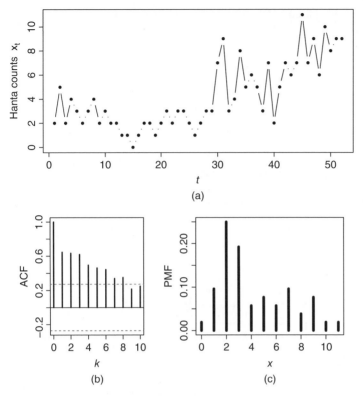

Figure 4.8 Hantavirus reports: (a) counts, (b) sample autocorrelation, (c) marginal frequencies. See Example 4.2.6.

The plot of the time series $x_1, \ldots, x_{52}$ in Figure 4.8a and the pmf plot in Figure 4.8c show that the counts do not exhaust the full range $\{0, \ldots, 38\}$: there are at most 11 territorial units in a week with new cases of hantavirus infections. The mean equals $\bar{x} \approx 4.212$, and the dispersion test (3.30) uncovers a significant degree of extra-binomial variation: $\hat{I}_{\text{Bin}} \approx 2.047$ with P value $< 10^{-4}$. The SACF in Figure 4.8b exhibits a medium autocorrelation level, with $\hat{\rho}(1) \approx 0.645$. Although the SPACF also shows a significant value at lag 2, we shall first see if an AR(1)-like model suffices to describe the data. So we fit the binomial INARCH(1) model from Example 4.2.5 to the data, and the (beta-)binomial AR(1) model discussed in Section 3.3 for comparison. Full ML estimates and the corresponding information criteria are summarized in Table 4.5.

The binomial and the beta-binomial AR(1) model not only perform worst in terms of AIC and BIC; an analysis of the respective Pearson residuals and the PIT histogram shows that these models are not adequate for the data.

Table 4.5 Hanta counts: ML estimates, AIC and BIC for different models.

Model	Parameter			AIC	BIC
	1	2	3		
Binomial AR(1)	0.112	0.539		226	230
(π, ρ)	(0.013)	(0.070)			
Beta-binomial AR(1)	0.114	0.570	0.027	221	227
(π, ρ, ϕ)	(0.017)	(0.073)	(0.015)		
Binomial INARCH(1)	0.030	0.734		215	219
(β, α)	(0.011)	(0.103)			

Figures in parentheses are standard errors.

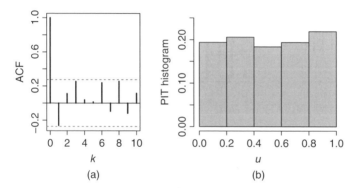

(a) (b)

Figure 4.9 Hanta counts, see Example 4.2.6: (a) SACF of Pearson residuals and (b) PIT histogram, both based on fitted binomial INARCH(1) model.

In contrast, the Pearson residuals of the fitted binomial INARCH(1) model (variance ≈ 1.096, SACF in Figure 4.9a) and its PIT histogram in Figure 4.9b show that this model does rather well. In particular, the residuals' SACF in Figure 4.9a does not suggest a need to use a higher-order model, although the SPACF of the original time series was significant at lag 2. The marginal distribution of the fitted binomial INARCH(1) model has mean 4.297 and binomial dispersion index 2.104, both being close to the empirical values. An important difference between the three fitted models becomes clear by looking at their conditional variances; see Figure 4.10. The binomial and the beta-binomial AR(1) model show increasing variance with increasing x_{t-1}, whereas the binomial INARCH(1) model has its largest conditional variances in the center of the range $\{0, \dots, 38\}$.

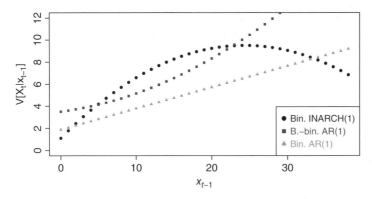

Figure 4.10 Hanta counts for Example 4.2.6: conditional variances of fitted models.

In this context, it is worth looking back to Figure 4.8a: it seems that the counts for $t \geq 30$, having reached a higher level, also show more variation. This phenomenon can be explained by the quadratic conditional variance (4.12); see Figure 4.10 as well as the detailed discussion in Weiß & Pollett (2014). A possible alternative for describing these data could be the *SET binomial INARCH model* as proposed by Möller (2016).

4.3 Multivariate INGARCH Models

While a lot of thinning-based models for multivariate counts have been proposed in the literature – see Section 3.4 for some of these models – little work has been done concerning multivariate extensions of the INGARCH model. A *bivariate Poisson INGARCH(1,1) model* is presented in Chapter 4 of Liu (2012); also see the works by Heinen & Rengifo (2003) and Andreassen (2013). Analogous to Definition 4.1.1, the bivariate counts X_t, conditioned on $X_{t-1}, X_{t-2}, \ldots$, are assumed to be bivariately Poisson distributed (Example A.3.1) according to MPoi($\lambda_0; \lambda_{1,t}, \lambda_{2,t}$), where the conditional mean $M_t := E[X_t \mid X_{t-1}, \ldots]$, with $M_{t,i} = \lambda_0 + \lambda_{i,t}$ for $i = 1, 2$, satisfies

$$M_t = b_0 + A\, X_{t-1} + B\, M_{t-1}, \tag{4.14}$$

where $b_0 \in (0; \infty)^2$, and where A, B are non-negative matrices. Liu (2012) shows that a unique stationary solution for $(M_t)_{\mathbb{Z}}$ given by (4.14) exists if the largest absolute eigenvalue of $A + B$ is smaller than 1, and if $\|A\|_p < 1$ for some $1 \leq p \leq \infty$. Here, the $\| \cdot \|_p$ denotes the induced norm corresponding to the conventional vector p-norm. To guarantee ergodicity, $\|A\|_q + 2^{1-1/q}\,\|B\|_q < 1$ for some $1 \leq q \leq \infty$ is also required. The stationary mean of X_t equals $(I - A - B)^{-1}\, b_0$, and formulae for variance and autocovariance are provided by Heinen & Rengifo (2003). The latter work mainly concentrates on an extension of the Poisson distribution, the so-called *double Poisson distribution*, and it deals with the general multivariate case. In addition, to allow for more flexible cross-correlation, a copula-based approach is presented; see also Andreassen (2013). A type of multivariate INARCH(1) model (expandable by trend and seasonal component) was proposed by Held et al. (2005).

An INARCH model for bivariate counts with range $\{0, \ldots, n\} :=$ $\{0, \ldots, n_1\} \times \{0, \ldots, n_2\}$ was proposed by Scotto et al. (2014). Analogous to (4.10), their *bivariate binomial INARCH(1) model* (BVB_{II}-INARCH(1)-INARCH) assumes the bivariate counts X_t, conditioned on $X_{t-1}, X_{t-2}, \ldots$, to be BVB_{II}-distributed (Example A.3.5) as

$$\text{BVB}_{\text{II}}\left(n_1, n_2, \min\{n_1, n_2\}; \ \beta_1 + \alpha_1 \frac{X_{t-1,1}}{n_1}, \ \beta_2 + \alpha_2 \frac{X_{t-1,2}}{n_2}, \ \phi\right),$$

(4.15)

where $\beta_i, \alpha_i + \beta_i \in (0; 1)$ for $i = 1, 2$, and where

$$\max\left\{-\sqrt{\frac{\beta_1 \beta_2}{(1-\beta_1)(1-\beta_2)}}, \ -\sqrt{\frac{(1-\beta_1-\alpha_1)(1-\beta_2-\alpha_2)}{(\beta_1+\alpha_1)(\beta_2+\alpha_2)}}\right\}$$

$$< \phi < \min\left\{\sqrt{\frac{\beta_1(1-\beta_2-\alpha_2)}{(1-\beta_1)(\beta_2+\alpha_2)}}, \ \sqrt{\frac{(1-\beta_1-\alpha_1)\beta_2}{(\beta_1+\alpha_1)(1-\beta_2)}}\right\}.$$

Scotto et al. (2014) showed that the BVB_{II}-INARCH(1) process constitutes a stationary, ergodic and ϕ-mixing Markov chain with the transition probabilities being determined by (4.15), where the components $(X_{t,i})_{\mathbb{Z}}$ for $i = 1, 2$ are just univariate binomial INARCH(1) processes with parameters α_i, β_i. The cross-covariance function has the form (Scotto et al., 2014):

$$Cov[X_{t,1}, X_{t,2}] = \frac{\phi \min\{n_1, n_2\}}{1 - \alpha_1 \alpha_2} \cdot$$

$$\frac{}{E\left[\sqrt{\prod_{i=1,2} \left(\beta_i + \alpha_i \frac{X_{t,i}}{n_i}\right)\left(1 - \beta_i - \alpha_i \frac{X_{t,i}}{n_i}\right)}\right]}$$

(4.16)

and may take also negative values, depending on the sign of ϕ.

5

Further Models for Count Time Series

The INARMA and INGARCH approaches described above have become very popular in recent years for the modeling of stationary and ARMA-like count processes. But a large number of other count time series models has also been proposed in the literature. Three of these alternatives are presented in this chapter: regression models in Section 5.1, hidden-Markov models in Section 5.2, and NDARMA models in Section 5.3.

5.1 Regression Models

A traditional approach for modeling count data (not just time series data) are *regression models*. The main advantage of regression models is their ability to incorporate covariate information (although also extensions of, for example, the INARMA models have been developed that include covariate information; see Remark 3.1.7). Here, we will review some of these regression models for count time series. Among others, we consider the observation-driven Markov models proposed by Zeger & Qaqish (1988) in Example 5.1.3, which had a groundbreaking effect for research on count time series similar to the work of McKenzie (1985) and Al-Osh & Alzaid (1987) on the thinning-based INAR(1) model (Section 2.1). It will become clear that the INGARCH model discussed in Chapter 4.1 can be understood as an instance of the family of count regression models. A much more detailed discussion of regression models for count time series can be found in Chapter 4 of the book by Kedem & Fokianos (2002); further recent references on this topic are provided by Fokianos (2011) and Tjøstheim (2012).

Let $(X_t)_{\mathbb{Z}}$ be a count process, and let $(Z_t)_{\mathbb{Z}}$ be a vector-valued covariate process (which might also be deterministic). To simplify the discussion, we shall mainly consider the case of a *conditional Poisson distribution* (Example A.1.1), although non-Poisson distributions like the negative binomial distribution (Example A.1.4) have also been considered in the literature. The conditional mean (as the parameter of the conditional Poisson distribution) is assumed

An Introduction to Discrete-Valued Time Series, First Edition. Christian H. Weiss.
© 2018 John Wiley & Sons Ltd. Published 2018 by John Wiley & Sons Ltd.
Companion website: www.wiley.com/go/weiss/discrete-valuedtimeseries

to be "linked" to a linear expression of the available information. Therefore, the considered models are commonly referred to as *generalized linear models* (GLMs) (Kedem & Fokianos, 2002).

Many of these Poisson regression models for count processes are *conditional regression models* in the sense of Fahrmeir & Tutz (2001); that is, they are defined by specifying the conditional distribution of the counts, given the available observations and covariates.

Definition 5.1.1 **(Conditional regression model)** Let $(Z_t)_{\mathbb{Z}}$ be a covariate process. The process $(X_t)_{\mathbb{Z}}$ follows a *conditional (Poisson) regression model* if

(i) X_t, conditioned on $X_{t-1}, \ldots$ and $Z_t, \ldots$, is Poisson distributed according to Poi(M_t), where
(ii) the conditional mean $M_t := E[X_t | X_{t-1}, \ldots, Z_t, \ldots]$ satisfies

$$g(M_t) = \boldsymbol{\theta}^\top \boldsymbol{V}_t$$

with a *link function g* and a parameter vector $\boldsymbol{\theta}$, where the design vector $\boldsymbol{V}_t$ is a function of $X_{t-1}, \ldots$ and $Z_t, \ldots$

The inverse of the link function, $h := g^{-1}$, is referred to as a *response function*: $M_t = h(\boldsymbol{\theta}^\top \boldsymbol{V}_t)$.

Part (i) specifies the *random component* of the model, while (ii) determines the *systematic component* (Kedem & Fokianos, 2002). Note that g (or $h = g^{-1}$, respectively) and the parameter range for $\boldsymbol{\theta}$ have to be chosen such that $h(\boldsymbol{\theta}^\top \boldsymbol{V}_t)$ always leads to a positive value, since the (conditional) mean of a count random variable is necessarily positive.[1] Choosing the *identity link* $g(u) = u$, M_t becomes a linear function in $\boldsymbol{V}_t$, with the INARCH models as discussed in Chapter 4.1 being instances of such conditional regression models with identity link. More generally, models of the form

$$g(M_t) = \boldsymbol{\gamma}^\top \boldsymbol{Z}_t + \sum_{i=1}^{p} \alpha_i \cdot \mathcal{A}(X_{t-i}, \boldsymbol{\gamma}^\top \boldsymbol{Z}_{t-i}) + \sum_{j=1}^{q} \beta_j \cdot \mathcal{M}(X_{t-j}, M_{t-j}) \qquad (5.1)$$

are referred to as *generalized autoregressive moving-average models* (GARMA models) of order (p, q), where $\mathcal{A}$ and $\mathcal{M}$ are functions representing the autoregressive and moving-average terms (Benjamin et al., 2003; Kedem & Fokianos, 2002). This approach not only includes the INGARCH model according to Definition 4.1.1, but many other important models, some of which are briefly presented below.

---

1 The situation would be even more restrictive if the counts X_t had a finite range, for example being described by a conditional binomial distribution. Then an approach like the one for the binary regression models discussed in Section 7.4 could be relevant.

Since the *canonical link* function (also *natural link*) of the Poisson distribution is the *log link* $g(u) = \ln u$, one often considers a *log-linear Poisson model* of the form

$$M_t = E[X_t \mid X_{t-1}, \ldots, Z_t, \ldots] = \exp(\theta^\top V_t), \tag{5.2}$$

that is, where the conditional mean is determined multiplicatively as $M_t = (e^{\theta_1})^{V_1} \cdot (e^{\theta_2})^{V_2} \cdots$. Note that the logarithm is a (bijective and strictly monotonic increasing) mapping between $(0; \infty)$ and $\mathbb{R}$; that is, the right-hand side of (5.2) will always produce a positive value, independent of the parameter range for θ.

Example 5.1.2 **(Poisson GLARMA model)** The *observation-driven Poisson model* proposed by Davis et al. (2003) is related to the GARMA approach (5.1). Let $Y_t := \ln M_t - \gamma^\top Z_t$ and $e_t := (X_t - M_t)/M_t^\lambda$ with $\lambda \in (0; 1]$ (note that $\lambda = \frac{1}{2}$ corresponds to the Pearson residuals). The authors then define the Poisson *GLARMA(p,q) process* (generalized linear ...) by the recursion

$$Y_t = \sum_{i=1}^{p} \alpha_i \, (Y_{t-i} + e_{t-i}) + \sum_{j=1}^{q} \beta_j \, e_{t-j},$$

where the corresponding characteristic polynomials $\alpha(z)$ and $\beta(z)$ are required to have their roots outside the unit circle (analogous to the basic ARMA model according to Definition B.3.5).

As a simple example, Davis et al. (2003) consider the case $\gamma^\top Z_t = \mu$ constantly and $(p, q) = (0, 1)$, that is,

$$\ln M_t = \mu + \beta_1 \, e_{t-1} = \mu + \beta_1 \frac{X_{t-1} - M_{t-1}}{M_{t-1}^\lambda}$$

with $\beta_1 \neq 0$. Davis et al. (2003) emphasize that $(\ln M_t)_{\mathbb{Z}}$ constitutes a Markov chain, while the observations process $(X_t)_{\mathbb{Z}}$ depends on the whole past. They also derive the formulae $E[\ln M_t] = \mu$ and $V[\ln M_t] = \beta_1^2 \, E[M_{t-1}^{1-2\lambda}]$, where the latter reduces to the constant term β_1^2 if $\lambda = \frac{1}{2}$. Davis et al. (2003) show that $(\ln M_t)_{\mathbb{Z}}$ has a stationary solution for $\lambda \in [\frac{1}{2}; 1]$, which is unique for $\lambda = 1$.

Possible properties of such Poisson GLARMA(0, 1) models can be recognized from Figure 5.1, where we set $\lambda = \frac{1}{2}$ (Pearson residuals). Parts (a) and (b) refer to a simulated sample path of length $T = 1000$ for the model parametrization $\mu = 0, \beta_1 = 0.75$. The plot in (a) shows that the model produces sporadic extreme observations ($\bar{x} = 1.40$, $s^2 = 6.64$), and the SACF in (b) is of MA(1)-type with a positive value for $\hat{\rho}(1)$. The SACF in (c), in contrast, refers to a sample path where β_1 is negative, $\beta_1 = -0.75$. The SACF is still of MA(1)-type ($\bar{x} = 1.25$, $s^2 = 1.81$), but with a negative value for $\hat{\rho}(1)$. This is a major difference to the INARMA models from Section 3.1 and

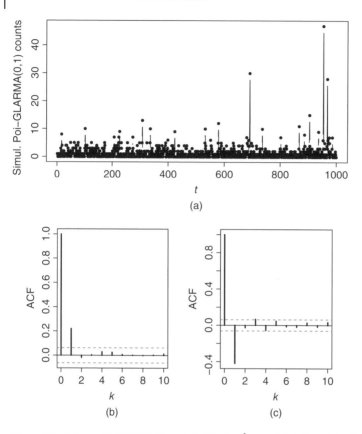

Figure 5.1 Poisson GLARMA(0, 1) model with $\lambda = \frac{1}{2}, \mu = 0$: (a) plot of simulated sample path ($\beta_1 = 0.75, T = 1000$), (b) the corresponding SACF, (c) SACF of another simulated sample path ($\beta_1 = -0.75, T = 1000$). See Example 5.1.2.

to the INGARCH models from Section 4.1, where the ACF can only take non-negative values.

Example 5.1.3 **(Log-linear Poisson autoregression)** Zeger & Qaqish (1988) consider *observation-driven Markov models* of the form

$$g(M_t) = \gamma^\top Z_t + \sum_{i=1}^{p} \alpha_i \cdot f_i(X_{t-1}, \dots, Z_t, \dots),$$

which constitute an instance of the GARMA(p, 0) model according to (5.1). As specific cases, they suggest two models, one following the recursion

$$\ln M_t = \gamma^\top Z_t + \sum_{i=1}^{p} \alpha_i \left(\ln (X_{t-i} + c) - \ln \left(\exp (\gamma^\top Z_{t-i}) + c \right) \right), \qquad (5.3)$$

the other following

$$\ln M_t = \gamma^\top Z_t + \sum_{i=1}^p \alpha_i \left(\ln \tilde{X}_{t-i} - \gamma^\top Z_{t-i}\right), \quad \text{where } \tilde{X}_t := \max\{X_t, c\}.$$

Here, $c > 0$ is a constant that avoids problems with the logarithm if $X_{t-i} = 0$. Both models might be understood as log-linear generalizations of the INARCH(p) model (4.3). This relation is further exploited by Fokianos & Tjøstheim (2011), who define a model by the *log-linear Poisson autoregression*

$$\ln M_t = \beta_0 + \alpha_1 \ln (X_{t-1} + c) + \beta_1 \ln M_{t-1} \quad \text{with } c > 0, \tag{5.4}$$

which constitutes a modification of the Poisson INGARCH(1, 1) model from Example 4.1.4. But while the INGARCH(1, 1) model has an additive structure, the one of the log-linear model (5.4) is multiplicative:

$$M_t = e^{\beta_0} (X_{t-1} + c)^{\alpha_1} M_{t-1}^{\beta_1}.$$

Although it has a feedback mechanism, the model is observation-driven, which follows arguments analogous to Example 4.1.4. The feedback mechanism constitutes a parametrically parsimonious way of creating a long memory.

Fokianos & Tjøstheim (2011) argue that the actual choice of c does not have a strong effect when fitting the model (5.4) to given data, so they recommend to simply set $c := 1$. To allow for consistent ML estimation, Fokianos & Tjøstheim (2011) show that the range of the real-valued parameters $\beta_0, \alpha_1, \beta_1$ has to be restricted by the requirement $|\alpha_1 + \beta_1| < 1$ if α_1, β_1 have the same sign, and by $\alpha_1^2 + \beta_1^2 < 1$ otherwise.

Possible features of the model (5.4) (with $c := 1$) can be studied based on simulated sample paths (of length $T = 10\ 000$). The SACF in Figure 5.2a,

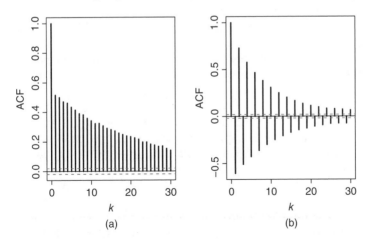

Figure 5.2 SACF for log-linear model (5.4): (a) $(\beta_0, \beta_1, \alpha_1) = (0.05, 0.72, 0.25)$, (b) $(\beta_0, \beta_1, \alpha_1) = (4, -0.25, -0.72)$. See Example 5.1.3.

which corresponds to the model $(\beta_0, \beta_1, \alpha_1) = (0.05, 0.72, 0.25)$ with $\bar{x} = 9.05$ and $s^2 = 14.81$ (the choice of the parameter values was motivated by Example 4.1.5), shows that a slowly decaying and positive-valued autocorrelation structure might be obtained, analogous to the Poisson INGARCH(1, 1) model from Example 4.1.4. But in contrast to this model, negative ACF values are also possible, as illustrated by the SACF shown in Figure 5.2b ($\bar{x} = 8.87$, $s^2 = 46.58$), which was generated based on the parametrization $(\beta_0, \beta_1, \alpha_1) = (4, -0.25, -0.72)$.

Example 5.1.4 **(Non-linear Poisson autoregression)** A further generalization of the Poisson INGARCH(1, 1) model is discussed by Neumann (2011), who defined the *non-linear Poisson autoregression*

$$M_t = f(X_{t-1}, M_{t-1}) \qquad \text{for some } f : \mathbb{N}_0 \times [0; \infty) \to [0; \infty).$$

Here, the function f has to satisfy the contractive condition

$$|f(x, m) - f(\tilde{x}, \tilde{m})| \leq \eta_1 |x - \tilde{x}| + \eta_2 |m - \tilde{m}|$$

to guarantee a stationary solution, where $\eta_1, \eta_2 \geq 0$ and $\eta_1 + \eta_2 < 1$. Furthermore, the process is α-mixing (actually, even β-mixing) with geometrically decreasing weights. An extension of the above approach allowing for non-Poisson conditional distributions is considered by Davis & Liu (2016); see also Christou & Fokianos (2014) for the particular case of a conditional negative binomial distribution.

The model by Neumann (2011) not only includes the INGARCH(1, 1) model as a special case, but also the *exponential autoregressive model*

$$M_t = \beta_0 + \alpha_1 X_{t-1} + \left(\beta_1 + a \exp\left(-b M_{t-1}^2\right)\right) M_{t-1}$$

with $\beta_0 \geq 0$ and $\alpha_1, \beta_1, a, b > 0$ as proposed by Fokianos et al. (2009) ($a = 0$ or $b = 0$ lead to the INGARCH(1, 1) model).

A similar *non-linear Poisson autoregression* is proposed by Fokianos & Tjøstheim (2012),

$$M_t = f_1(X_{t-1}) + f_2(M_{t-1}) \qquad \text{for some } f_1, f_2 : [0; \infty) \to [0; \infty),$$

where f_1, f_2 have to satisfy the regularity conditions given in Fokianos & Tjøstheim (2012) in view of, for example, consistent ML estimation. A particular instance is given by $f_1(x) = \alpha_1 x$ and $f_2(m) = \beta_1 m + \beta_0/(1 + m)^a$, where $a = 0$ leads to the INGARCH(1, 1) model. For $a > 0$, the model becomes truly non-linear.

Remark 5.1.5 **(Partial likelihood)** To estimate the parameters of a conditional regression model by a full likelihood approach (Remark B.2.1.2), knowledge about the distribution of the covariate process would be required, except for the case where the covariates are purely deterministic. A way of circumventing this issue is the *partial likelihood* approach introduced by Cox (1975); see also Wong (1986), Kedem & Fokianos (2002) and Fokianos & Kedem (2004). The idea is to factorize the full likelihood function $L(\theta)$ as

$$L(\theta) = \overbrace{P(X_1 \mid \mathbf{Z}_1, \theta) \cdot \prod_{t=2}^{T} P(X_t \mid X_{t-1}, \dots, \mathbf{Z}_t, \dots, \theta)}^{=: PL(\theta)}$$

$$\cdot \; P(\mathbf{Z}_1 \mid \theta) \cdot \prod_{t=2}^{T} P(\mathbf{Z}_t \mid X_{t-1}, \dots, \mathbf{Z}_{t-1}, \dots, \theta).$$

The first factor, $PL(\theta)$, is referred to as the *partial likelihood function*, and estimates are obtained by maximizing just this partial likelihood. Note that the tth factor of $PL(\theta)$ depends only on data being readily available at time t.

The conditional approach according to Definition 5.1.1 assumes that the count at time t can be explained by the past observations and the covariate information up to time t. For a *marginal (Poisson) regression model* in the sense of Fahrmeir & Tutz (2001), the past observations are without explanatory power provided that the current covariates are given. So the marginal distribution of the counts can be modeled directly. In its basic form, a marginal Poisson regression model requires X_t, conditioned on $\mathbf{Z}_t$, to be Poisson distributed according to $\mathrm{Poi}(M_t)$, where the mean $M_t := E[X_t \mid \mathbf{Z}_t]$ satisfies

$$g(M_t) = \theta^{\top} \mathbf{V}_t \tag{5.5}$$

with the design vector $\mathbf{V}_t$ now being a function of only $\mathbf{Z}_t$. A typical example is the *seasonal log-linear model* being used by Höhle & Paul (2008) for epidemic counts, defined by

$$\ln(M_t) = \gamma_0 + \gamma_1 \, t + \sum_{s=1}^{S} \left(\gamma_{2s} \, \cos(s \, \omega t) + \gamma_{2s+1} \, \sin(s \, \omega t) \right), \tag{5.6}$$

where $\omega = 2\pi/\mathrm{p}$ with period p.

Example 5.1.6 **(Legionnaires' disease infections)** Legionnaires' disease, which often leads to pneumonia for infected persons, with a mortality rate of up to 15%, is caused by Legionella bacteria, which can be found in hot water systems. Infections happen by inhaling droplets of contaminated water; Legionnaires' disease is not spread from person to person.[2] In an analogous way to Example 4.2.6, we consider a count time series obtained from the database *SurvStat@RKI 2.0* of the Robert-Koch-Institut (2016) (data as at 22 January 2016). The counts $x_1, \dots, x_T$ provide the weekly numbers of new infections with Legionnaires' disease in Germany, for the period 2002–2008 ($T = 365$).

---

2 A factsheet for the general public by the European Centre for Disease Prevention and Control is available from: http://ecdc.europa.eu/en/healthtopics/legionnaires_disease/.

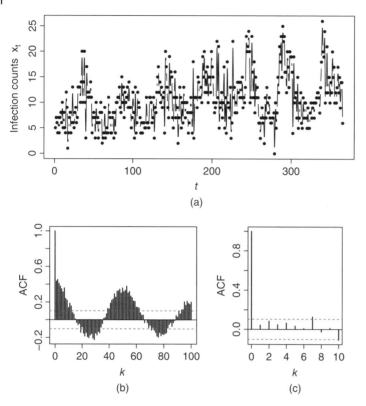

Figure 5.3 Legionnaires' disease: (a) counts, (b) sample autocorrelation, (c) Pearson residuals' SACF (w. r. t. NB model). See Example 5.1.6.

A plot of the data is shown in Figure 5.3a. A seasonal behavior is obvious, but also a slightly increasing trend. The seasonality is also apparent from the ACF in Figure 5.3b, and further inspection of a periodogram confirms that period p = 52 is dominant (referring to the ≈ 52 weeks per year). So we start by fitting a marginal regression model to the data: the seasonal log-linear Poisson model (5.6), where $S = 1$ is chosen (larger values of S did not lead to significant estimates). For improved estimation, the original linear term $\gamma_1 t$ in (5.6) was reparametrized as $\gamma_1 t/T$. The obtained ML estimates are shown in Table 5.1. To check the model adequacy, Pearson residuals are computed. While the residuals' ACF confirms the fitted marginal regression model, their variance of 1.396 indicates overdispersion, thus voting against the Poisson model.

For ease of presentation, we have concentrated on *Poisson* regression models until now, but any other count distribution could also be used for constructing a regression model, for example a zero-inflated Poisson distribution, as in

Table 5.1 Legionnaires' disease counts: ML estimates, AIC and BIC for different models.

Model	Parameter					AIC	BIC
	γ_0	γ_1	γ_2	γ_3	n		
Poi. SLL	2.069	0.478	−0.142	−0.322		2014	2029
$(\gamma_0, ..., \gamma_3)$	(0.035)	(0.057)	(0.023)	(0.024)			
NB SLL	2.068	0.480	−0.138	−0.322	27.957	1995	2014
$(\gamma_0, ..., \gamma_3, n)$	(0.041)	(0.067)	(0.027)	(0.028)	(7.721)		

Figures in parentheses are standard errors.

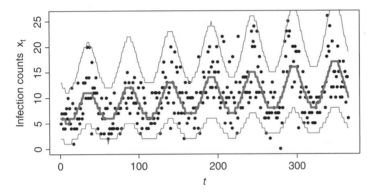

Figure 5.4 Plot of the Legionnaires' disease counts, together with median and 2.5%-/97.5%-quantiles of the fitted NB(n, π_t) distribution; see Example 5.1.6.

Yang et al. (2013). In view of the overdispersion observed for the data, we shall now consider a modified seasonal log-linear model, based on the *negative binomial* distribution (Example A.1.4). Analogous to the NB-INGARCH model by Zhu (2011) (see Example 4.2.1, Höhle & Paul (2008), Davis et al. (2009) and Christou & Fokianos (2014)), we consider the approach $n \frac{1-\pi_t}{\pi_t} := M_t$ (with M_t still following (5.6)); that is, π_t varies with time according to M_t, whereas the NB-parameter n is an additional parameter. As a consequence, the variance at time t equals $M_t (1 + M_t/n)$. The ML-fitted seasonal log-linear NB model in Table 5.1 not only leads to improved AIC and BIC, but also the Pearson residuals confirm the model adequacy (variance 1.021, ACF shown in Figure 5.3c). Note the significantly positive estimate for γ_1; according to the fitted NB model, the mean number of Legionnaires' disease infections increases in time with factor exp $(\frac{0.480}{T} t) \approx 1.001316^t$, corresponding to about a 7.1% increase per year. This is also visible from Figure 5.4, where the median as well as the 2.5%- and 97.5%-quantiles of the fitted model with time-varying marginal distribution NB(n, π_t) are shown.

Example 5.1.7 **(Cryptosporidiosis infections)** As in Example 5.1.6, we consider a time series $x_1, \ldots, x_T$ of weekly counts of new infections in Germany (period 2002–2008, $T = 365$), again taken from the *SurvStat@RKI 2.0* database (Robert-Koch-Institut, 2016, data as at 22 January 2016). But now these counts refer to cryptosporidiosis infections, which cause watery diarrhoea. Cryptosporidiosis is commonly transmitted by infected water or food, but in contrast to Legionnaires' disease, it can also be passed from person to person by direct contact.[3] The plot in Figure 5.5a shows a strong seasonal pattern (again period p = 52) such that the seasonal log-linear Poisson model (5.6) (with linear term $\gamma_1 \, t/T$) is again a reasonable first candidate for the data (now we use $S = 2$; that is, we allow for a half-year effect). An analysis of the corresponding Pearson residuals, however, not only indicates strong overdispersion (variance 3.030), but this time, the residuals also exhibit significant autocorrelations:

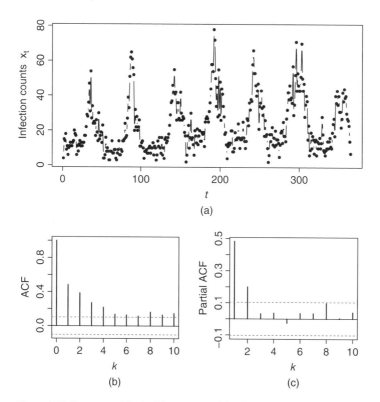

Figure 5.5 Cryptosporidiosis: (a) counts, and for the Pearson residuals' w. r. t. marginal model: (b) the SACF and (c) the SPACF. See Example 5.1.6.

---

3 Factsheet for the general public by the European Centre for Disease Prevention and Control (ECDC): http://ecdc.europa.eu/en/healthtopics/cryptosporidiosis/.

the plots of SACF and SPACF in Figures 5.5b and 5.5c, respectively, indicate an AR(2)-like autocorrelation structure. Therefore, a conditional regression model with an additional autoregressive component appears to be more appropriate for the data. This might be obtained by adding the autoregressive terms directly to the means M_t from (5.6) (as for INARCH models) as suggested by Held et al. (2006). Alternatively, the log-linear autoregressive model (5.3) of Zeger & Qaqish (1988) (Example 5.1.3) can be used. We shall follow the latter approach here; that is, we consider the second-order model defined by

$$
\ln M_t = \overbrace{\gamma_0 + \gamma_1\, t/T + \sum_{s=1}^{2}\, (\gamma_{2s} \cos(s\,\omega t) + \gamma_{2s+1} \sin(s\,\omega t))}^{=:\ \ln \mu_{0,t}(\gamma)}
$$
$$
+ \sum_{i=1}^{2} \alpha_i \left(\ln \left(X_{t-i} + 1 \right) - \ln \left(\mu_{0,t-i}(\gamma) + 1 \right) \right).
$$

In view of the overdispersion, a negative binomial conditional distribution with parametrization $n\ \frac{1-\pi_t}{\pi_t} := M_t$ is used. The obtained CML estimates and corresponding standard errors are displayed in Table 5.2.

Table 5.2 Cryptosporidiosis counts: ML estimates for NB SLL model with additional AR(2) part.

γ_0	γ_1	γ_2	γ_3	γ_4	γ_5	n	α_1	α_2
2.803	0.451	−0.101	−0.630	−0.185	0.039	18.023	0.418	0.130
(0.077)	(0.129)	(0.051)	(0.050)	(0.047)	(0.047)	(2.654)	(0.052)	(0.050)

Figures in parentheses are standard errors.

The Pearson residuals computed for this negative binomial log-linear autoregressive model have variance 1.023, and their SACF indicates an adequate autocorrelation structure. The plot in Figure 5.6, where a graph for the conditional means M_t has been added, shows that the model explains the data quite well.

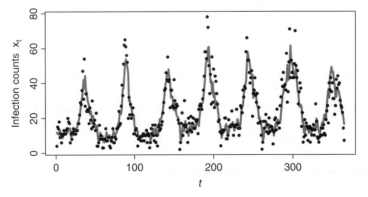

Figure 5.6 Plot of the Cryptosporidiosis counts together with conditional means; see Example 5.1.7.

Similar to the Legionnaires' disease counts from Example 5.1.6, the linear term is significantly positive; that is, the mean number of cryptosporidiosis infections also increases with time according to the model. But the main difference between the model for Legionnaires' disease and the one for cryptosporidiosis infections is the additional autoregressive part (of order 2) for the latter. A possible explanation could be that cryptosporidiosis may spread from person to person (the disease may last up to two weeks); that is, the autoregressive part serves as the "epidemic component" (Held et al., 2006).

An immediate extension of the above marginal models towards parameter-driven models is obtained by assuming an additional latent process – one may also assume that a part of the covariate process is unobservable – say, $(\epsilon_t)_{\mathbb{Z}}$. Then the conditional means defined by $M_t := E[X_t \mid \mathbf{Z}_t, \epsilon_t, \ldots]$ are modeled by the approach in (5.5).

Example 5.1.8 **(Parameter-driven regression models)** A famous instance of such a marginal model is the *parameter-driven regression model* of Zeger (1988). Let $(\epsilon_t)_{\mathbb{Z}}$ be a positive real-valued and weakly stationary process with $E[\epsilon_t] = 1$ and $Cov[\epsilon_t, \epsilon_{t-k}] = \sigma_\epsilon^2 \cdot \rho_\epsilon(k)$. Conditioned on the latent process $(\epsilon_t)_{\mathbb{Z}}$ (and possibly on deterministic covariate information), $(X_t)_{\mathbb{Z}}$ is assumed to be a process of independent counts with

$$E[X_t \mid \mathbf{z}_t, \epsilon_t, \epsilon_{t-1}, \ldots] = V[X_t \mid \mathbf{z}_t, \epsilon_t, \epsilon_{t-1}, \ldots] = \epsilon_t \cdot \exp\left(\boldsymbol{\gamma}^\top \mathbf{z}_t\right),$$

where $\mathbf{z}_t$ is the covariate known at time t. Hence, the time-varying marginal mean and variance follow as

$$\mu_t := E[X_t] = \exp\left(\boldsymbol{\gamma}^\top \mathbf{z}_t\right), \qquad V[X_t] = \mu_t + \sigma_\epsilon^2 \, \mu_t^2,$$

see Zeger (1988), while the ACF is obtained as

$$\rho_t(k) := Corr[X_t, X_{t-k}] = \frac{\rho_\epsilon(k)}{\sqrt{(1 + (\mu_t \, \sigma_\epsilon^2)^{-1})(1 + (\mu_{t-k} \, \sigma_\epsilon^2)^{-1})}}.$$

The model of Zeger (1988) for the case of a conditional Poisson distribution is further discussed in Davis et al. (2000), while a related approach using a conditional negative binomial distribution is proposed by Davis et al. (2009). In particular, Chan & Ledolter (1995) and Davis et al. (2000) consider the special case where ϵ_t follows a lognormal distribution (for identifiability reasons, Davis et al. (2000) set the mean of ϵ_t equal to 1). Defining $\tilde{\epsilon}_t := \ln \epsilon_t \sim N(-\sigma_{\tilde{\epsilon}}^2, \sigma_{\tilde{\epsilon}}^2)$, the model recursion can then be rewritten as

$$X_t \mid \mathbf{z}_t, \epsilon_t, \epsilon_{t-1}, \ldots \sim \text{Poi}\left(\exp\left(\tilde{\epsilon}_t + \boldsymbol{\gamma}^\top \mathbf{z}_t\right)\right).$$

It holds that $\gamma_\epsilon(k) = \exp\left(\gamma_{\tilde{\epsilon}}(k)\right) - 1$; see Davis et al. (2000).

While the stochastic properties of the parameter-driven model of Zeger (1988) are easily obtained, parameter estimation is much more demanding; possible approaches are presented by Zeger (1988) and Davis et al. (2000, 2009).

Parameter-driven models for *multivariate* count processes have been proposed by Jørgensen et al. (1999) and Jung et al. (2011). The state space model by Jørgensen et al. (1999) uses a conditional Poisson distribution and assumes the latent process to be a type of gamma Markov process; the covariate information is embedded with a log-link approach. The dynamic factor model by Jung et al. (2011) generates the components of the d-dimensional counts from conditionally independent Poisson distributions. The corresponding d-dimensional vectors of Poisson means constitute a latent process and are determined by three latent factors (log-linear model), which themselves are assumed to be independent Gaussian AR(1) processes. More information on these and further models for multivariate count time series can be found in the survey by Karlis (2016).

5.2   Hidden-Markov Models

A very popular type of parameter-driven model for count processes is the *hidden-Markov model* (HMM); actually, such HMMs can be defined for any kind of range, even for categorical processes; see Section 7.3. According to Ephraim & Merhav (2002), the first paper about HMMs was the one by Baum & Petrie (1966), who referred to them as "probabilistic functions of Markov chains"; they in fact focussed on the categorical case discussed in Section 7.3. This section gives an introduction to these models (for the count data case), while a much more comprehensive treatment of HMMs is provided by the book by Zucchini & MacDonald (2009); also the survey article by Ephraim & Merhav (2002) is recommended for further reading.

HMMs assume a bivariate process $(X_t, Q_t)_{\mathbb{N}_0}$, where the X_t are the *observable random variables*, whereas the Q_t are the *hidden states* (latent states) with range $Q = \{0, \ldots, d_Q\}$ where $d_Q \in \mathbb{N}$. Note that the numbers $0, \ldots, d_Q$ just constitute a numerical coding of the hidden states, which are assumed to be of categorical nature (possibly not even ordinal; see Chapter 6 for more details). Possible choices for the observations' range are discussed below. The (categorical) state process $(Q_t)_{\mathbb{N}_0}$ is assumed to be a homogeneous Markov chain (Appendix B.2). Given the state process, the observation process $(X_t)_{\mathbb{N}_0}$ is serially independent with its pmf being solely determined through the current state Q_t (in this sense, we are concerned with a "probabilistic function" of a Markov chain; see Baum & Petrie (1966)). A common graphical representation of this data-generating mechanism is shown in Figure 5.7.

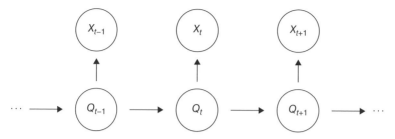

Figure 5.7 Graphical representation of the data-generating mechanism of an HMM.

Remark 5.2.1 (State space models) HMMs are special types of *state space models*. Let $(X_t, Q_t)_{\mathbb{N}_0}$ be a bivariate and discrete-valued process. Then the tth conditional probability splits into

$$P(X_t, Q_t \mid X_{t-1}, \dots, Q_{t-1}, \dots)$$
$$= P(X_t \mid X_{t-1}, \dots, Q_t, \dots) \cdot P(Q_t \mid X_{t-1}, \dots, Q_{t-1}, \dots). \tag{5.7}$$

The process $(X_t, Q_t)_{\mathbb{N}_0}$ follows a *(generalized) state space model* (Brockwell & Davis, 2016, Section 9.8) if the first conditional probability in (5.7) is assumed to simplify to

$$P(X_t \mid X_{t-1}, \dots, Q_t, \dots) = P(X_t \mid Q_t) \qquad \text{for all } t \in \mathbb{N}_0. \tag{5.8}$$

So the conditional distribution of X_t is completely determined by the current state Q_t (as also assumed for HMMs); Equation 5.8 is referred to as the *observation equation*.

Next, consider the second conditional probability in (5.7). According to Cox (1981), $(X_t, Q_t)_{\mathbb{N}_0}$ is classified as being *parameter-driven* if the following *state equation* holds:

$$P(Q_t \mid X_{t-1}, \dots, Q_{t-1}, \dots) = P(Q_t \mid Q_{t-1}, \dots), \tag{5.9}$$

that is, together with (5.8), the distribution of the observation X_t is determined by the latent states; see also Example 5.1.8. Equation 5.9 also holds for HMMs, where it is further assumed that $P(Q_t \mid Q_{t-1}, \dots) = P(Q_t \mid Q_{t-1})$; that is, the state process $(Q_t)_{\mathbb{N}_0}$ is simply assumed to be a Markov chain (Appendix B.2).

In contrast, $(X_t, Q_t)_{\mathbb{N}_0}$ is classified as being *observation-driven* (Cox, 1981) if the state equation equals

$$P(Q_t \mid X_{t-1}, \dots, Q_{t-1}, \dots) = P(Q_t \mid X_{t-1}, \dots), \tag{5.10}$$

that is, summing out Q_t in (5.7) and (5.8), the distribution of the observation X_t would be solely determined by the past observations $X_{t-1}, \dots$

Let us return to HMMs. These models are defined by two sets of parameters: one determining the distribution of the state process $(Q_t)_{\mathbb{N}_0}$, and another concerning the conditional distribution of the observation X_t given the current state Q_t (*state-dependent distribution*). The state process $(Q_t)_{\mathbb{N}_0}$ is assumed to satisfy the above *state equation* (5.9) and, in addition, to be a homogeneous Markov chain with the *state transition probabilities* being given by

$$P(Q_t = q \mid Q_{t-1} = r) = a_{q|r} \quad \text{for all } q, r \in Q = \{0, \ldots, d_Q\}. \tag{5.11}$$

Let $\mathbf{A} = (a_{q|r})_{q,r}$ denote the corresponding transition matrix. The initial distribution $\boldsymbol{p}_0$ of Q_0 either leads to additional model parameters, or it is determined by a stationarity assumption; that is, $\boldsymbol{p}_0 := \boldsymbol{\pi}$, where $\boldsymbol{\pi}$ satisfies the invariance equation $\mathbf{A}\,\boldsymbol{\pi} = \boldsymbol{\pi}$ (see (B.4)). We shall restrict ourselves to *stationary* HMMs here; that is, $P(Q_t = q) = \pi_q$ for all $q \in Q$ and all $t \in \mathbb{N}_0$.

Concerning the observations, the *observation equation* (5.8) has to hold; that is,

$$P(X_t \mid X_{t-1}, \ldots, Q_t, \ldots) = P(X_t \mid Q_t) \quad \text{for all } t \in \mathbb{N}.$$

These state-dependent distributions are also assumed to be time-homogeneous, say $p(\cdot|q)$ for the states $q \in Q$. So $P(X_t = x \mid Q_t = q) = p(x|q)$ for all t.

In applications, parametric distributions are assumed for the $p(\cdot|q)$. For illustration, we shall mainly focus on the *Poisson HMM*, but any other count model could be used as well, or even different models for different states. As mentioned before, HMMs might be adapted to any kind of range for the observations (whereas the states are always categorical), for example, to continuous-valued cases like $\mathbb{R}$ or to purely categorical cases, as discussed in Section 7.3. The Poisson HMM assumes the distribution of X_t, conditioned on $Q_t = q$, to be the Poisson distribution $\text{Poi}(\lambda_q)$; that is,

$$P(X_t = x \mid Q_t = q) = e^{-\lambda_q} \cdot \frac{\lambda_q^x}{x!}, \tag{5.12}$$

and consequently $E[X_t \mid Q_t = q] = \lambda_q = V[X_t \mid Q_t = q]$. The complete set of model parameters is given by $\lambda_0, \ldots, \lambda_{d_Q} > 0$ and $a_{0|0}, a_{0|1}, \ldots, a_{d_Q|d_Q} \in [0; 1]$, where $\sum_{q=0}^{d_Q} a_{q|r} = 1$ has to hold for all $r \in Q$.

Let us look at some stochastic properties of the resulting observation process $(X_t)_{\mathbb{N}_0}$ (Zucchini & MacDonald, 2009, Section 2.2). Let $\mathbf{P}(x)$, as a function of $x \in \mathbb{N}_0$, denote the diagonal matrices $\mathbf{P}(x) := \text{diag}(p(x|0), \ldots, p(x|d_Q)) \in [0; 1]^{(d_Q+1) \times (d_Q+1)}$. Then the marginal pmf and the bivariate probabilities are computed as

$$P(X_t = x) = \sum_{q \in Q} p(x|q)\, \pi_q = \mathbf{1}^\top \mathbf{P}(x)\, \boldsymbol{\pi},$$

$$P(X_t = x, X_{t-k} = y) = \mathbf{1}^\top \mathbf{P}(x)\, \mathbf{A}^k\, \mathbf{P}(y)\, \boldsymbol{\pi}, \tag{5.13}$$

where $\mathbf{1}$ denotes the vector of ones. To express mean $\mu = E[X_t]$ and variance $\sigma^2 = V[X_t]$, let us introduce the notation $\boldsymbol{\mu} = (\mu_0, \ldots, \mu_{d_Q})^\top$ with

$\mu_q := E[X_t \mid Q_t = q]$, and $\sigma_q^2 := V[X_t \mid Q_t = q]$. Then it follows that

$$\mu = \sum_{q \in Q} \pi_q \, \mu_q = \pi^\top \mu,$$

$$\sigma^2 = \sum_{q \in Q} \pi_q \, (\sigma_q^2 + \mu_q^2) - \mu^2. \tag{5.14}$$

The autocovariance $\gamma(k) = Cov[X_t, X_{t-k}]$ equals

$$\gamma(k) = \sum_{q,r \in Q} (\mathbf{A}^k)_{q,r} \pi_r \, \mu_q \mu_r - \mu^2. \tag{5.15}$$

For the limiting behavior of $\mathbf{A}^k$ for $k \to \infty$, see Remark B.2.2.1 on the Perron–Frobenius theorem.

Example 5.2.2 **(Poisson HMM)** For the *Poisson HMM* (5.12), Equations 5.13–5.15 further simplify. Equation 5.13 implies that the marginal pmf of a Poisson HMM is a mixture of Poisson distributions. In particular, (5.14) simplifies to

$$\mu = \sum_{q \in Q} \pi_q \, \lambda_q, \qquad \sigma^2 = \mu + \sum_{q \in Q} \pi_q \, \lambda_q^2 - \mu^2 \geq \mu,$$

that is, the Poisson HMM is marginally overdispersed; the more diverse the mixed Poisson distributions, the stronger the overdispersion. This is illustrated by Figure 5.8a, where the pmfs of two stationary three-state Poisson HMMs are shown: both have the same state transition matrix (also see Example 7.1.1),

$$\mathbf{A} = \begin{pmatrix} 0.90 & 0.05 & 0.25 \\ 0.05 & 0.80 & 0.05 \\ 0.05 & 0.15 & 0.70 \end{pmatrix} \quad \text{with} \quad \pi = \begin{pmatrix} 0.6 \\ 0.2 \\ 0.2 \end{pmatrix},$$

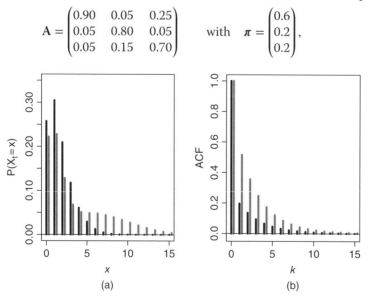

Figure 5.8 Pmf (a) and ACF (b) of two three-state Poisson HMMs; see Example 5.2.2.

but different state-dependent Poisson distributions. The black pmf corresponds to the Poisson means $\lambda_0 = 1, \lambda_1 = 2, \lambda_2 = 3$, and it has mean 1.6 and dispersion ratio 1.4. The gray pmf, in contrast, has the much more distant Poisson means $\lambda_0 = 1, \lambda_1 = 5, \lambda_2 = 9$, which lead to mean 3.4 and to a very strong level of overdispersion, at about 4.012.

For the special case of a *two-state* Poisson HMM (that is, with $d_Q = 1$), mean and variance further simplify to

$$\mu = \pi_0 \lambda_0 + \pi_1 \lambda_1, \qquad \sigma^2 = \mu + \pi_0 \pi_1 (\lambda_0 - \lambda_1)^2.$$

It is also known that the state transition probabilities $a_{q|r}$ for $q, r \in \{0, 1\}$ can be rewritten in the form $a_{q|r} = (1 - \rho) \pi_q + \rho \delta_{q,r}$ with $\pi_0 = 1 - \pi_1$ and with a $\rho \in$ $(\max\{-\frac{\pi_1}{\pi_0}, -\frac{\pi_0}{\pi_1}\}; 1)$; see (7.6). Furthermore, the powers of $\mathbf{A}$ become $(\mathbf{A}^k)_{q,r} = (1 - \rho^k) \pi_q + \rho^k \delta_{q,r}$. Hence, the autocovariance function (5.15) simplifies to

$$\gamma(k) = \rho^k \sum_{q \in Q} \pi_q \lambda_q^2 - \rho^k \mu^2 = \rho^k (\sigma^2 - \mu).$$

So the ACF $\rho(k) = (1 - \mu/\sigma^2) \rho^k$ is exponentially decaying, and the damping effect of the factor $(1 - \mu/\sigma^2)$ decreases with increasing overdispersion (an analogous conclusion holds for the ACFs of the three-state Poisson HMMs shown in Figure 5.8b). Note that the formulae for $\gamma(k)$ and $\rho(k)$ also hold for an arbitrary number $d_Q + 1$ of states, provided that the states follow a DAR(1) model (see Example 7.2.2).

Next, we turn to the question of parameter estimation. A widely used approach is the *Baum–Welch algorithm*, which is an instance of the *expectation-maximization* (EM) algorithm; see Chapter 4 in Zucchini & MacDonald (2009) for a detailed description. Alternatively, a direct (numerical) maximization of the likelihood function can be performed. Provided that accurate starting values have been selected, the latter approach usually converges much faster than the Baum–Welch algorithm; see Bulla & Berzel (2008). Also, MacDonald (2014) concludes that the direct maximization of the likelihood is often advantageous. Therefore, we shall concentrate on this latter approach here.

Remark 5.2.3 **(Likelihood estimation)** The *likelihood function* of a HMM with parameter vector θ (see also Remark B.2.1.2), given the observations $x_1, \dots, x_T$, can be computed as (Zucchini & MacDonald, 2009, Chapter 3)

$$L(\theta) = \mathbf{1}^\top \mathbf{P}(x_T) \, \mathbf{A} \mathbf{P}(x_{T-1}) \, \mathbf{A} \cdots \mathbf{P}(x_1) \, \boldsymbol{\pi}.$$

Obviously, the following recursive scheme holds:

$$\boldsymbol{\alpha}_1 = \mathbf{P}(x_1) \, \boldsymbol{\pi}, \qquad \boldsymbol{\alpha}_t = \mathbf{P}(x_t) \, \mathbf{A} \, \boldsymbol{\alpha}_{t-1}, \qquad L(\theta) = \mathbf{1}^\top \boldsymbol{\alpha}_T.$$

Here, $\boldsymbol{\alpha}_t$ is the vector of *forward probabilities* at time t:

$$\alpha_{t,q} = P(Q_t = q, \ X_t = x_t, \dots, X_1 = x_1) \qquad \text{for all } q \in Q. \tag{5.16}$$

Later, we shall also need the *backward probabilities* (Zucchini & MacDonald, 2009, Section 4.1)

$$\beta_{t,q} = P(Q_t = q, \ X_{t+1} = x_{t+1}, \dots, X_T = x_T) \qquad \text{for all } q \in Q, \tag{5.17}$$

which follow recursively from $\boldsymbol{\beta}_T = \mathbf{1}$ and $\boldsymbol{\beta}_t^\top = \boldsymbol{\beta}_{t+1}^\top \, \mathbf{P}(x_{t+1}) \, \mathbf{A}$.

Once $L(\boldsymbol{\theta})$ has been implemented according to the above scheme, it can be maximized by using a numerical optimization routine; see also the discussion in Remark B.2.1.2. Common asymptotic approximations for the standard errors and distribution of the ML estimator, however, have to be treated with caution. For example, a very large sample size might be required to obtain a reasonable approximation. See the discussion in Section 3.6 of Zucchini & MacDonald (2009).

It should be pointed out that for large T, one may experience numerical underflow when computing $\boldsymbol{\alpha}_t$. For such a case, Zucchini & MacDonald (2009) recommend computing $(w_t/w_{t-1}, \boldsymbol{\phi}_t)$ instead of $\boldsymbol{\alpha}_t$, where $w_t := \mathbf{1}^\top \boldsymbol{\alpha}_t$ and $\boldsymbol{\phi}_t := \boldsymbol{\alpha}_t/w_t$. Note that the recursive scheme $w_t/w_{t-1} \cdot \boldsymbol{\phi}_t = \mathbf{P}(x_t) \, \mathbf{A} \, \boldsymbol{\phi}_{t-1}$ holds. So one computes

$$w_1 = \mathbf{1}^\top \boldsymbol{\alpha}_1, \quad \boldsymbol{\phi}_1 = \boldsymbol{\alpha}_1/w_1;$$

$$\boldsymbol{u}_t := \mathbf{P}(x_t) \, \mathbf{A} \, \boldsymbol{\phi}_{t-1}, \quad \frac{w_t}{w_{t-1}} = \mathbf{1}^\top \boldsymbol{u}_t, \quad \boldsymbol{\phi}_t = \boldsymbol{u}_t \Big/ \frac{w_t}{w_{t-1}} \tag{5.18}$$

for $t = 2, 3, \dots$ The likelihood function is obtained as

$$L(\boldsymbol{\theta}) = w_T = \frac{w_T}{w_{T-1}} \cdots \frac{w_2}{w_1} \cdot w_1.$$

The forward probabilities defined in the Remark 5.2.3 are not only useful in view of likelihood computation, but also for forecasting future observations. The observations' h-step-ahead *forecasting distribution*, given the observations $x_1, \dots, x_t$, is computed as (Zucchini & MacDonald, 2009, Section 5.2):

$$p_{t,h}(x) := P(X_{t+h} = x \mid x_t, \dots, x_1) = \frac{\mathbf{1}^\top \mathbf{P}(x) \, \mathbf{A}^h \, \boldsymbol{\alpha}_t}{\mathbf{1}^\top \boldsymbol{\alpha}_t}. \tag{5.19}$$

Note that these probabilities are easily updated for increasing t according to the recursive scheme in (5.16). Such an updating is also required if residuals are to be computed for the fitted model. While the *forecast pseudo-residuals* (Zucchini & MacDonald, 2009, Section 6.2.3) can be computed exactly using (5.19) with $h = 1$, the *standardized Pearson residuals* (Section 2.4) need to be approximated by computing $E[X_{t+1}^k \mid x_t, \dots, x_1] \approx \sum_{x=0}^{M} x^k \, p_{t,1}(x)$, with M being sufficiently large.

In some applications, it might also be necessary to predict a future state of the HMM; in this case,

$$P(Q_{t+h} = q \mid x_t, \dots, x_1) = \frac{e_q^\top \mathbf{A}^h \, \boldsymbol{\alpha}_t}{\mathbf{1}^\top \, \boldsymbol{\alpha}_t} \tag{5.20}$$

should be used, where e_q is the qth unit vector (Example A.3.3).

Remark 5.2.4 **(Decoding the hidden states)** Schemes for identifying the hidden states, a task commonly referred to as *decoding*, are derived in Section 5.3 of Zucchini & MacDonald (2009). Here, we present brief summaries of these schemes.

Local decoding refers to the identification of the *single* hidden state Q_t, given the observations $x_1, \dots, x_T$. The following approach is based on the forward probabilities (5.16) and backward probabilities (5.17) defined in Remark 5.2.3. Since $\alpha_{t,q} \cdot \beta_{t,q} = P(Q_t = q, X_T, \dots, X_1)$ can be shown to hold, the "most plausible" state at time t is given by

$$\hat{q}_t := \arg\max_q P(Q_t = q \mid x_T, \dots, x_1) = \arg\max_q \frac{\alpha_{t,q} \cdot \beta_{t,q}}{\mathbf{1}^\top \, \boldsymbol{\alpha}_T}.$$

Global decoding refers to the identification of the *complete* series of hidden states $q_1, \dots, q_T$. To find the sequence of states maximizing $P(Q_T = q_T, \dots, Q_1 = q_1 \mid x_T, \dots, x_1)$, the *Viterbi algorithm* can be used. For all $q \in Q$, define the probabilities

$$m_{1,q} := P(Q_1 = q, \, X_1 = x_1) = p(x_1 \mid q) \, \pi_q,$$
$$m_{t+1,q} := \max_{q_1, \dots, q_t} P(Q_{t+1} = q, Q_t = q_t, \dots, Q_1 = q_1,$$
$$X_{t+1} = x_{t+1}, \dots, X_1 = x_1),$$

which are computed recursively as

$$m_{t+1,q} = p(x_{t+1} \mid q) \cdot \max_r \{ m_{t,r} \cdot a_{q \mid r} \} \qquad \text{for } t \geq 1.$$

Then the decoded states are obtained as

$$\hat{q}_T := \arg\max_q \{ m_{T,q} \},$$
$$\hat{q}_t := \arg\max_q \{ m_{t,q} \cdot a_{\hat{q}_{t+1} \mid q} \} \qquad \text{for } t = T - 1, \dots, 1.$$

Note that the results of local and global decoding might differ from each other.

Example 5.2.5 **(Download counts)** Let us pick up again the time series of download counts, as discussed in Section 2.5 and Example 3.2.1. These serially dependent and overdispersed counts have been shown to be reasonably described by an NB-INAR(1) or NB-RCINAR(1) model. Now we shall investigate if a Poisson HMM might also be appropriate for these data. Since the number of parameters increases quadratically in the number of states $((d_Q + 1)^2$

parameters if there are $d_Q + 1$ states), we shall first try a two-state Poisson HMM (that is, with $d_Q = 1$). In this case ($\ell_{\max} \approx -548.9$), the ML-estimated Markov model for the state process is

$$\hat{\mathbf{A}}_{\mathrm{ML}} \approx \begin{pmatrix} 0.803 & 0.395 \\ 0.197 & 0.605 \end{pmatrix}, \qquad \hat{\boldsymbol{\pi}}_{\mathrm{ML}} \approx \begin{pmatrix} 0.668 \\ 0.332 \end{pmatrix}.$$

So, for example, the overall probability of being in state 0 is about 66.8%, and the probability for remaining in state 0 even equals about 80.3%. In each of the states 0 and 1, we have a conditional Poisson model, with the estimated means being about 0.995 (std. err. 0.127) in state 0 and 5.267 (std. err. 0.409) in state 1, respectively. So the predominant state 0 corresponds to "low download activity", while the less frequent state 1 might be interpreted as "high download activity".

To check if this simple two-state Poisson HMM is adequate for the data, let us first look at some properties of the fitted model. While mean (≈ 2.415) and ACF ($\approx 0.256, 0.105, \ldots$) of the fitted model are reasonably close to the corresponding sample values (Section 2.5), the marginal variance appears to be slightly too small (dispersion index 2.676 vs. 3.127). A similar observation (now with respect to the conditional variance) is made if looking at the Pearson residuals; these have a variance of about 1.113. Hence, in contrast to the above NB-INAR(1) or NB-RCINAR(1) models, the fitted two-state Poisson HMM is not able to fully explain the observed (conditional) variance. Therefore, let us fit a three-state Poisson HMM ($d_Q = 2$; that is, with nine model parameters) to the data. For the state process, this leads to

$$\hat{\mathbf{A}}_{\mathrm{ML}} \approx \begin{pmatrix} 0.713 & 0.282 & 0.158 \\ 0.245 & 0.599 & 0.502 \\ 0.042 & 0.118 & 0.341 \end{pmatrix}, \qquad \hat{\boldsymbol{\pi}}_{\mathrm{ML}} \approx \begin{pmatrix} 0.473 \\ 0.421 \\ 0.106 \end{pmatrix},$$

where the second and third column do not sum up to exactly 1 because of rounding. For the corresponding state-dependent Poisson distributions, we obtain the estimates $\hat{\boldsymbol{\lambda}} \approx (0.631, 2.979, 8.230)^{\top}$ with approximate standard errors $(0.151, 0.426, 0.917)^{\top}$. So from a practical point of view, we now distinguish between phases of low (state 0), medium (state 1) and high (state 2) download activity. While states 0 and 1 are "inert" in the sense that the respective conditional probabilities $\hat{a}_{0|0}, \hat{a}_{1|1}$ for remaining in the present state are largest among all conditional probabilities $\hat{a}_{q|0}, \hat{a}_{q|1}$, state 2 will most likely change to state 1. As a consequence, we can only expect rather short periods of continuously high download activity. Another major difference is the marginal probabilities $\hat{\boldsymbol{\pi}}_{\mathrm{ML}}$ for these states, with state 2 happening only rarely ($\approx 10.6\%$) and with states 0 and 1 being roughly equiprobable.

The fitted three-state Poisson HMM allows for a refined explanation of the time series, but how does it perform compared to the other models? First, it can be observed that it leads to the maximal log-likelihood, namely $\ell_{\max} \approx -536.2$

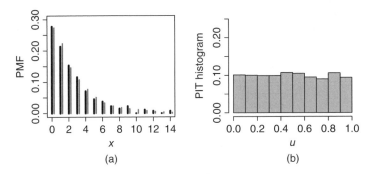

Figure 5.9 Download counts, see Example 5.2.5: (a) marginal frequencies (black) together with PMF of fitted three-state Poisson HMM (gray), and (b) PIT histogram based on this fitted model.

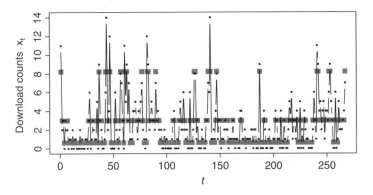

Figure 5.10 Plot of the download counts together with globally decoded states of fitted three-state Poisson HMM (mean $\hat{\lambda}_{q, \text{ML}}$ is shown in gray if state equals q). See Example 5.2.5.

compared to $\ell_{\max} \approx -543.0$ for the NB-INAR(1) model and to $\ell_{\max} \approx -539.4$ for the NB-RCINAR(1) model. Due to the large number of model parameters, however, it has worse values for AIC and BIC: 1090 and 1123, respectively. Let us further investigate the properties of the fitted model. Its marginal pmf is rather close to the empirical one (Figure 5.9a), having the mean ≈ 2.422 and dispersion index ≈ 3.150, and the ACF of the fitted model, $0.274, 0.120, \ldots$, is also reasonable. Analyzing the Pearson residuals, we find no significant auto-correlations, and their variance ≈ 0.969 is close to 1 this time. Finally, the PIT histogram in Figure 5.9b is close to uniformity (see Figure 2.9 for comparison). So, in summary, besides the drawback of a large number of parameters, the three-state Poisson HMM does rather well for the download counts, and it is also easy to interpret.

Related to this last aspect, let us decode the hidden states as described in Remark 5.2.4. The result of a global decoding (using the Viterbi algorithm) is shown in Figure 5.10, where the qth state is represented by the corresponding Poisson mean $\hat{\lambda}_{q,\text{ML}}$. As already conjectured from the estimated transition matrix $\hat{\mathbf{A}}_{\text{ML}}$, we observe long runs of the states 0 and 1, but only short runs or sporadic occurrences of the state 2. If we had done a local decoding for all $t = 1, \ldots, 267$ instead, only three states would have been decoded differently: at times $t = 142, 144, 145$, we would have obtained the state 2 instead of state 1.

Remark 5.2.6 (**Further extensions**) There are a number of ways of generalizing the basic HMM; see Chapter 8 in Zucchini & MacDonald (2009) for a detailed survey. As an example, one may allow the state process $(Q_t)_{\mathbb{N}_0}$ to follow a *higher-order Markov model*, possibly with additional parametric assumptions like those discussed in Section 7.1 (for example, the MTD(p) or DAR(p) models). Other options are the inclusion of covariate information or additional dependencies at the observation level, for example analogues of *Markov-switching autoregressive models*; see also Remark 7.3.5.

Also the more general *state space approach* discussed in Remark 5.2.1 offers a way of obtaining further models for count processes. While we considered the case of a finite and discrete state space for the ease of presentation, one can also allow for, say, a continuous-valued range; see Section 9.8 in Brockwell & Davis (2016) for details. Doing this, the parameter-driven model of Chan & Ledolter (1995) and Davis et al. (2000) according to Example 5.1.8 belongs to the class of generalized state space models, with the lognormally distributed ϵ_t as the latent state at time t. An observation-driven example is the Poisson model proposed by Harvey & Fernandes (1989), where the states are conditionally gamma distributed, given the previous observations.

5.3 Discrete ARMA Models

The "new" *discrete ARMA* (NDARMA) models were proposed by Jacobs & Lewis (1983). They generate an ARMA-like dependence structure through some kind of random mixture. There are several ways of formulating these models, for example through a backshift mechanism, as in Jacobs & Lewis (1983), or by using Pegram's operator, as in Biswas & Song (2009). Here, we follow the approach of Weiß & Göb (2008) to give a representation close to the conventional ARMA recursion.

Definition 5.3.1 (**NDARMA model for counts**) Let the observations $(X_t)_{\mathbb{Z}}$ and the innovations $(\epsilon_t)_{\mathbb{Z}}$ be count processes, where $(\epsilon_t)_{\mathbb{Z}}$ is i.i.d. with $P(\epsilon_t = i) = \pi_i$, and where ϵ_t is independent of $(X_s)_{s<t}$. The random mixture is obtained through the i.i.d. multinomial random vectors

$$(\alpha_{t,1}, \ldots, \alpha_{t,\text{p}}, \beta_{t,0}, \ldots, \beta_{t,q}) \quad \sim \quad \text{MULT}(1; \phi_1, \ldots, \phi_\text{p}, \varphi_0, \ldots, \varphi_\text{q}),$$

which are independent of $(\epsilon_t)_{\mathbb{Z}}$ and of $(X_s)_{s<t}$. Then $(X_t)_{\mathbb{Z}}$ is said to be an *NDARMA*(p, q) *process* if it follows the recursion

$$X_t = \alpha_{t,1} \cdot X_{t-1} + \ldots + \alpha_{t,p} \cdot X_{t-p} + \beta_{t,0} \cdot \epsilon_t + \ldots + \beta_{t,q} \cdot \epsilon_{t-q}. \tag{5.21}$$

The cases q = 0 and p = 0 are referred to as a *DAR(p) process* and *DMA(q) process*, respectively.

Note that exactly one out of $\alpha_{t,1}, \ldots, \beta_{t,q}$ becomes 1; all others are equal to 0. Hence the NDARMA recursion (5.21) implies that each observation X_t chooses either one of the past observations $X_{t-1}, \ldots, X_{t-p}$ or one of the past (unobservable) innovations $\epsilon_t, \ldots, \epsilon_{t-q}$. Because of this mechanism, the stationary marginal distribution of X_t is identical to that of ϵ_t; that is, $P(X_t = i) = \pi_i = P(\epsilon_t = i)$, and we always have

$$P(X_t = i \mid X_{t-k} = j) = \pi_i \cdot (1 - \rho(k)) + \delta_{i,j} \cdot \rho(k).$$

The autocorrelations are non-negative and can be determined from the Yule–Walker equations (Jacobs & Lewis, 1983)

$$\rho(k) = \sum_{j=1}^{p} \phi_j \cdot \rho(|k - j|) + \sum_{i=0}^{q-k} \varphi_{i+k} \cdot r(i) \qquad \text{for } k \geq 1, \tag{5.22}$$

where the $r(i)$ satisfy

$$r(i) = \sum_{j=\max\{0,i-p\}}^{i-1} \phi_{i-j} \cdot r(j) + \varphi_i \mathbb{1}(0 \leq i \leq q),$$

which implies $r(i) = 0$ for $i < 0$, and $r(0) = \varphi_0$. While these properties might suggest that the NDARMA models should be very attractive in practice for ARMA-like count processes, they have an important limitation: the sample paths generated by NDARMA processes tend to show long runs (constant segments) of a certain count value. This is illustrated by Figure 5.11, where the plotted sample path differs markedly from the corresponding INAR(1) path in Figure 2.5b and the INARCH(1) path in Figure 4.3. Since these long runs and large jumps between them are a rather uncommon pattern in real count time series, the NDARMA models are rarely used in the count data context, although we shall see in Section 7.2 that they are quite useful when considering *categorical* time series. An important exception is the modeling of *video traffic data* (Tanwir & Perros, 2014), as briefly sketched in the following example.

Example 5.3.2   (Video traffic modeling)   A video can be understood as a sequence of frames, where each frame is displayed for, say, 1/30 of a second. To reduce the amount of data corresponding to a video sequence, many different video compression schemes have been developed, making use of (among other approaches) the fact that successive frames belonging to the same scene

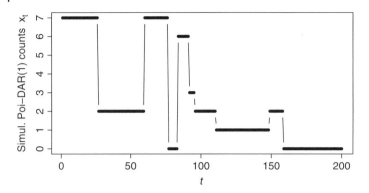

Figure 5.11 Simulated sample path of Poisson DAR(1) process with $\mu = 3$ and $\phi_1 = 0.95$.

are usually very similar to each other. The resulting sequence of frame sizes (say, the number of bytes per frame) is typically characterized by large variation and a strong autocorrelation (also by seasonality caused by regular patterns of so-called I-, B- and P-frames); see Tanwir & Perros (2014) for more details.

Videoconference traffic data is often characterized by high autocorrelation and low motion. As shown by Heyman et al. (1992) and Lazaris & Koutsakis (2010), a DAR(1) model with a (truncated) negative binomial marginal distribution is well-suited to describing *multiplexed* videoconference traffic data (with separate models for I-, B- and P-frames), but may not be appropriate for single traces (according to Lazaris & Koutsakis (2010), at least five traces need to be superpositioned), and not for video sequences with frequent scene changes (Heyman & Lakshman, 1996; Tanwir & Perros, 2014). These results appear plausible in view of the characteristic feature of NDARMA models to produce runs of certain values.

Part II

Categorical Time Series

The time series discussed in Part I had discrete and quantitative ranges, which allowed us to apply standard analytic tools used for real-valued time series analysis: the time series plot, the autocorrelation function, and many more. In the second part of this book, we consider another type of discrete-valued time series where we skip the second of the aforementioned assumptions. In other words, the time series now exhibit a qualitative range consisting of a finite number of categories (including the special case of a binary time series). In some applications, the categorical range exhibits at least a natural ordering; that is, it is ordinal. Otherwise, if not even such an inherent ordering exists, the range is said to be *nominal*. In particular, a nominal range implies a number of difficulties when trying to analyze the time series: completely different measures of dispersion or serial dependence have to be developed, and a visualization of the time series is quite demanding; see Chapter 6.

In addition, the *modeling* of categorical processes requires new approaches; see Chapter 7. The previously discussed INARMA and INGARCH models cannot be applied, but it is possible to adapt NDARMA models to the categorical case, thus offering some kind of counterpart to conventional ARMA models. The serial dependence structure of these discrete ARMA models, which cannot be expressed in terms of the autocorrelation function (since the range is qualitative), shows an ARMA-like behavior if an appropriate measure of serial dependence is taken as a basis. Like the NDARMA models, hidden-Markov models are also easily applied to categorical processes, while regression models require more extensive modifications. Besides these adjustments of models from Chapter 5, tailor-made solutions, such as parsimoniously parametrized Markov models, are also surveyed.

An Introduction to Discrete-Valued Time Series, First Edition. Christian H. Weiss.
© 2018 John Wiley & Sons Ltd. Published 2018 by John Wiley & Sons Ltd.
Companion website: www.wiley.com/go/weiss/discrete-valuedtimeseries

6

Analyzing Categorical Time Series

In Part II of this book, we shall be concerned with *categorical processes* $(X_t)_\mathbb{N}$; that is, the range S of X_t is not only assumed to be discrete, but also to be qualitative, consisting of a finite number $d + 1$ of categories, with $d \in \mathbb{N}$ (*state space*). The time series $x_1, \ldots, x_T$ stemming from this kind of process are referred to as *categorical time series*. In some applications, the range of $(X_t)_\mathbb{N}$ exhibits at least a natural ordering; it is then referred to as an *ordinal* range. In other cases, not even such an inherent order exists (a *nominal* range). In Brenčič et al. (2015), for instance, time series about atmospheric circulation patterns are analyzed. Each day is assigned 1 out of 41 categories, called elementary circulation mechanisms (ECMs); although there are some relationships (similarities) between these categories (for example, they can be arranged in four groups), the categories do not exhibit an inherent ordering; that is, we are concerned with a nominal range. In contrast, Chang et al. (1984) consider time series for daily precipitation and distinguish between dry days, days with medium or with strong precipitation, thus leading to an ordinal time series. Another example of nominal "time" series are nucleotide sequences (a range of four DNA bases) and protein sequences (twenty amino acids) (Churchill, 1989; Krogh et al., 1994; Dehnert et al., 2003), although again similarities exist within these types of nominal range (Taylor, 1986). The time series of electroencephalographic (EEG) sleep states (per minute), as analyzed by Stoffer et al. (2000), are also ordinal time series.

Here, unless stated otherwise, we shall consider the more general case of a nominal range. So even if there is some ordering, we do not make use of it but assume that each random variable X_t takes one of a finite number of *unordered* categories. To simplify notation, we adapt the convention from Appendix B.2 and assume the possible outcomes to be arranged in a certain *lexicographical order*, $S = \{s_0, s_1, \ldots, s_d\}$.

As discussed in the context of Example A.3.3, a categorical random variable X_t can be represented equivalently as a binary random vector $\boldsymbol{Y}_t$, with the range consisting of the unit vectors $\boldsymbol{e}_0, \ldots, \boldsymbol{e}_d \in \{0,1\}^{d+1}$, by defining $\boldsymbol{Y}_t = \boldsymbol{e}_j$

An Introduction to Discrete-Valued Time Series, First Edition. Christian H. Weiss.
© 2018 John Wiley & Sons Ltd. Published 2018 by John Wiley & Sons Ltd.
Companion website: www.wiley.com/go/weiss/discrete-valuedtimeseries

if $X_t = s_j$. We shall sometimes switch to this kind of representation, referred to as a *binarization*, if it allows us to simplify expressions.

6.1 Introduction to Categorical Time Series Analysis

For (stationary) *real-valued* time series, a huge toolbox for analysis and modeling is readily available and well known to a broad audience. To highlight a few basic approaches, the time series are visualized by simply plotting the observed values against time, the marginal properties such as location and dispersion may be measured in terms of mean/median and variance/quartile range, respectively, and serial dependence is commonly quantified in terms of autocorrelation; see also Section 2.3.

Things change if the time series is *categorical*. As an example, since the elementary mathematical operations are not applicable for such a qualitative range, moments like the mean or the autocovariance can no longer be computed. In the *ordinal* case, at least a few methods can be preserved. For example, a time series plot is still feasible by arranging the possible outcomes in their natural ordering along the Y-axis, and the location can be measured by the median (more generally, quantiles and cdf are defined for ordinal data). But in the purely nominal case (as mainly considered here), not even these basic analytic tools are applicable. Therefore, tailor-made solutions are required for visualizing such time series, or for quantifying location, dispersion and serial dependence.

Example 6.1.1 (Ordinal vs. nominal data) In Table 1 of Stoffer et al. (2000)[1] we find a categorical time series of length $T = 107$, which expresses the EEG sleep state (per minute) for an infant 24–36 hours after birth. The range is ordinal, with the six possible states (in their natural ordering):

- 'qt' (quiet sleep, trace alternant)
- 'qh' (quiet sleep, high voltage)
- 'tr' (transitional sleep)
- 'al' (active sleep, low voltage)
- 'ah' (active sleep, high voltage)
- 'aw' (awake).

The observed frequencies are as shown in Table 6.1.

So the median equals 'al', while the mode (most frequent category) is given by 'qt'. The time series plot shown in Figure 6.1 has a meaningful interpretation, since the six states are arranged in their natural ordering along the *y*-axis. For a purely nominal range, in contrast, any arrangement of the states along the

---

1 See www.stat.pitt.edu/stoffer/specrev.pdf.

Table 6.1 Frequency table of infant EEG sleep states data.

State	qt	qh	tr	al	ah	aw
Absolute frequency	33	3	12	27	32	0

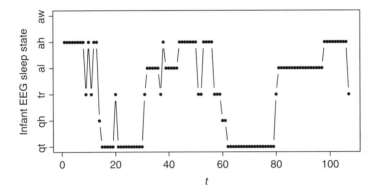

Figure 6.1 Time series plot of infant EEG sleep states (per minute); see Example 6.1.1.

ordinate would be arbitrary and hence misleading. Different approaches are required for a visual representation.

In the sequel, when calling a categorical process $(X_t)_\mathbb{N}$ stationary, we refer to the concept of *strict stationarity* according to Definition B.1.3. While specific models for such stationary categorical processes are discussed in Chapter 7, the particular instance of an i.i.d. categorical process will be of importance here, since it constitutes the benchmark when trying to uncover serial dependence.

Example 6.1.2 (Rate evolution graph) As already emphasized in Example 6.1.1, the widely-used time series plot cannot be applied to a purely nominal time series in a meaningful way. There are several proposals for a *visual analysis* of a nominal time series; see the survey by Weiß (2008d). Although none of them seems to be a perfect substitute for the time series plot, the *rate evolution graph* as suggested by Ribler (1997) is at least an easily implemented visual tool that can be used for stationarity analysis. If $y_1, \ldots, y_T$ denotes the binarization of the available time series, then component-wise graphs of the cumulated sums $c_t := \sum_{s=1}^{t} y_s$ – that is, the component series $c_{t,i}$ for $i = 0, \ldots, d$ – are plotted in one graph against time t. The slope of the graphs is an estimate for the corresponding marginal probability. If the process is stationary, then the graphs should be approximately linear in t, while visible violations of linearity indicate non-stationarity.

Table 6.2 Frequency table of wood pewee data.

State	1	2	3
Absolute frequency	691	357	279

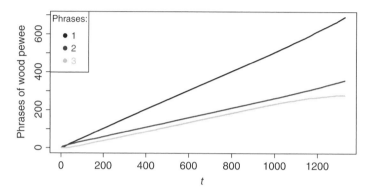

Figure 6.2 Rate evolution graph of wood pewee data; see Example 6.1.2.

For illustration, let us consider a time series referring to the morning twilight song of the wood pewee, a North American song bird famous for its great vocal abilities. The time series data (length $T = 1327$) are printed in Table 12 of Raftery & Tavaré (1994).[2] The data date back to Craig (1943) – apart from a few deviations, they correspond to Record 9 given there – and were analyzed afterwards by several authors, including Raftery & Tavaré (1994) and Berchtold (2002). The wood pewee song is composed of three different phrases, labeled '1', '2' and '3' (Craig, 1943, p. 21):

- The gliding phrases:
  '1' "pee-ah-wee"
  '2' "pee-oh"
- The rhythmic phrase:
  '3' "ah-di-dee".

So the range of the time series is of size $d + 1 = 3$. The observed frequencies are as shown in Table 6.2: Hence, the mode (the most frequent phrase) is given by '1'. The rate evolution graph shown in Figure 6.2 indicates a stationary behavior (at least with respect to the marginal distribution), since the three graphs appear roughly linear. Their slopes are computed (via linear regression) at about 0.509 for state '1', 0.265 for '2' and 0.210 for '3', respectively, expressing

---

2 See www.stat.washington.edu/raftery/Research/PDF/tavare1994.pdf.

the overall "rates" (estimated marginal probabilities) for the wood pewee's phrases.

An application leading to a visibly non-linear rate evolution graph is presented by Brenčič et al. (2015), who analyzed a time series about atmospheric circulation patterns. Other tools for visually analyzing a categorical time series, such as the IFS (iterated function systems) circle transformation (Weiß, 2008d), look for the occurrence of patterns; that is, the occurrence of tuples ("strings") $s \in S^r$ or of sets of such tuples. A comprehensive survey of tools for visualizing time series data in general (not restricted to the categorical case) is provided by Aigner et al. (2011).

Remark 6.1.3 (Frequency-domain analysis) At this point, it is worth mentioning the so-called *spectral envelope* developed by Stoffer et al. (1993, 2000), see also Section 7.9 in the textbook by Shumway & Stoffer (2011). The idea is to look at different numerical codings (called *scalings*) of the categorical process: for $\beta \in \mathbb{R}^{d+1}$, $\beta^\top Y_t$ represents the coding of the X_t range S by the numbers $\beta_0, \ldots, \beta_d \in \mathbb{R}$. As a simple example, $\beta = (1, 1, 0, \ldots, 0)^\top$ implies that $(X_t)_\mathbb{N}$ is mapped onto a binary process, where '1' occurs if either s_0 or s_1 are observed, and '0' otherwise.

Depending on the particular coding, certain periodicities might be observed in the time series. For a given frequency ω, the idea is now to determine the "most striking" $\beta = \beta(\omega)$ (in some sense). For this purpose, Stoffer et al. (1993, 2000) apply a Fourier transform[3] and compute the spectral density $f(\omega; \beta)$, or a sample version of it for given time series data. If $\sigma^2(\beta)$ denotes the variance of $\beta^\top Y_t$, then $\beta(\omega)$ is chosen to maximize $f(\omega; \beta)/\sigma^2(\beta)$. The corresponding maximal value, or

$$\lambda(\omega) = \sup_{\beta \neq \alpha \cdot 1} \frac{f(\omega; \beta)}{\sigma^2(\beta)}$$

to be more precise, is called the *spectral envelope* of the process $(X_t)_\mathbb{N}$. So $\lambda(\omega)$ expresses the maximal proportion of the variance that can be explained by the frequency ω, and this maximal proportion is reached if the *optimal scaling* $\beta(\omega)$ is used. More details on the computation of $\lambda(\omega)$ and on corresponding sample versions $\hat{\lambda}(\omega)$ can be found in Stoffer et al. (1993, 2000) and Shumway & Stoffer (2011). If $\hat{\lambda}(\omega)$ is plotted against ω, a visual frequency analysis of the categorical time series is possible. For illustration, Figure 6.3 shows the spectral envelope of the wood pewee data from Example 6.1.2. The plot was created by adapting Examples 7.17 and 7.18 in Shumway & Stoffer (2011). It can be seen that frequencies around 1/4 and 1/2 are dominant, an observation that will also be plausible in view of our analyses in Example 6.3.1 below.

---

3 A related approach based on the Walsh–Fourier transform was proposed by Stoffer (1987).

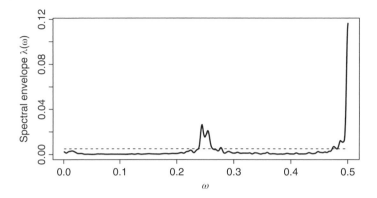

Figure 6.3 Spectral envelope of wood pewee data; see Remark 6.1.3.

6.2 Marginal Properties of Categorical Time Series

Let $(X_t)_{\mathbb{N}}$ be a stationary categorical process with marginal distribution $\pi = (\pi_{s_0}, \dots, \pi_{s_d})^{\top}$. Given the segment $X_1, \dots, X_T$ from this process, we estimate π by the vector $\hat{\pi}$ of relative frequencies computed from $X_1, \dots, X_T$, which is also expressed as $\hat{\pi} = \frac{1}{T} \sum_{t=1}^{T} Y_t$ by using the above binarization of the process. Especially if d is large, the complete (estimated) marginal distribution might be difficult to interpret. So, as with real-valued data, it is necessary in practice to reduce the full information about the marginal distribution into a few metrics that concentrate on features such as location and dispersion.

Measuring the location of a categorical random variable X (or to estimate it from $X_1, \dots, X_T$) is rather straightforward; see also Examples 6.1.1 and 6.1.2. In any case, it is possible to compute "the" (sample) mode, although such a mode is sometimes not uniquely determined. If X is even ordinal, then the median (or any other quantile) can be used to express the "center" of π or $\hat{\pi}$, respectively.

Categorical dispersion is not that obvious in the beginning. Even in the ordinal case, a quantile-based dispersion measure such as the inter quartile range (IQR) is not applicable, since a difference between categories is not defined (one might use the number of categories between the quartiles as a substitute). Therefore, let us first think about the intuitive meaning of dispersion. For a real-valued random variable Z, measures such as variance or IQR ultimately aim at expressing uncertainty. The smaller the dispersion of Z, the better we can predict the outcome of Z. Adapting this intuitive understanding of dispersion to the categorical case, we have maximal dispersion if all probabilities π_j are equal to each other, because then, every outcome is equally probable and a reasonable prediction is impossible. So a *uniform distribution* in S constitutes one extreme of categorical dispersion. At the other extreme, if $\pi_j = 1$ for one

$j \in S$ and 0 otherwise (*one-point distribution*, so π equals one of the unit vectors $e_0, \ldots, e_d$), then we are able to perfectly predict the outcome of X, so X has minimal dispersion in this sense.

Remark 6.2.1 (Categorical dispersion) Note that categorical dispersion as characterized before is just the opposite phenomenon to the concentration of a categorical distribution. Furthermore, this concept also has a meaningful interpretation in the ordinal case. Nevertheless, in the ordinal case, a different approach is also possible. Here, it would also be reasonable to define maximal dispersion as an extreme two-point distribution, where the minimal and the maximal value of the range have probability 0.5 each; see Kiesl (2003) for instance. But we shall not consider the latter approach here.

Now that the extremes of categorical dispersion are known, we can think of dispersion measures v that map these extremes at the extremes of their range. In fact, several measures for this purpose are readily available in the literature; see the survey in Appendix A of Weiß & Göb (2008), for instance. Furthermore, any concentration index can be used as a measure of dispersion.

For the sake of simplicity, we consider measures v with range $[0; 1]$, where 0 refers to minimal dispersion, and 1 to maximal dispersion. Two popular (and, in the author's opinion, quite useful) measures of categorical dispersion are the Gini index and entropy. We define the (sample) *Gini index* as

$$v_{\mathrm{G}} = \frac{d+1}{d}\left(1 - \sum_{j \in S}\pi_j^2\right) \quad \text{and} \quad \hat{v}_{\mathrm{G}} = \frac{d+1}{d}\left(1 - \sum_{j \in S}\hat{\pi}_j^2\right), \qquad (6.1)$$

respectively. The theoretical Gini index v_{G} has range $[0; 1]$, where increasing values indicate increasing dispersion, with the extremes $v_{\mathrm{G}} = 0$ iff X_t has a one-point distribution, and $v_{\mathrm{G}} = 1$ iff X_t has a uniform distribution. The sample Gini index $\hat{v}_{\mathrm{G}}$ is asymptotically normally distributed in the i.i.d. case, and the variance is approximated by $\frac{4}{T}\left(\frac{d+1}{d}\right)^2\left(\sum_{j \in S}\pi_j^3 - \left(\sum_{j \in S}\pi_j^2\right)^2\right)$. Furthermore, although it is a biased estimator of v_{G}, its bias is easily corrected in the i.i.d. case by considering $\frac{T}{T-1}\hat{v}_{\mathrm{G}}$ instead (Weiß, 2011a).

As an alternative, we define the (sample) *entropy* as

$$v_{\mathrm{E}} = \frac{-1}{\ln(d+1)}\sum_{j \in S}\pi_j \ln \pi_j \quad \text{and} \quad \hat{v}_{\mathrm{E}} = \frac{-1}{\ln(d+1)}\sum_{j \in S}\hat{\pi}_j \ln \hat{\pi}_j, \qquad (6.2)$$

respectively, where we always use the convention $0 \cdot \ln 0 := 0$. v_{E} has the same properties as mentioned for the theoretical Gini index. In the i.i.d. case, $\hat{v}_{\mathrm{E}}$ is also asymptotically normally distributed, now with approximate variance $\frac{1}{T}\frac{1}{\ln(d+1)^2}\left(\sum_{j \in S}\pi_j\left(\ln \pi_j\right)^2 - \left(\sum_{j \in S}\pi_j \ln \pi_j\right)^2\right)$, but there is no simple way to exactly correct the bias of $\hat{v}_{\mathrm{E}}$ (Weiß, 2013b).

Example 6.2.2 **(Marginal properties)**   For the (nominal) wood pewee data from Example 6.1.2, we get the point estimates 0.918 and 0.929 for the Gini index and entropy, respectively, both indicating a rather large degree of dispersion. In fact, the estimated marginal distribution $(0.521, 0.269, 0.210)^\top$ is reasonably close to a uniform distribution. For the (ordinal) sleep states data from Example 6.2.2, in contrast, the corresponding estimates 0.886 and 0.791, respectively, indicate less dispersion, which is again also visible from the estimated marginal distribution $(0.308, 0.028, 0.112, 0.252, 0.299, 0.000)^\top$.

If we would like to do a bias correction or compute confidence intervals for the dispersion measures in Example 6.2.2, we would first need to further investigate the serial dependence structure of the available time series, say to establish a possible i.i.d.-behavior such that the above asymptotics could be used. Corresponding tools for measuring serial dependence are presented in the next section.

6.3 Serial Dependence of Categorical Time Series

For the count time series considered in Part I, we simply used the well-known autocorrelation function to analyze the serial dependence structure; see Section 2.3. But this function is not defined in the categorical case (neither nominal nor ordinal), so different approaches are required. Before presenting particular measures, let us again start with some more general thoughts. As for the autocorrelation function, we shall look at pairs (X_t, X_{t-k}) with $k \geq 1$ from the underlying stationary categorical process. If, after having observed X_{t-k}, it is possible to perfectly predict X_t, then it would be plausible to refer to X_t and X_{t-k} as perfectly dependent. If, in contrast, knowledge about X_{t-k} would not help in these respects, then X_t and X_{t-k} would seem to be independent.

To translate this intuition into formulae, let us introduce the notation $p_{ij}(k) := P(X_t = i, X_{t-k} = j)$ with $i, j \in S$ for the lagged bivariate probabilities, with the sample counterpart $\hat{p}_{ij}(k)$ being the relative frequency of (i, j) within the pairs $(X_{k+1}, X_1), \ldots, (X_T, X_{T-k})$. Using the binarization, we can express the latter as $(\hat{p}_{ij}(k))_{i,j \in S} = \frac{1}{T-k} \sum_{t=k+1}^{T} Y_t\, Y_{t-k}^\top$; that is, $\hat{p}_{s_i s_j}(k) = \frac{1}{T-k} \sum_{t=k+1}^{T} Y_{t,i}\, Y_{t-k,j}$. The corresponding conditional bivariate probabilities are denoted as $p_{i|j}(k) := P(X_t = i \mid X_{t-k} = j) = p_{ij}(k)/\pi_j$ for $i, j \in S$; see also Appendix B.2. To avoid computational difficulties, we assume that all marginal probabilities are truly positive ($\pi_j > 0$ for all $j \in S$); otherwise, we would first have to reduce the state space.

Following Weiß & Göb (2008), we now say that:

- we have perfect *(unsigned) serial dependence* at lag $k \in \mathbb{N}$ iff for any $j \in S$, the conditional distribution $p_{\cdot|j}(k)$ is a one-point distribution

- we have perfect *serial independence* at lag $k \in \mathbb{N}$ iff $p_{ij}(k) = \pi_i \pi_j$ for any $i, j \in S$ (or, equivalently, if $p_{i|j}(k) = \pi_i$).

The term "unsigned" was used above for the following reason: the autocorrelation function may take positive or negative values, hence being a signed measure, and positive autocorrelation implies, amongst other things, that large values tend to be followed by large values (and vice versa). This motivates us to introduce an analogous concept of *signed* categorical dependence, where positive dependence implies that the process tends to stay in the state it has reached (and vice versa). So again following Weiß & Göb (2008), and given that we have already established perfect serial dependence at lag k (in the unsigned sense above), we now say that

- we even have perfect *positive* serial dependence iff all $p_{i|i}(k) = 1$, or
- we even have perfect *negative* serial dependence iff all $p_{i|i}(k) = 0$.

The latter implies that X_t necessarily has to take a state other than X_{t-k}. A number of measures of *unsigned* serial dependence have been proposed in the literature so far (Dehnert et al., 2003; Weiß & Göb, 2008; Biswas & Song, 2009; Weiß, 2013b). We shall consider one such measure here, namely *Cramer's v*, where the selection is motivated by the attractive properties of the theoretical v as well as of the sample version $\hat{v}$ of this measure. It is defined by

$$v(k) := \sqrt{\frac{1}{d} \sum_{i,j \in S} \frac{(p_{ij}(k) - \pi_i \pi_j)^2}{\pi_i \pi_j}} \quad \text{and}$$

$$\hat{v}(k) := \sqrt{\frac{1}{d} \sum_{i,j \in S} \frac{(\hat{p}_{ij}(k) - \hat{\pi}_i \hat{\pi}_j)^2}{\hat{\pi}_i \hat{\pi}_j}}, \quad \text{respectively.} \tag{6.3}$$

$v(k)$ has the range $[0; 1]$, where the boundaries 0 and 1 are reached iff we have perfect serial independence/dependence at lag k. The distribution of its sample counterpart $\hat{v}(k)$, in the case of an underlying i.i.d. process, is asymptotically approximated by a χ^2-distribution (Weiß, 2013b): $T d \, \hat{v}^2(k) \underset{a}{\sim} \chi_{d^2}^2$.

This relationship is quite useful in practice, since it allows us to uncover significant serial dependence. If the null of serial independence at lag k is to be tested on (approximate) level α, and if $\chi_{d^2;\,1-\alpha}^2$ denotes the $(1 - \alpha)$-quantile of the $\chi_{d^2}^2$-distribution, then we will reject the null if $\hat{v}(k) > \sqrt{\frac{1}{T d} \chi_{d^2;\,1-\alpha}^2}$. This critical value can also be plotted into a graph of $\hat{v}(k)$ against k, as a substitute for the ACF plot familiar from real-valued time series analysis; see also Remark 2.3.1. On the other hand, this asymptotic result also shows that $\hat{v}(k)$ is generally a biased estimator of $v(k)$.

As a measure of *signed* serial dependence, we consider the (sample) *Cohen's κ*

$$\kappa(k) = \frac{\sum_{j \in S} \left(p_{jj}(k) - \pi_j^2 \right)}{1 - \sum_{j \in S} \pi_j^2} \quad \text{and} \quad \hat{\kappa}(k) := \frac{\sum_{j \in S} \left(\hat{p}_{jj}(k) - \hat{\pi}_j^2 \right)}{1 - \sum_{j \in S} \hat{\pi}_j^2}. \quad (6.4)$$

The range of $\kappa(k)$ is given by $\left[-\frac{\sum_{j \in S} \pi_j^2}{1 - \sum_{j \in S} \pi_j^2} ; 1 \right]$, where 0 corresponds to serial independence, with positive (negative) values indicating positive (negative) serial dependence at lag k. For the i.i.d. case, Weiß (2011a) showed that $\hat{\kappa}(k)$ is asymptotically normally distributed, with approximate mean $-1/T$ and variance $\frac{1}{T} (1 - (1 + 2 \sum_{j \in S} \pi_j^3 - 3 \sum_{j \in S} \pi_j^2)/(1 - \sum_{j \in S} \pi_j^2)^2)$. So there is only a small negative bias, which is easily corrected by adding $1/T$ to $\hat{\kappa}(k)$, and the asymptotic result can again be applied to test for significant dependence.

Example 6.3.1 **(Serial dependence plots)** Let us have a look again at the wood pewee time series from Example 6.1.2. On level $\alpha = 0.05$, we want to test for serial dependence in the data. Instead of evaluating the results numerically, we will draw a "serial dependence plot" analogous to the common plots of the sample ACF. With $T = 1327$ and $d = 2$, it follows that the critical value for $\hat{v}(k)$ equals about 0.060. For $\hat{\kappa}(k)$, we also require information about the marginal distribution to compute the asymptotic distribution. So we plug $\hat{\pi}$ instead of π into the above asymptotic formula and obtain the two-sided critical values -0.040 and 0.038. The resulting plots are shown in Figure 6.4. Both indicate strong significant serial dependencies. As a consequence, returning to the discussion at the end of Section 6.2, confidence intervals for v_G or v_E could not be constructed with the asymptotics given there, since these rely on an i.i.d. assumption. Besides the serial dependencies being very strong, regular patterns can also be observed. $\hat{v}(k)$ shows larger values at even lags k than at the adjacent odd lags. An even more complex pattern is observed for $\hat{\kappa}(k)$, with negative values at odd lags, positive values at even lags, and with much larger values at lags of the form $k = 4l$ than $k = 4l - 2$. These periodicities appear plausible in view of our earlier discussion in Remark 6.1.3. They also match the analyses of Raftery & Tavaré (1994) and Berchtold (2002), who emphasize the repeated occurrence of certain patterns, especially the pattern "1312".

At this point, it is also illuminating to look at the "partial versions" of $\hat{v}(k)$ and $\hat{\kappa}(k)$; that is, at a *partial Cramer's v* and a *partial Cohen's κ*, defined by exactly the same relation as in Theorem B.3.4, but replacing ρ by v or κ; see also the discussion in Section 7.2. For both partial measures, a rather abrupt decline after lag 4 can be observed (see Table 6.3).

This indicates some kind of fourth-order autoregressive dependence, which is in line with a result in Berchtold (2002), where a fourth-order Markov model performed quite well. Many details about Markov models are provided in Section 7.1.

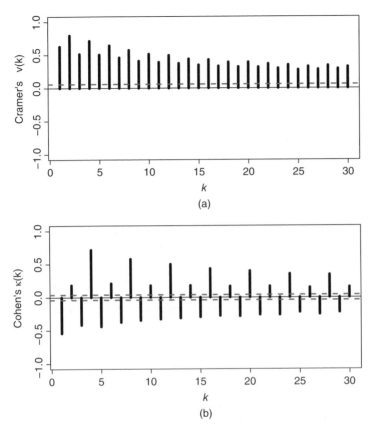

Figure 6.4 Serial dependence plots of wood pewee data based on (a) Cramer's $\hat{v}(k)$, (b) Cohen's $\hat{\kappa}(k)$. See Example 6.3.1.

For completeness, the serial dependence plots for the EEG sleep state data (Example 6.1.1) are also shown, in Figure 6.5, although these data are even ordinal. This specific application illustrates a possible issue with Cramer's v: for short time series (here, we have $T = 107$), it may be that some states are not observed (the state 'aw' in this case). To circumvent division by zero when

Table 6.3 Partial Cramer's v and partial Cohen's κ for wood pewee data.

k	1	2	3	4	5	6	7
$\hat{v}_{part}(k)$	0.626	0.665	−0.207	0.315	0.053	−0.027	0.024
$\hat{\kappa}_{part}(k)$	−0.542	−0.157	−0.564	0.431	0.143	0.137	−0.041

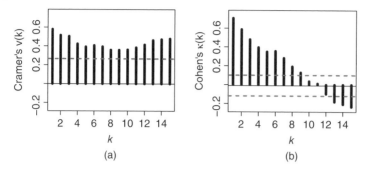

Figure 6.5 Serial dependence plots of EEG sleep state data, based on (a) Cramer's $\hat{v}(k)$, (b) Cohen's $\hat{\kappa}(k)$; see Example 6.1.1.

computing (6.3), all summands related to 'aw' have been dropped while computing $\hat{v}(k)$. The dependence measure $\hat{\kappa}(k)$, in contrast, is robust with respect to zero frequencies.

Let us conclude this section with the special case of a binary process (that is, $d = 1$). If the range of X_t is coded by 0 and 1, then a quantitative interpretation is possible, since each X_t then simply follows a Bernoulli distribution with $\pi_1 = \pi$ and $\pi_0 = 1 - \pi$; see Example A.2.1.

Example 6.3.2 (**Serial dependence for binary processes**) Computing the Gini index (6.1), we obtain $v_G = 2\,(1 - (1 - \pi)^2 - \pi^2) = 4\pi(1 - \pi)$, so except the factor 4, this coincides with the variance of X_t.

For the autocovariance function, in turn, we get $\gamma(k) = E[X_t\,X_{t-k}] - E[X_t]^2 = p_{11}(k) - \pi^2$. Furthermore, $p_{00}(k) = 1 - \pi - p_{01}(k) = 1 - \pi - \pi + p_{11}(k)$, so $p_{00}(k) - (1 - \pi)^2$ equals $p_{11}(k) - \pi^2$. Altogether, this shows that

$$\rho(k) = \frac{\gamma(k)}{\gamma(0)} = \frac{\frac{1}{2}\sum_{j\in\{0,1\}}\left(p_{jj}(k) - \pi_j^2\right)}{\frac{1}{2}\sum_{j\in\{0,1\}}\left(\pi_j - \pi_j^2\right)} = \kappa(k);$$

that is, autocorrelation function and Cohen's κ are identical to each other in the binary case. Analogously, it follows that $v(k) = |\rho(k)|$ in this case.

7

Models for Categorical Time Series

As in Part I, Markov models are very attractive for categorical processes, because of the ease of interpreting the model, making likelihood inferences (see Remark B.2.1.2 in the appendix) and making forecasts. However, without further restrictions concerning the conditional distributions, the number of model parameters becomes quite large. Therefore, Section 7.1 presents approaches for defining parsimoniously parametrized Markov models. One of these approaches is linked to a family of discrete ARMA models, which exhibit an ARMA-like serial dependence structure and allow also for non-Markovian forms of dependence; see Section 7.2. Note that the count data version of this model was discussed in Section 5.3. Two other approaches from Chapter 5, namely hidden-Markov models and regression models, can also be adapted to the categorical case, as described in Sections 7.3 and 7.4, respectively.

7.1   Parsimoniously Parametrized Markov Models

Perhaps the most obvious approach to model a categorical process $(X_t)_{\mathbb{Z}}$ with state space $S = \{s_0, \ldots, s_d\}$ is to use a (homogeneous) pth-order *Markov model*; see (B.1) for the definition:

$$P(X_t = x \mid X_{t-1} = x_{-1}, \ldots) = p(x \mid x_{-1}, \ldots, x_{-p}).$$

The idea of having a limited memory is often plausible in practice, and the Markov assumption is also advantageous, say for parameter estimation (see Remark B.2.1.2) or for forecasting. Concerning the latter application, it is important to note that a pth-order Markov process can always be transformed into a first-order one (for p = 1, we speak of a *Markov chain*; see Appendix B.2) by considering the vector-valued process $(\boldsymbol{X}_t)_{\mathbb{Z}}$ with $\boldsymbol{X}_t := (X_t, \ldots, X_{t-p+1})^{\top}$. And for a Markov chain with transition matrix $\mathbf{P}$, in turn, h-step-ahead transition probabilities are obtained as the entries of the matrix $\mathbf{P}^h$. So, to summarize, h-step-ahead conditional distributions of the form $P(X_{T+h} \mid x_T, \ldots, x_{T-p+1})$ are calculated with relatively little effort for a pth-order Markov process.

An Introduction to Discrete-Valued Time Series, First Edition. Christian H. Weiss.
© 2018 John Wiley & Sons Ltd. Published 2018 by John Wiley & Sons Ltd.
Companion website: www.wiley.com/go/weiss/discrete-valuedtimeseries

Then, *point forecasts* are computed as a mode of this conditional distribution, or, for an ordinal process, as the median. In the ordinal case, it is obvious how to define a *prediction region* on level $\geq 1 - \alpha$: for a two-sided region, the limits are defined based on the $\alpha/2$- and the $(1 - \alpha/2)$-quantile of the conditional distribution, while one uses the α- or $(1 - \alpha)$-quantile, respectively, in case of a one-sided region ("best/worst-case scenario"). If α is very small, however, it may be that one ends up with a trivial region (say, the full range S). The same problem may occur in the nominal case, but here, in addition, a reasonable definition of a prediction region is more difficult because of the lack of a natural ordering. Commonly, one defines a prediction set for X_{T+h} to consist of the most frequent states in S (most frequent with respect to the h-step-ahead conditional distribution) such that the required level $\geq 1 - \alpha$ is ensured.

Example 7.1.1   (Three-state Markov chain)   Let us consider two examples of a three-state Markov chain ($d = 2$, $S = \{s_0, s_1, s_2\}$), which are defined by their transition matrices

$$\mathbf{P}_1 = \begin{pmatrix} 0.90 & 0.05 & 0.25 \\ 0.05 & 0.80 & 0.05 \\ 0.05 & 0.15 & 0.70 \end{pmatrix}, \qquad \mathbf{P}_2 = \begin{pmatrix} 0.10 & 0.05 & 0.60 \\ 0.85 & 0.05 & 0.25 \\ 0.05 & 0.90 & 0.15 \end{pmatrix}.$$

Solving the invariance equation (B.4), we obtain the corresponding stationary marginal distributions as

$$\pi_1 = (0.6, 0.2, 0.2)^\top, \qquad \pi_2 \approx (0.273, 0.345, 0.381)^\top,$$

where the latter does not sum exactly to 1 because of rounding. π_2 is closer to a uniform distribution than π_1; this manifests itself in the Gini dispersions $v_{G,1} = 0.84$ and $v_{G,2} \approx 0.991$, respectively.

The transition matrix $\mathbf{P}_1$ concentrates its probability mass along the diagonal; that is, each state tends to be followed by itself (positive dependence). This is illustrated by the directed graph in Figure 7.1a, where the thickness of the edges represents the size of the respective transition probabilities, and where the positive dependence causes the loops to be dominant. Different behavior is observed for the Markov chain $\mathbf{P}_2$, where the diagonal probabilities are close to zero (negative dependence) and hence the loops in Figure 7.1b are rather thin. The most probable rules are "$s_0 \Rightarrow s_1$", "$s_1 \Rightarrow s_2$" and "$s_2 \Rightarrow s_0$", as can be seen from the dominant edges in Figure 7.1b.

The respective extent of serial dependence is illustrated by the serial dependence plots shown in Figure 7.1, where only Cohen's κ is able to distinguish between the positive and negative dependencies. The positive values of $\kappa(k)$ for $k = 3, 6, \ldots$ in part (d) are reasonable, since the above rules imply a probable return to the starting state after three time units.

The computation of the serial dependence measures (6.3), (6.4) for a Markov chain with transition matrix $\mathbf{P}$ and stationary marginal distribution π, as in

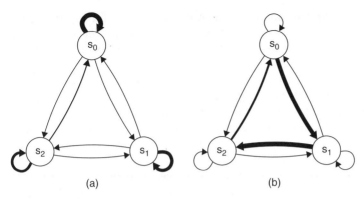

Figure 7.1 Visual representation of two three-state Markov chains (a) $\mathbf{P}_1$ and (b) $\mathbf{P}_2$; see Example 7.1.1.

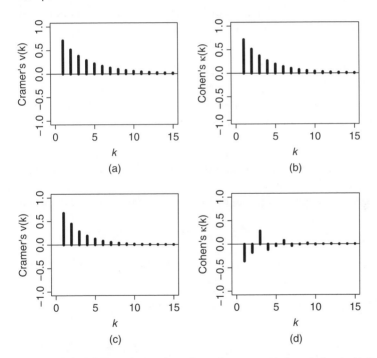

Figure 7.2 Serial dependence plots of two three-state Markov chains: (a, b) $\mathbf{P}_1$ and (c, d) $\mathbf{P}_2$; see Example 7.1.1.

Example 7.1.1, is done by first computing the matrices $\mathbf{P}(k) := (p_{ij}(k))_{i,j}$ of bivariate *joint* probabilities via $\mathbf{P}(k) = \mathbf{P}^k \operatorname{diag}(\pi_0, \ldots, \pi_d)$.

While a full Markov model might be a feasible approach if p is very small, higher orders p will cause a practical problem: a general pth-order Markov process (that is, where the conditional probabilities are not further restricted by parametric assumptions) has a huge number of model parameters, $d (d + 1)^p$, which increases exponentially in p. This problem becomes visible in Figure 7.3, where the possible paths in the past for increasing p are illustrated. Each path $x_{-1}, \dots, x_{-p}$ (out of $(d + 1)^p$ such paths) requires d parameters to be specified to obtain the conditional distribution $p(\cdot \mid x_{-1}, \dots, x_{-p})$, thus leading to altogether $d (d + 1)^p$ parameters. For the EEG sleep state data from Example 6.1.1, for instance, we have $d + 1 = 6$ states such that a first-order Markov model would already require 30 parameters to be estimated (from only $T = 107$ observations available).

To make Markov models more useful in practice, the number of parameters has to be reduced. One suggestion in this direction is due to Bühlmann & Wyner (1999): the *variable-length Markov model* (VLMM). Here, the "embedding Markov model" possibly has a rather large order p, but then several branches of the tree in Figure 7.3 are "cut" by assuming identical conditional distributions; that is, by assuming shortened tuples $(x_{-1}, \dots, x_{-c}) \in S^c$ with $c < p$ such that

$$P(X_t = x \mid X_{t-1} = x_{-1}, \dots, X_{t-p} = x_{-p}) = p(x \mid x_{-1}, \dots, x_{-c}) \qquad (7.1)$$

for all $(x_{-c-1}, \dots, x_{-p}) \in S^{p-c}$ and all $x \in S$. Hence, if the last c observations have been equal to $x_{-1}, \dots, x_{-c}$, the remaining past is negligible. So the order p of the embedding Markov model determines the maximal length of the required memory, but depending on the observed past, a shorter memory may also suffice, which explains the name "variable-length" Markov model. While this kind of parameter reduction is reasonable at a first glance, the number of possible VLMMs is very large, so the model choice is non-trivial. Algorithms for constructing VLMMs have been proposed by Ron et al. (1996).

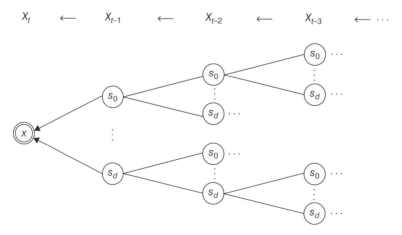

Figure 7.3 Tree structure of Markov model: past observations influencing current outcome.

Another proposal to reduce the number of parameters of the full Markov model aims at introducing parametric relations between the conditional probabilities: the pth-order *mixture transition distribution* (MTD(p)) model of Raftery (1985). The idea is to start with a (full) Markov chain with *transition matrix* $Q = (q_{i|j})_{i,j \in S}$ (Appendix B.2), and to define the pth-order conditional probabilities as a mixture of these transition probabilities:

$$p(x \mid x_{-1}, \dots, x_{-p}) := \sum_{j=1}^{p} \lambda_j \, q_{x|x_{-j}} \qquad \text{for all } x, x_{-1}, \dots, x_{-p} \in S. \qquad (7.2)$$

Here, $\sum_{j=1}^{p} \lambda_j = 1$ is required. A common further restriction is to assume all $\lambda_j \geq 0$, but it would even be possible to allow some λ_j to be negative (Raftery, 1985).

While the general pth-order Markov model has $d\,(d+1)^p$ parameters, this number is now reduced to $d\,(d+1) + p - 1$; that is, we have the number of parameters for a Markov chain (the ones from Q) plus only one additional parameter for each increment of p. Since the conditional probabilities are just some kind of weighted mean of transition probabilities from Q, Raftery (1985) showed that the stationary marginal distribution π for positive Q is simply obtained as the solution of

$$Q \, \pi = \pi, \qquad (7.3)$$

see also (B.4). Furthermore, Raftery (1985) showed that the bivariate probabilities $p_{ij}(k)$ satisfy a set of Yule–Walker-type equations. Denoting $P(k) := (p_{ij}(k))_{i,j}$ for $k \in \mathbb{N}$, and $P(0) := \text{diag}(\pi_0, \dots, \pi_d)$, then

$$P(k) = \sum_{r=1}^{p} \lambda_r \, Q \, P(|k - r|). \qquad (7.4)$$

For $p = 2$, for instance, we immediately obtain that $P(1) = (I - \lambda_2 Q)^{-1} \lambda_1 Q P(0)$, and the remaining $P(k)$s are computed via (7.4). In this way, (7.4) can be used to compute serial dependence measures such as Cramer's $v(k)$ from (6.3) or Cohen's $\kappa(k)$ from (6.4), although simple closed-form results will usually not be available due to the general form of Q. Note that because of (7.3), Equation 7.4 can also be rewritten in a centered version as

$$P(k) - \pi \, \pi^\top = \sum_{r=1}^{p} \lambda_r \, Q \, (P(|k - r|) - \pi \, \pi^\top).$$

Approaches for parameter estimation are discussed by Raftery & Tavaré (1994). For a survey of results for MTD models, see Berchtold & Raftery (2001).

Example 7.1.2 **(Three-state MTD(2) processes)** Let us pick up Example 7.1.1. We define two three-state MTD(2) models by setting $Q_1 := P_1$ and $Q_2 := P_2$, so the stationary marginal distributions remain π_1, π_2 as before;

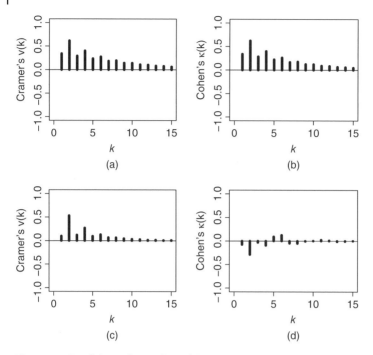

Figure 7.4 Serial dependence plots of three-state MTD(2) models : (a, b) $(\mathbf{Q}_1, \lambda)$ and (c, d) $(\mathbf{Q}_2, \lambda)$; see Example 7.1.2.

see (7.3). We choose $\lambda_1 := \lambda := 0.2$ and $\lambda_2 := 1 - \lambda = 0.8$ such that lag 2 gets a higher weight than lag 1. Comparing the resulting serial dependence plots in Figure 7.4 with those of the corresponding Markov chains (the MTD(1) model) in Figure 7.2, we see increased values for even lags k but reduced ones for odd lags, as to be expected from the chosen weights. Furthermore, especially for model 1, we generally observe increased values for higher lags; that is, the second-order model indeed leads to a longer memory.

A further reduction in the number of parameters is obtained by assuming parametric relations for $\mathbf{Q}$. Jacobs & Lewis (1978c) and Pegram (1980) suggest defining $q_{i|j} := (1 - \lambda)\,\pi_i + \lambda\,\delta_{i,j}$ with $\lambda \in (0; 1)$ such that

$$p(x \mid x_{-1}, \ldots, x_{-p}) = (1 - \lambda)\,\pi_x + \lambda \sum_{j=1}^{p} \lambda_j\,\delta_{x,x_{-j}}. \tag{7.5}$$

Jacobs & Lewis (1978c) require $\lambda_j \geq 0$ and refer to these models as the *discrete autoregressive models* of order p (DAR(p)), while Pegram (1980) even allows some λ_j to be negative (see above). This model has only $d + p$ parameters. We shall provide a more detailed discussion of it in Section 7.2 when considering the extension to a full discrete ARMA model. Also, the discussion of another type of parsimoniously parametrized Markov model, namely the autoregressive logit model, is postponed until later; see Example 7.4.4.

Example 7.1.3 (**Binary Markov chain**) To conclude this section, let us have a look at the special case of a *binary Markov process* (that is, $d = 1$), where we use the coding $S = \{0, 1\}$ for the state space to simplify the notation, as in Example 6.3.2. If the Markov order equals $p = 1$ (so a *binary Markov chain*; see Appendix B.2), then we have only two model parameters. A useful parametrization in the stationary case is

$$\mathbf{P} = \begin{pmatrix} p_{0|0} & p_{0|1} \\ p_{1|0} & p_{1|1} \end{pmatrix} = \begin{pmatrix} (1 - \pi)(1 - \rho) + \rho & (1 - \pi)(1 - \rho) \\ \pi(1 - \rho) & \pi(1 - \rho) + \rho \end{pmatrix}. \quad (7.6)$$

Then $\pi \in (0; 1)$ determines the marginal distribution, $P(X_t = 1) = \pi = 1 - P(X_t = 0)$, and $\rho \in (\max\{\frac{-\pi}{1-\pi}, \frac{1-\pi}{-\pi}\}; 1)$ controls the serial dependence structure, as we have an AR(1)-like autocorrelation function $\rho(k) = \rho^k$. Denoting $\pi_1 := \pi$ and $\pi_0 := 1 - \pi$, the transition probabilities in (7.6) can also be written as $\pi_{i|j} = (1 - \rho)\,\pi_i + \rho\,\delta_{i,j}$, which shows that a binary Markov chain has a DAR(1)-like dependence structure; see (7.5).

With increasing p, not only the number of parameters of a binary Markov model increases rapidly (2^p); in general, we also do not have an AR(p)-like autocorrelation structure anymore. An exception is the binary version of the MTD(p) model. If $\mathbf{Q}$ is assumed to be parametrized as in (7.6) – that is, $q_{i|j} = (1 - \rho)\,\pi_i + \rho\,\delta_{i,j}$ – then it immediately follows that (7.5) holds. This implies an AR(p)-like autocorrelation structure; see Section 7.2 below.

A closely related approach to the binary MTD(p) and DAR(p) models is the binary AR(p) model of Kanter (1975), which was extended to a full binary ARMA(p, q) model by McKenzie (1981) and Weiß (2009d). The model recursion looks somewhat artificial, as it uses addition modulo 2. On the other hand, this definition also allows for negative autocorrelations; see the cited references for further details.

7.2 Discrete ARMA Models

Based on preliminary works (Jacobs & Lewis, 1978a–1978c) on discrete counterparts to the ARMA(1, q) and AR(p) model, respectively, Jacobs & Lewis (1983) introduced two types of discrete counterparts to the full ARMA(p, q) model. In particular the second of these models, referred to as the *NDARMA model* by Jacobs & Lewis (1983), is quite attractive, since its definition and serial dependence structure are very close to those of a conventional ARMA model; see also our discussion of the count data version of this model in

Section 5.3. Let us pick up Definition 5.3.1 and adapt it to the categorical case (Weiß & Göb, 2008).

Definition 7.2.1 **(NDARMA model)** Let the observations $(X_t)_{\mathbb{Z}}$ and the innovations $(\epsilon_t)_{\mathbb{Z}}$ be categorical processes with state space S, where $(\epsilon_t)_{\mathbb{Z}}$ is i.i.d. with marginal distribution π, and where ϵ_t is independent of $(X_s)_{s<t}$. The random mixture is obtained through the i.i.d. multinomial random vectors

$$(\alpha_{t,1}, \ldots, \alpha_{t,p}, \beta_{t,0}, \ldots, \beta_{t,q}) \sim \text{MULT}(1; \phi_1, \ldots, \phi_p, \varphi_0, \ldots, \varphi_q),$$

which are independent of $(\epsilon_t)_{\mathbb{Z}}$ and of $(X_s)_{s<t}$. Then $(X_t)_{\mathbb{Z}}$ is said to be an *NDARMA(p, q) process* (and the cases q = 0 and p = 0 are referred to as a *DAR(p) process* and a *DMA(q) process*, respectively) if it follows the recursion

$$X_t = \alpha_{t,1} \cdot X_{t-1} + \ldots + \alpha_{t,p} \cdot X_{t-p} + \beta_{t,0} \cdot \epsilon_t + \ldots + \beta_{t,q} \cdot \epsilon_{t-q}. \tag{7.7}$$

Here, if the state space S is not numerically coded, we assume $0 \cdot s = 0$, $1 \cdot s = s$ and $s + 0 = s$ for each $s \in S$.

Note that the NDARMA(p, q) model has only $d + p + q$ parameters. Although its recursion is written down in an "ARMA" style, X_t does nothing but choose the state of either X_{t-1} or … or ϵ_{t-q}. So X_t is generated as a random choice from $X_{t-1}, \ldots, \epsilon_{t-q}$ (a random mixture). But as we shall see in (7.9), the ARMA-like notation is indeed adequate and reasonable for NDARMA processes.

With the same argument as in Section 5.3, it follows that X_t and ϵ_t have the same stationary marginal distribution; that is, $P(X_t = i) = \pi_i = P(\epsilon_t = i)$ for all $i \in S$. It is useful to know that an NDARMA model can always be represented by a (p + q)-dimensional finite Markov chain with a primitive transition matrix (Jacobs & Lewis, 1983; Weiß, 2013b). Using this Markov representation, Weiß (2013b) showed that an NDARMA process is ϕ-mixing with exponentially decreasing weights (see Definition B.1.5) such that the central limit theorem (CLT) in Billingsley (1999, p. 200) is applicable. Conditional distributions of an NDARMA process are determined by

$$\begin{aligned} P(X_t = i_0 \mid X_{t-1} = i_1, \ldots, \epsilon_{t-1} = j_1, \ldots) \\ = \varphi_0 \, \pi_{i_0} + \sum_{r=1}^{p} \delta_{i_0, i_r} \, \phi_r + \sum_{s=1}^{q} \delta_{i_0, j_s} \, \varphi_s, \end{aligned} \tag{7.8}$$

which reduces to an expression like the one in (7.5) for the DAR(p) case (q = 0). Only in the latter case, we have Markov dependence (of order p), while an NDARMA(p, q) process $(X_t)_{\mathbb{Z}}$ with $q \geq 1$ is not Markovian, although it can be represented as a (p + q)-dimensional Markov chain.

Let us now investigate the serial dependence structure of an NDARMA (p, q) process $(X_t)_{\mathbb{Z}}$ using the concepts described in Section 6.3. Only positive dependence is possible (implying that NDARMA processes tend to show long runs of their states); that is, $\kappa(k) \geq 0$, and it even holds that $\kappa(k) = v(k)$ (Weiß &

Göb, 2008). These properties can be utilized to identify if an NDARMA model might be appropriate for a set of time series data (by looking at the sample versions $\hat{\kappa}(k)$, $\hat{v}(k)$). For example, the serial dependence measures for the wood pewee time series shown in Figure 6.4 clearly deviate from NDARMA behavior, with negative dependencies and strong deviation between $\hat{v}(k)$ and $\hat{\kappa}(k)$.

Since both measures lead to identical results anyway, let us focus on $\kappa(k)$ in the following. $\kappa(k)$ itself is obtained from a set of Yule–Walker-type equations (Weiß & Göb, 2008):

$$\kappa(k) = \sum_{j=1}^{p} \phi_j \cdot \kappa(|k-j|) + \sum_{i=0}^{q-k} \varphi_{i+k} \cdot r(i) \qquad \text{for } k \geq 1, \tag{7.9}$$

where the $r(i)$ satisfy

$$r(i) = \sum_{j=\max\{0,i-p\}}^{i-1} \phi_{i-j} \cdot r(j) + \varphi_i \; \mathbb{1}(0 \leq i \leq q),$$

which implies $r(i) = 0$ for $i < 0$, and $r(0) = \varphi_0$. Note the analogy to (5.22) for the autocorrelation function in the count data case. The relation between $\kappa(k)$ and the bivariate distributions with lag k is given by

$$p_{i|j}(k) = \pi_i \cdot (1 - \kappa(k)) + \delta_{i,j} \cdot \kappa(k), \tag{7.10}$$

see Weiß & Göb (2008). The equations (7.9) can now be applied in an analogous way to the use of the original Yule–Walker equations for an ARMA process; see Appendix B.3. For a DMA(q) model (p = 0), we have $r(i) = \varphi_i$. Hence, $\kappa(k) = \sum_{i=0}^{q-k} \varphi_i \, \varphi_{i+k}$ such that $\kappa(k)$ vanishes after lag q; see (B.11) for the corresponding MA(q) result. So the model order q can be estimated using $\hat{\kappa}(k)$ or $\hat{v}(k)$, as described in Appendix B.3.

For a DAR(p) model (q = 0), in turn, we have $\kappa(k) = \sum_{j=1}^{p} \phi_j \cdot \kappa(|k-j|)$, which corresponds exactly to the AR(p) result (B.13). Hence, defining the *partial Cohen's* κ (or *partial Cramer's* v) with exactly the same relation as in Theorem B.3.4 (just replacing ρ by κ or v), we obtain a tool for identifying the autoregressive order p: $\kappa_{\text{part}}(k) = v_{\text{part}}(k) = 0$ for lags $k > p$. These results also apply to the binary MTD(p) model; see Section 7.1, which was shown to have an DAR(p) dependence structure, and where $\rho(k) = \kappa(k)$ according to Example 6.3.1.

Example 7.2.2 **(DAR(1) model)** The DAR(1) model constitutes a parsimoniously parametrized Markov chain with $\kappa(k) = \phi_1^k$. Its transition matrix is given by

$$\mathbf{P} = \begin{pmatrix} \pi_0(1-\phi_1)+\phi_1 & \pi_0(1-\phi_1) & \cdots & \pi_0(1-\phi_1) \\ \pi_1(1-\phi_1) & \pi_1(1-\phi_1)+\phi_1 & & \vdots \\ \vdots & & \ddots & \\ \pi_d(1-\phi_1) & \pi_d(1-\phi_1) & \cdots & \pi_d(1-\phi_1)+\phi_1 \end{pmatrix}, \tag{7.11}$$

see (7.8); that is, the diagonal probabilities are increased by ϕ_1 compared to the non-diagonal ones, thus increasing the tendency to stay in the current state (positive dependence). This structure is completely analogous to the transition matrix of a binary Markov chain; see (7.6). Note that the h-step-ahead transition probabilities do not need to be computed from $\mathbf{P}^h$, but they are directly available from (7.10) as $p_{i|j}(h) = \pi_i\,(1 - \phi_1^h) + \delta_{i,j}\,\phi_1^h$. This shows that $p_{i|j}(h) \to \pi_i$ geometrically fast, confirming the ergodicity of the DAR(1) process (Appendix B.2.2).

Remark 7.2.3 (Derived processes) If $(X_t)_{\mathbb{Z}}$ is a process exhibiting a certain serial dependence structure, and if $(Z_t)_{\mathbb{Z}}$ is derived from it by applying a deterministic function f; that is, $Z_t := f(X_t)$ – for example Z_t equals one component of the binarization Y_t of X_t – then, generally, the serial dependence structure is not preserved. For instance, if $(X_t)_{\mathbb{Z}}$ is a Markov chain and $Z_t := Y_{t,j}$ for some $j \in S$, then $(Z_t)_{\mathbb{Z}}$ is usually not a Markov chain anymore. An exception is the NDARMA process, where, due to the construction (7.7), any binarization $(Y_{t,j})_{\mathbb{Z}}$ shows the same serial dependence structure as the parent process $(X_t)_{\mathbb{Z}}$. An analogous statement also holds for hidden-Markov processes, as discussed in Section 7.3.

Now let us look at the sample measures of dispersion from Section 6.2 and the sample measures of serial dependence from Section 6.3. There, we presented the asymptotic properties of these measures if applied to i.i.d. categorical processes. Now let us analyze what changes if these measures are applied to an underlying NDARMA process.

The asymptotic results for the Gini index (6.1) and the entropy (6.2) are easily adapted. As shown by Weiß (2013b), they are still asymptotically normally distributed, and the i.i.d. variance just has to be inflated by the factor

$$c := 1 + 2 \cdot \sum_{k=1}^{\infty} \kappa(k) < \infty. \tag{7.12}$$

While $c = 1$ in the i.i.d. case, it is given by $c = (1 + \phi_1)/(1 - \phi_1)$ for a DAR(1) model, as an example. So altogether, we have

$$V[\hat{v}_G] \approx \frac{4c}{T} \left(\frac{d+1}{d}\right)^2 \left(\sum_{j \in S} \pi_j^3 - \left(\sum_{j \in S} \pi_j^2\right)^2\right),$$

$$V[\hat{v}_E] \approx \frac{c}{T} \frac{1}{\ln(d+1)^2} \left(\sum_{j \in S} \pi_j\,(\ln \pi_j)^2 - \left(\sum_{j \in S} \pi_j \cdot \ln \pi_j\right)^2\right). \tag{7.13}$$

At least for the sample Gini index, an exact bias correction is possible. As shown by Weiß (2013b),

$$E[\hat{v}_\mathrm{G}] = v_\mathrm{G}\left(1 - \frac{1}{T}\cdot\left(1 + 2\sum_{k=1}^{T-1}(1 - \frac{k}{T})\cdot\kappa(k)\right)\right) \geq v_\mathrm{G}\,\frac{T-c}{T}. \quad (7.14)$$

The latter formula provides a simple way to obtain an approximate bias correction, and it is exact in the i.i.d. case.

For the sample serial dependence measures from Section 6.3, asymptotic normality can also be established, and explicit expressions for the asymptotic variances can be derived. But these expressions are much more complex than in the i.i.d. case, so we refer the reader to Weiß (2013b) for further details.

Let us now look at applications of discrete ARMA models. An application of count time series to video traffic count data was sketched in Example 5.3.2. For categorical time series, Chang et al. (1984) and Delleur et al. (1989) used discrete ARMA models for modeling daily precipitation. While Chang et al. (1984) used a three-state model (that is, $d = 2$) representing either dry days, days with medium or with strong precipitation, Delleur et al. (1989) used a two-state model (that is, $d = 1$) to just distinguish between wet and dry days. If the rainfall quantity also has to be considered, Delleur et al. (1989) propose using a state-dependent exponential distribution, analogous to the corresponding approach with HMMs (Sections 5.2 and 7.3). Another field of application of discrete ARMA models is DNA sequence data; see, for example, Dehnert et al. (2003). Returning to this application, let us consider the DNA sequence corresponding to the bovine leukemia virus.

Example 7.2.4 **(Bovine leukemia DNA data)** We consider a "time" series that was analyzed in Weiß & Göb (2008) and Weiß (2013b): the DNA sequence of the bovine leukemia virus, which was published by the National Center for Biotechnology Information (NCBI).[1] It is of length $T = 8419$, and its range consists of the $d + 1 = 4$ DNA bases 'a', 'c', 'g' and 't' (adenine, cytosine, guanine and thymine, respectively). Certainly, such a biological sequence cannot be assumed to be a realization of a stochastic process, but corresponding stochastic models are commonly applied in practice as a tool for summarizing properties of the considered sequence; see Churchill (1989) and Dehnert et al. (2003), for instance.

For the bovine leukemia DNA data, the rate evolution graph shown in Figure 7.5a indicates stationary behavior. The estimated marginal distribution $\hat{\pi} = (0.220, 0.331, 0.210, 0.239)^\top$ is quite close to a uniform distribution (with the mode being 'c'). So it is not surprising that the point estimates 0.988 and 0.987 for Gini index and entropy, respectively, are close to 1, indicating a strong degree of dispersion.

---

1 See http://www.ncbi.nlm.nih.gov/nuccore/NC_001414?%3Fdb=nucleotide.

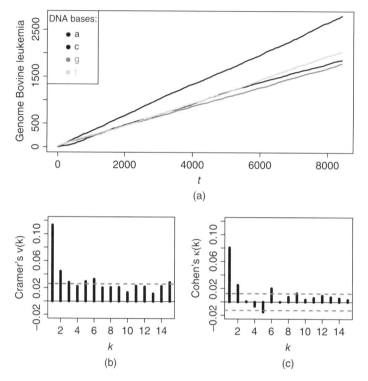

Figure 7.5 Bovine leukemia DNA data: (a) rate evolution graph; serial dependence plots based on (b) Cramer's $\hat{v}(k)$ and (c) Cohen's $\hat{\kappa}(k)$. See Example 7.2.4.

The serial dependence plots shown in Figure 7.5 both exhibit significant serial dependencies, especially for lags 1 and 2, although the absolute extent is rather small, even at lag 1. This, together with an analysis of the corresponding partial $\hat{v}$ and $\hat{\kappa}$, indicates that an AR(1)- or AR(2)-like model might be appropriate for the series. The values of $\hat{v}(k)$ and $\hat{\kappa}(k)$ are also relatively close to each other, and the ones of $\hat{\kappa}(k)$ are mainly positive. So we shall try to fit the DAR(1) and DAR(2) models to the data, as well as the MTD(1) (full first-order Markov model) and MTD(2) model as further candidate models. In any case, final estimates are obtained using the conditional maximum likelihood (CML) approach (see Remark B.2.1.2), and they are computed with a numerical optimization routine.

Let us start with the DAR(1) and DAR(2) models. After obtaining initial estimates based on $\hat{\pi}$ and $\hat{\kappa}(k)$ (the "method of moments"), the CML estimates are computed by maximizing the respective conditional log-likelihood functions $\ell(\theta \mid x_p, \ldots, x_1) = \sum_{t=p+1}^{T} \ln p(x_t \mid x_{t-1}, \ldots, x_{t-p})$ from (B.6). The required transition probabilities are given by (7.8), with p = 1, 2 and q = 0, leading to the CML estimates shown in Table 7.1.

Table 7.1 Bovine leukemia DNA data: CML estimates for DAR(p) models, together with maximized log-likelihood and BIC.

DAR	$\hat{\pi}_{a,\mathrm{CML}}$	$\hat{\pi}_{c,\mathrm{CML}}$	$\hat{\pi}_{g,\mathrm{CML}}$	$\hat{\pi}_{t,\mathrm{CML}}$	$\hat{\phi}_{1,\mathrm{CML}}$	$\hat{\phi}_{2,\mathrm{CML}}$	$\frac{T}{T-p}\ell_{max}$	BIC
$p = 1$	0.220	0.331	0.208	0.241	0.081		$-11\,446$	$22\,927$
$p = 2$	0.219	0.331	0.209	0.241	0.079	0.020	$-11\,440$	$22\,926$

It can be seen that there is just a little difference between the fitted DAR(1) and DAR(2) models, so the more parsimonious DAR(1) model appears to be preferable. It is a Markov chain, the transition matrix of which is easily computed; see Example 7.2.2. We just have to multiply the marginal probabilities $\hat{\pi}_{\cdot,\mathrm{CML}}$ by $1 - \hat{\phi}_{1,\mathrm{CML}}$ and to increase the diagonal elements by $\hat{\phi}_{1,\mathrm{CML}}$, leading to

$$\hat{\mathbf{P}}_{\mathrm{DAR}(1)} \approx \begin{pmatrix} 0.283 & 0.202 & 0.202 & 0.202 \\ 0.304 & 0.385 & 0.304 & 0.304 \\ 0.192 & 0.192 & 0.272 & 0.192 \\ 0.222 & 0.222 & 0.222 & 0.302 \end{pmatrix},$$

where some columns do not sum up to exactly 1 because of rounding. The columns of $\hat{\mathbf{P}}_{\mathrm{DAR}(1)}$ are the 1-step-ahead conditional distributions and might be applied to determine the conditional modes as the point forecasts. Note that 'c' is always the most probable state; that is, 'c' is the 1-step-ahead point forecast independent of the previous observation. The h-step-ahead forecast distributions for $h \geq 2$ are computed in the same way but by replacing $\hat{\phi}_{1,\mathrm{CML}}$ by $\hat{\phi}_{1,\mathrm{CML}}^{h}$; see Example 7.2.2. However, since $\hat{\phi}_{1,\mathrm{CML}} \approx 0.081$ is already rather close to 0, they barely differ from the marginal distribution.

For the MTD models, it turned out that the MTD(2) model does not lead to a visible improvement compared to the MTD(1) model, the parameter estimate for λ_2 is nearly equal to 0. So we concentrate on the fitted MTD(1) model; that is, a standard Markov chain model with 12 parameters (transition probabilities). CML estimation leads to the following transition matrix and stationary marginal distribution (the latter being computed from the invariance equation (B.4)):

$$\hat{\mathbf{P}}_{\mathrm{MTD}(1)} \approx \begin{pmatrix} 0.282 & 0.202 & 0.233 & 0.170 \\ 0.268 & 0.392 & 0.297 & 0.336 \\ 0.226 & 0.134 & 0.298 & 0.225 \\ 0.224 & 0.272 & 0.172 & 0.270 \end{pmatrix}, \quad \hat{\pi}_{\mathrm{MTD}(1)} \approx \begin{pmatrix} 0.218 \\ 0.331 \\ 0.210 \\ 0.240 \end{pmatrix},$$

where again some columns do not sum exactly to 1 because of rounding. Considering the conditional modes as the point forecasts, we obtain the rules "a $\Rightarrow$ a", "c $\Rightarrow$ c", "g $\Rightarrow$ g", and "t $\Rightarrow$ c", which differs from the DAR(1) rule, which always predicts 'c'. But the 2-step-ahead conditional distributions – that is, the columns of $\hat{\mathbf{P}}_{\mathrm{MTD}(1)}^2$ – are again very close to $\hat{\pi}_{\mathrm{MTD}(1)}$, so the

Table 7.2 Gini index and entropy for Bovine leukemia DNA data, and for fitted models.

	Observed	DAR(1)	Full Markov
v_G	0.98767	0.98763	0.98757
v_E	0.98736	0.98728	0.98725

2-step-ahead mode forecast is always equal to 'c'. This is not surprising in view of the Perron–Frobenius theorem (Remark B.2.2.1), since the second largest eigenvalue is rather small, at about 0.156.

In terms of the BIC ($\approx 22\,825$), the full Markov model should be preferred to the DAR(1) model. Comparing the marginal properties of the fitted models to the observed ones, the DAR(1) model does reproduce the observed dispersion slightly better (see Table 7.2): while the full Markov model does better in terms of the serial dependence structure. For instance, it has $\kappa(1) \approx 0.0803$ and $v(1) \approx 0.1131$ (observed values 0.0804 and 0.1134, respectively), while $\kappa(k) = v(k) \approx 0.0806^k$ for the fitted DAR(1) model. For computing these serial dependence measures, we make use of the fact that the matrix $(p_{ij}(k))$ of bivariate probabilities of a Markov chain is computed as $\mathbf{P}^k \, \mathrm{diag}(\boldsymbol{\pi})$; see Appendix B.2.1.

Altogether, the full Markov model seems preferable. Nevertheless, let us conclude this example with the following exercise. Using the fitted DAR(1) model, the constant c from (7.12) becomes ≈ 1.175. Following (7.14), we compute the bias-corrected point estimate of the Gini index as $\frac{T}{T-c} \, \hat{v}_G \approx 0.98780$. The approximate standard errors are obtained from (7.13) as 0.00141 (Gini) and 0.00163 (entropy), so approximate 95% confidence intervals follow as [0.98504; 0.99057] and [0.98417; 0.99055], respectively.

7.3 Hidden-Markov Models

Another important type of model for categorical processes with a non-Markovian serial dependence structure is the *hidden-Markov model* (HMM), which we are already familiar with from the count data case; see Section 5.2 as well as the book by Zucchini & MacDonald (2009) and the survey article by Ephraim & Merhav (2002). HMMs refer to a bivariate process $(X_t, Q_t)_{\mathbb{N}_0}$, in which the *hidden states* Q_t (latent states) constitute a homogeneous Markov chain (Appendix B.2) with range $Q = \{0, \dots, d_Q\}$ and $d_Q \in \mathbb{N}$, and where the *observable random variables* X_t (now also categorical with state space $S = \{s_0, s_1, \dots, s_d\}$) are generated conditionally independently, given the state process; see Figure 5.7 for an illustration of the data-generating mechanism.

As mentioned in Section 5.2, we may interpret a HMM as a "probabilistic function of a Markov chain" (Baum & Petrie (1966); see also Remark 7.3.3). While we shall again concentrate on HMMs for stationary processes, these models can also be used for non-stationary processes by, for example, including covariate information (Remark 5.2.6). HMMs for categorical processes have been applied in many contexts: in biological sequence analysis (Churchill, 1989; Krogh et al., 1994) and especially in fields related to natural languages:

- speech recognition (Rabiner, 1989): transforming spoken into written text
- text recognition (Makhoul et al., 1994; Natarajan et al., 2001): handwriting recognition or optical character recognition
- part-of-speech tagging (Cutting et al., 1992; Thede & Harper, 1999): where each word in a text is assigned its correct part of speech.

As described in Section 5.2, an HMM for $(X_t, Q_t)_{\mathbb{N}_0}$ is defined based on three assumptions:

- the *observation equation* (5.8):
  $$P(X_t \mid X_{t-1}, \ldots, Q_t, \ldots) = P(X_t \mid Q_t) \qquad \text{for all } t \in \mathbb{N}_0,$$
- the *state equation* (5.9):
  $$P(Q_t \mid X_{t-1}, \ldots, Q_{t-1}, \ldots) = P(Q_t \mid Q_{t-1}, \ldots),$$
- the Markov assumption with *state transition probabilities* (5.11):
  $$P(Q_t = q \mid Q_{t-1} = r) = a_{q|r} \qquad \text{for all } q, r \in Q = \{0, \ldots, d_Q\}.$$

As before, we denote the hidden states' transition matrix by $\mathbf{A} = (a_{q|r})_{q,r}$. The initial distribution $\boldsymbol{p}_0$ of Q_0 is assumed to be determined by the stationarity assumption; that is, $\boldsymbol{p}_0 := \boldsymbol{\pi}$, where $\boldsymbol{\pi}$ satisfies the invariance equation $\mathbf{A}\,\boldsymbol{\pi} = \boldsymbol{\pi}$ (see (B.4)). In contrast to Section 5.2, the time-homogeneous *state-dependent distributions* $p(\cdot|q)$ – that is, $P(X_t = x \mid Q_t = q) = p(x|q)$ for all t – are *categorical* distributions, each having d parameters. So altogether, we have $d_Q\,(d_Q + 1)$ parameters for the hidden states, plus $d\,(d_Q + 1)$ parameters related to the observations.

Most of the stochastic properties discussed in Section 5.2 directly carry over to the categorical case (certainly except those referring to moments, since the latter do not exist for categorical random variables). For instance, defining again the diagonal matrices $\mathbf{P}(x) := \mathrm{diag}(p(x|0), \ldots, p(x|d_Q)) \in [0;1]^{(d_Q+1) \times (d_Q+1)}$ for $x \in S$, the marginal pmf and the bivariate probabilities are given by (5.13); that is, by

$$P(X_t = x) = \sum_{q \in Q} p(x|q)\,\pi_q = \mathbf{1}^\top \mathbf{P}(x)\,\boldsymbol{\pi},$$

$$\tag{7.15}$$

$$P(X_t = x, X_{t-k} = y) = \mathbf{1}^\top \mathbf{P}(x)\,\mathbf{A}^k\,\mathbf{P}(y)\,\boldsymbol{\pi}.$$

Furthermore, maximum likelihood (ML) estimation is still possible, as described in Remark 5.2.3, the forecast distributions (5.19) and (5.20) remain valid, and both decoding schemes from Remark 5.2.4 are also applicable in the categorical case.

As the main difference, the serial dependence structure of the HMM's observations can no longer be described in terms of the ACF, and measures such as *Cramer's v* (6.3) or *Cohen's κ* (6.4) have to be computed using (7.15). Measures of categorical dispersion (Section 6.2), such as the *Gini index* (6.1) or *entropy* (6.2), are also available in this way.

Example 7.3.1 **(Two-state HMM)** Analogous to Example 5.2.2, let us consider the special case of a categorical HMM with only two hidden states; that is, with $d_Q = 1$. From (7.6), we know that the state transition probabilities $a_{q|r}$ for $q, r \in \{0, 1\}$ can be rewritten in the form $a_{q|r} = (1 - \rho) \pi_q + \rho \delta_{q,r}$ with $\pi_0 = 1 - \pi_1$ and with a $\rho \in (\max\{-\frac{\pi_1}{\pi_0}, -\frac{\pi_0}{\pi_1}\}; 1)$. Furthermore, the powers of **A** become $(\mathbf{A}^k)_{q,r} = (1 - \rho^k) \pi_q + \rho^k \delta_{q,r}$.

As a result, it follows that

$$P(X_t = x, X_{t-k} = y) - P(X_t = x) P(X_{t-k} = y)$$
$$\overset{(7.15)}{=} \mathbf{1}^\top \mathbf{P}(x) (\mathbf{A}^k - \boldsymbol{\pi} \mathbf{1}^\top) \mathbf{P}(y) \boldsymbol{\pi} = \mathbf{1}^\top \mathbf{P}(x) (\rho^k (\delta_{q,r} - \pi_q))_{q,r} \mathbf{P}(y) \boldsymbol{\pi},$$

which simplifies to

$$\rho^k \cdot \pi_0 \pi_1 (p(x|0) - p(x|1)) (p(y|0) - p(y|1)).$$

So both Cramer's $v(k)$ (6.3) and Cohen's $\kappa(k)$ (6.4) are proportional to ρ^k, analogous to the count data case from Example 5.2.2, where the ACF was proportional to ρ^k. This serial dependence caused by the underlying Markov chain, however, becomes damped if the state-dependent distributions $p(\cdot|0)$ and $p(\cdot|1)$ do not differ much, since then $(p(x|0) - p(x|1)) (p(y|0) - p(y|1))$ is close to zero. See Example 5.2.2 for an analogous conclusion in the count data case.

Example 7.3.2 **(Three-state HMM)** We define two stationary three-state HMMs using the same state transition matrix **A** as in Example 5.2.2 (see also Example 7.1.1), but with categorical observations having the state space $S = \{s_0, s_1, s_2, s_3\}$ ($d = 3$) and the following state-dependent distributions:

	HMM$_1$;	$x =$			HMM$_2$;	$x =$		
$p(x|q)$	s_0	s_1	s_2	s_3	s_0	s_1	s_2	s_3
$q = 0$	0.4	0.2	0.2	0.2	0.80	0.10	0.05	0.05
1	0.2	0.4	0.3	0.1	0.05	0.80	0.10	0.05
2	0.1	0.2	0.4	0.3	0.05	0.05	0.40	0.50

So for both models, the hidden state '0' tends to observation 's_0', '1' to observation 's_1', and '2' to observations 's_2' or 's_3'. But while this tendency is rather weak for model 1 (here, the different state-dependent distributions are quite close to each other), it is very pronounced for model 2. As expected from Example 7.3.1, model 1 causes a strong damping of the underlying Markov chain's serial dependence structure (see Figure 7.2a,b):

	HMM$_1$; $k=$				HMM$_2$; $k=$			
	1	2	3	4	1	2	3	4
$\kappa(k)$	0.031	0.022	0.016	0.011	0.293	0.213	0.156	0.114
$v(k)$	0.038	0.028	0.021	0.016	0.291	0.212	0.157	0.119

The marginal distributions of the observations (7.15) are computed as $(0.30, 0.24, 0.26, 0.20)^\top$ and $(0.50, 0.23, 0.13, 0.14)^\top$, respectively, so the first one shows more dispersion (Gini index 0.993 vs. 0.881).

Picking up Remark 7.2.3, it is obvious by model construction that any binarization $(Y_{t,j})_{\mathbb{N}_0}$ of the HMM's observation process $(X_t)_{\mathbb{N}_0}$, with $Y_{t,j} = \delta_{X_t, j}$ for a $j \in S$, follows an HMM again, now with state-dependent probabilities $P(Y_{t,j} = 1 \mid Q_t = q) = p(j|q)$ and $P(Y_{t,j} = 0 \mid Q_t = q) = 1 - p(j|q)$. For such a binary HMM, in turn, moments and hence the ACF will be well-defined, where the relationship between ACF and v, κ has already been investigated; see Example 6.3.2.

Remark 7.3.3 (**Markov representation**) In Baum & Petrie (1966), an HMM is introduced as a "probabilistic function of a Markov chain". But in the same article, it is also shown that an HMM can be expressed as a deterministic function of (another) Markov chain; see also Remark 7.2.3. The idea is quite simple: define the bivariate process $(Z_t)_{\mathbb{N}_0}$ by $Z_t := (X_t, Q_t)^\top$, then $(Z_t)_{\mathbb{N}_0}$ is a *finite* Markov chain with transition probabilities

$$P(Z_t = (x, q)^\top \mid Z_{t-1} = (y, r)^\top)$$
$$= P(X_t = x \mid Q_t = q)\, P(Q_t = q \mid Q_{t-1} = r) = p(x|q)\, a_{q|r}.$$

The observable random variables X_t are obtained from the Markovian variables Z_t by applying the deterministic function $f(z) := z_1$. As a benefit of this representation, if the finite Markov chain $(Z_t)_{\mathbb{N}_0}$ can be shown to satisfy, say, some mixing properties (see Definition B.1.5 and Appendix B.2.2), these carry over to the observations process $(X_t)_{\mathbb{N}_0}$, since this is obtained by just applying a deterministic function to $(Z_t)_{\mathbb{N}_0}$.

Example 7.3.4 (**Bovine leukemia DNA data**) Let us continue Example 7.2.4, where we modeled the bovine leukemia DNA series with its $d + 1 = 4$ observable states 'a', 'c', 'g' and 't'. As an alternative to the models considered before, we shall now try to fit a two-state HMM. Since the time series is rather long ($T = 8419$), we are faced with the numerical issues mentioned at the end of Remark 5.2.3, so the likelihood computation has to be based on $(w_t/w_{t-1}, \boldsymbol{\phi}_t)$ instead of $\boldsymbol{\alpha}_t$; see formula (5.18).

The (full) ML estimates of the two-state HMM are

$$\hat{\mathbf{A}}_{\mathrm{ML}} \approx \begin{pmatrix} 0.549 & 0.349 \\ 0.451 & 0.651 \end{pmatrix} \quad \text{and} \quad \hat{\boldsymbol{\pi}}_{\mathrm{ML}} \approx \begin{pmatrix} 0.436 \\ 0.564 \end{pmatrix}$$

for the hidden states' stationary Markov model, and the state-dependent distributions are estimated as

$\hat{p}_{\mathrm{ML}}(x\|q)$	$x =$ 'a'	'c'	'g'	't'
$q = 0$	0.449	0.000	0.481	0.070
1	0.043	0.589	0.000	0.368

The maximal log-likelihood equals $\approx -11\,412$, which is less than for the full Markov model from Example 7.2.4 ($\frac{T}{T-1}\,\ell_{\max} \approx -11\,358$), but better than for the DAR models fitted there. At this point, it should also be mentioned that a three-state HMM does not lead to a visible improvement in terms of model performance (see also the discussion below): the full Markov model remains the preferred choice. However, it is interesting to further interpret and analyze the fitted two-state HMM.

Looking at the state-dependent distributions $\hat{p}_{\mathrm{ML}}(\cdot|q)$, it becomes clear that the hidden state 0 mainly leads to either observation 'a' or 'g', while state 1 goes along with 'c' and 't'. Such a separation is plausible, since the nucleotides 'a' and 'g' form the group of purines, while 'c' and 't' are the pyrimidines. So state 0 might be interpreted as a "purine state", while state 1 constitutes a "pyrimidine state". The corresponding transition matrix $\hat{\mathbf{A}}_{\mathrm{ML}}$ has maximal entries on the diagonal, so a purine tends to be followed by a purine again, for example, but the probability for changing between both groups is also rather large. Overall, $\hat{\boldsymbol{\pi}}_{\mathrm{ML}}$ shows that the "pyrimidine state" is dominant.

The observations' marginal distribution within the fitted model, given by equation (7.15), equals about $(0.220, 0.331, 0.210, 0.239)^{\mathsf{T}}$ and thus agrees with the marginal frequencies up to three decimal places. There is, however, a visible discrepancy between the sample dependence measures $\hat{v}(k), \hat{\kappa}(k)$ on the one hand (as plotted in Figure 7.5), and the theoretical $v(k), \kappa(k)$ within the fitted model on the other hand. For the latter, we compute $(v(k))_k = (0.092, 0.018, \ldots)$ and $(\kappa(k))_k = (0.055, 0.011, \ldots)$, respectively, both being lower than the corresponding sample values. This confirms our earlier conclusion that the two-state HMM is not the optimal choice for the bovine leukemia DNA series.

Remark 7.3.5 **(Higher-order HMM)** As pointed out in Remark 5.2.6, there are several ways of extending the basic HMM to higher-order models, for example by using a higher-order Markov model (such as an MTD or DAR model) for the hidden states (Zucchini & MacDonald, 2009, Section 8.3). With respect to categorical processes, the *double-chain Markov model* (DCMM), as proposed by Berchtold (1999), is particularly worth mentioning. Here, the current observation X_t is influenced by both the current state Q_t and the past observation X_{t-1}. So the *observation equation* (5.8) is modified to

$$P(X_t \mid X_{t-1}, \dots, Q_t, \dots) = P(X_t \mid X_{t-1}, Q_t) \qquad \text{for all } t \in \mathbb{N}_0.$$

A further extension of this model was developed by Berchtold (2002), who allows for pth-order dependence with respect to past observations, and for qth-order Markov dependence concerning the hidden states; that is, (5.8) and (5.9) become

$$P(X_t \mid X_{t-1}, \dots, Q_t, \dots) = P(X_t \mid X_{t-1}, \dots, X_{t-p}, Q_t) \quad \text{and}$$
$$P(Q_t \mid X_{t-1}, \dots, Q_{t-1}, \dots) = P(Q_t \mid Q_{t-1}, \dots, Q_{t-q}).$$

This DCMM(p, q) model was applied by Berchtold (2002) to a categorical time series representing the song of a wood pewee (as presented in Examples 6.1.2, 6.2.2 and 6.3.1), and to a categorical time series expressing the behavior of young rhesus monkeys, with four behaviors: "passive", "explore", "fear/disturb" and "play").

7.4 Regression Models

In Section 5.1, we introduced *regression models* for time series of counts, the main advantage of which is their ability to easily incorporate covariate information, the latter being represented by the (possibly deterministic) vector-valued covariate process $(\boldsymbol{Z}_t)_{\mathbb{Z}}$. In our discussion, we focussed on *generalized linear models* (GLMs), where the observations' conditional mean is linked to a linear expression of the "available information". The available information is not necessarily limited to the covariate information, but it may also include past observations of the process, as in the case of the *conditional regression models* according to Definition 5.1.1. If, however, only the current covariate is required for "explaining" the current observation, then we referred to such a situation as a *marginal regression model* (5.5) (Fahrmeir & Tutz, 2001). In the sequel, we shall see that the regression approach can also be adapted to the case of the observations process $(X_t)_{\mathbb{Z}}$ being categorical. A much more detailed discussion of such categorical regression models together with several real-data examples is provided in Chapters 2 and 3 of the book by Kedem & Fokianos (2002).

Before turning to the general categorical case, let us first look at the special situation, where $(X_t)_{\mathbb{Z}}$ is a *binary process* with the state space being coded as $S = \{0, 1\}$ (see Example 6.3.2); that is, where the X_t are Bernoulli random variables (Example A.2.1). The parameter of the Bernoulli distribution is its "success probability", which is also equal to its mean. Hence, it lends itself to proceed in the same way as in Section 5.1; that is, to define a GLM with respect to this mean parameter. So Definition 5.1.1 now reads as follows.

Definition 7.4.1 (**Binary regression model**)   Let $(Z_t)_\mathbb{Z}$ be a covariate process. The binary process $(X_t)_\mathbb{Z}$ follows a *conditional binary regression model* if

(i) X_t, conditioned on $X_{t-1}, \ldots$ and $Z_t, \ldots$, is Bernoulli distributed according to Bin$(1, \pi_t)$, where
(ii) the conditional mean $\pi_t := E[X_t \mid X_{t-1}, \ldots, Z_t, \ldots]$ satisfies

$$g(\pi_t) = \boldsymbol{\theta}^\top V_t$$

with a *link function* g and a parameter vector $\boldsymbol{\theta}$, where the design vector V_t is a function of $X_{t-1}, \ldots$ and $Z_t, \ldots$

The inverse of the link function, $h := g^{-1}$, is referred to as a *response function*: $\pi_t = h(\boldsymbol{\theta}^\top V_t)$.

The definition (5.5) of a marginal regression model is adapted accordingly. Note that π_t is expressed equivalently as $\pi_t = P(X_t = 1 \mid X_{t-1}, \ldots, Z_t, \ldots)$; this allows for a simplified notation of the (partial) likelihood function (Remark 5.1.5), since $P(X_t \mid X_{t-1}, \ldots, Z_t, \ldots, \boldsymbol{\theta})$ now equals $\pi_t^{X_t}(1 - \pi_t)^{1-X_t}$. A detailed discussion of likelihood estimation for binary regression models is provided by Slud & Kedem (1994).

While in the count data case, we had to ensure that $h(\boldsymbol{\theta}^\top V_t)$ always produces a positive value, here, we even have to ensure that the value of $h(\boldsymbol{\theta}^\top V_t)$ is in the interval $(0; 1)$. If one wants to avoid severe restrictions concerning the parameter range for $\boldsymbol{\theta}$, it is recommended to use response functions h with range $(0; 1)$; for example, a (strictly monotonic increasing) cdf. The most common choice is the cdf of the standard logistic distribution, leading to a *logit model*.

Example 7.4.2 (**Binary logit model**)   The *logit link*, which is the canonical link function of the Bernoulli distribution, is given by

$$g(u) = \ln \frac{u}{1 - u} \qquad \text{and} \qquad h(u) = \frac{e^u}{1 + e^u} = \frac{1}{1 + e^{-u}},$$

Looking at the definition of g, we may interpret a logit GLM as a log-linear model with respect to the odds $\pi_t/(1 - \pi_t)$; the quantity $\ln\left(\pi_t/(1 - \pi_t)\right)$ is also referred to as the *log-odds*. In particular, the conditional odds are again determined multiplicatively as $\pi_t/(1 - \pi_t) = (e^{\theta_1})^{V_1} \cdot (e^{\theta_2})^{V_2} \ldots$ Further motivating arguments for the particular choice of a logit model are presented by Slud & Kedem (1994) and in Section 2.1.1 of Kedem & Fokianos (2002).

A simple *autoregressive logit model* was defined by Kedem & Fokianos (2002) and Fokianos & Kedem (2004)

$$\ln\left(\frac{\pi_t}{1-\pi_t}\right) = \beta_0 + \alpha_1\, X_{t-1} + \dots + \alpha_p\, X_{t-p}. \tag{7.16}$$

Binary logit models with a *feedback* mechanism (analogous to the INGARCH(1, 1) model from Example 4.1.4) are discussed by Moysiadis & Fokianos (2014); for example, the basic model

$$\lambda_t := \ln\left(\frac{\pi_t}{1-\pi_t}\right) = \beta_0 + \alpha_1\, X_{t-1} + \beta_1\, \lambda_{t-1}, \tag{7.17}$$

where the condition $|\alpha_1| + 4\,|\beta_1| < 1$ ensures a stationary solution.

Possible alternatives to the logit approach are the *probit model* (based on the cdf of the standard normal distribution), the *log–log model* (cdf of standard maximum extreme value distribution), or the *complementary log–log model* (cdf of standard minimum extreme value distribution); see Section 2.1 in Kedem & Fokianos (2002) for further details.

To extend the methods described before to the general *categorical* case – that is, where X_t has the state space $S = \{s_0, s_1, \dots, s_d\}$ – it is helpful to look at the binarization Y_t of X_t, defined by $Y_t = e_j$ if $X_t = s_j$, with $e_0, \dots, e_d \in \{0, 1\}^{d+1}$ being the unit vectors (see Example A.3.3). If X_t is distributed according to $\pi_t = (\pi_{t,0}, \dots, \pi_{t,d})^\top \in \mathbb{S}_{d+1}$ (unit simplex, see Remark A.3.4), then Y_t is multinomially distributed according to MULT$(1; \pi_{t,0}, \dots, \pi_{t,d})$, and its jth component $Y_{t,j}$ follows the Bernoulli distribution Bin$(1, \pi_{t,j})$. The basic idea in the sequel is to apply the above binary approaches (especially the logit approach) to these components $Y_{t,j}$.

Because of the sum constraint $Y_{t,0} + \dots + Y_{t,d} = 1$, it is reasonable to concentrate on the reduced vectors $Y_t^* := (Y_{t,1}, \dots, Y_{t,d})^\top$ (then $Y_{t,0} = 1 - Y_{t,1} - \dots - Y_{t,d}$), the distribution of which is denoted by MULT$^*(1; \pi_{t,1}, \dots, \pi_{t,d})$ according to Example A.3.3. To further simplify the notation, we define the *open d-part unit simplex*

$$\mathbb{S}_d^* := \{u \in (0;1)^d \mid u_1 + \dots + u_d < 1\};$$

then $\pi_t^* = (\pi_{t,1}, \dots, \pi_{t,d})^\top$ satisfies $\pi_t^* \in \mathbb{S}_d^*$, and we just write $Y_t^* \sim$ MULT* $(1; \pi_t^*)$.

Following Fahrmeir & Kaufmann (1987), Kedem & Fokianos (2002) and Fokianos & Kedem (2003), we now extend Definition 7.4.1 to a conditional categorical regression model.

Definition 7.4.3 **(Categorical regression model)** Let $(Z_t)_{\mathbb{Z}}$ be a covariate process, and represent the categorical process $(X_t)_{\mathbb{Z}}$ as $(Y_t^*)_{\mathbb{Z}}$. The process $(Y_t^*)_{\mathbb{Z}}$ follows a *conditional categorical regression model* if

(i) Y_t^*, conditioned on $Y_{t-1}^*, \dots$ and $Z_t, \dots$, is multinomially distributed according to MULT$^*(1; \pi_t^*)$, where

(ii) the conditional mean $\pi_t^* := E[Y_t^* \mid Y_{t-1}^*, \dots, Z_t, \dots]$ satisfies

$$\pi_t^* = h(V_t^\top \theta)$$

with a *response function* $h : \mathbb{R}^d \to \mathbb{S}_d^*$ (inverse link function) and a parameter vector θ, where the design matrix V_t has d columns and is a function of $Y_{t-1}^*, \dots$ and $Z_t, \dots$.

Analogous to the binary case, the probabilities $P(X_t \mid X_{t-1}, \dots, Z_t, \dots, \theta)$ are expressed in a simple way, namely as $\prod_{j=0}^d \pi_{t,j}^{Y_{t,j}}$ (the component $\pi_{t,j}$ of π_t expresses $P(X_t = s_j \mid \dots)$). These probabilities are required for (partial) likelihood computation. Likelihood estimation for categorical regression models is discussed by Fahrmeir & Kaufmann (1987) and Fokianos & Kedem (2003) as well as in the book by Kedem & Fokianos (2002).

To illustrate Definition 7.4.3, let us consider a particular instance of a categorical GLM: the categorical logit model as discussed by Kedem & Fokianos (2002) and Fokianos & Kedem (2003).

Example 7.4.4 (Categorical logit model) Let us assume that V_t is a k-dimensional design vector (as a function of $Y_{t-1}^*, \dots$ and $Z_t, \dots$), and let $\theta_1, \dots, \theta_d$ be k-dimensional parameter vectors. A categorical GLM is obtained by defining $\theta \in \mathbb{R}^{kd}$ and V_t with dimension $kd \times d$ via

$$\theta = \begin{pmatrix} \theta_1 \\ \vdots \\ \theta_d \end{pmatrix}, \qquad V_t = \begin{pmatrix} V_t & 0 & \cdots & 0 \\ 0 & V_t & \ddots & \vdots \\ \vdots & \ddots & \ddots & 0 \\ 0 & \cdots & 0 & V_t \end{pmatrix}.$$

Then $V_t^\top \theta$ is a d-dimensional column vector, the jth component of which just equals $V_t^\top \theta_j = \theta_j^\top V_t$.

A common choice for the response function $h : \mathbb{R}^d \to \mathbb{S}_d^*$ is the *multinomial logit*, where the jth component $h_j(u)$ of $h(u)$ equals

$$h_j(z) = \frac{\exp z_j}{1 + \sum_{i=1}^d \exp z_i} \qquad \text{for } j = 1, \dots, d.$$

Then the systematic component of the model becomes

$$\pi_{t,j} = \frac{\exp(\theta_j^\top V_t)}{1 + \sum_{i=1}^d \exp(\theta_i^\top V_t)} \qquad \text{for } j = 1, \dots, d, \tag{7.18}$$

and $\pi_{t,0}$ follows as $\pi_{t,0} = 1/(1 + \sum_{i=1}^d \exp(\theta_i^\top V_t))$.

It should be noted that this model can be rewritten by considering odds with respect to the 0th category:

$$\frac{\pi_{t,j}}{\pi_{t,0}} = \exp\left(\boldsymbol{\theta}_j^\top \boldsymbol{V}_t\right) \qquad \text{or} \qquad \ln\left(\frac{\pi_{t,j}}{\pi_{t,0}}\right) = \boldsymbol{\theta}_j^\top \boldsymbol{V}_t,$$

respectively, for $j = 1, \ldots, d$. If comparing the ith and jth category with $i, j = 1, \ldots, d$, in contrast, it follows that

$$\frac{\pi_{t,j}}{\pi_{t,i}} = \exp\left((\boldsymbol{\theta}_j - \boldsymbol{\theta}_i)^\top \boldsymbol{V}_t\right) \qquad \text{and} \qquad \ln\left(\frac{\pi_{t,j}}{\pi_{t,i}}\right) = (\boldsymbol{\theta}_j - \boldsymbol{\theta}_i)^\top \boldsymbol{V}_t.$$

A particular example of the categorical logit model described in Example 7.4.4 is a pth-order autoregressive model (analogous to (7.16); see also Kedem & Fokianos (2002) and Fokianos & Kedem (2003)), where V_t is composed of $1, Y_{t-1}^*, \ldots, Y_{t-p}^*$; that is, of dimension $d\,p + 1$. Hence, the total number of model parameters is given by $d\,(d\,p + 1)$, with the components' parameter vectors $\boldsymbol{\theta}_1, \ldots, \boldsymbol{\theta}_d$ being of dimension $d\,p + 1$. So the pth-order autoregressive logit model can be understood as a parsimoniously parametrized pth-order Markov model; see also Section 7.1. Partitioning the components of $\boldsymbol{\theta}_j$ as $\boldsymbol{\theta}_j = (\beta_{j,0}, \boldsymbol{\alpha}_{j,1}, \ldots, \boldsymbol{\alpha}_{j,p})^\top$ with the $\boldsymbol{\alpha}_{j,i}$ consisting of d parameters, we can rewrite the *autoregressive logit model*'s recursion as

$$\boldsymbol{\theta}_j^\top \boldsymbol{V}_t = \beta_{j,0} + \boldsymbol{\alpha}_{j,1}^\top \boldsymbol{Y}_{t-1}^* + \ldots + \boldsymbol{\alpha}_{j,p}^\top \boldsymbol{Y}_{t-p}^*. \tag{7.19}$$

A non-Markovian categorical regression model with an additional feedback component, analogous to equation (7.17), was developed by Moysiadis & Fokianos (2014).

Example 7.4.5 **(Bovine leukemia DNA data)** Let us continue Examples 7.2.4 and 7.3.4 about the bovine leukemia DNA series (length $T = 8419$) with state space $S = \{\text{'a', 'c', 'g', 't'}\}$ of size $d + 1 = 4$. In view of the AR-like serial dependence structure of these data, we try to fit the autoregressive logit model as described in (7.19). For model order p, it has $d\,(d\,p + 1)$ parameters. The $d = 3$ components of Y_t^* refer to the states $s_1 = \text{'c'}$, $s_2 = \text{'g'}$ and $s_3 = \text{'t'}$, and $X_t = \text{'a'}$ is represented by $Y_t^* = \mathbf{0}$.

For model order p = 1, we have $d\,(d + 1) = 12$ parameters, which is exactly the same number as for a full Markov chain model; see Section 7.1. In fact, the first-order autoregressive logit model and a full Markov chain are equivalent to each other; the logit model just constitutes a reparametrization of the Markov chain. This reparametrization, however, is quite useful in practice, because it avoids constraints for the model parameters and thus simplifies their estimation. We also consider the model orders 2 and 3, leading to 21 and 30 model parameters, respectively. These numbers are already quite large, but much lower than for a full second- or third-order Markov model (48 and 192,

Table 7.3 Bovine leukemia DNA data: maximized log-likelihood, AIC and BIC for AR(p) logit models.

$\frac{T}{T-p}\ell_{\max}$			AIC			BIC		
p = 1	p = 2	p = 3	p = 1	p = 2	p = 3	p = 1	p = 2	p = 3
−11 358	−11 330	−11 323	22 740	22 703	22 706	22 824	22 851	22 917

respectively). The obtained (rounded) values for the maximized (conditional) log-likelihood as well as the AIC and BIC are shown in Table 7.3.

So the BIC prefers the first-order logit model (full MC), while the AIC prefers the second-order one. Note that the values obtained for the first-order logit model slightly deviate from those in Example 7.2.4 for the Markov chain. This is caused by the use of different numerical optimization routines, namely unconstrained vs. constrained optimization.

In view of the diffuse picture caused by the information criteria, let us compare some stochastic properties of the fitted first- and second-order logit models with the corresponding observed ones. The required transition probabilities $p(k|l)$ and $p(k|l, m)$ are computed from (7.18) and (7.19), respectively, by inserting the estimated parameter values (the latter are not shown here to save space). Marginal distribution and serial dependence measures for the first-order logit model (full MC) have already been checked in Example 7.2.4.

For the second-order logit model, it would again be possible to compute the considered characteristics exactly, by transforming the model into a bivariate Markov chain (see the discussion below (B.1) in Appendix B.1). But for the sake of simplicity, the fitted model was used to simulate a time series of length 10 million, the sample properties of which serve as an approximation to the true model's values. For the marginal distribution, we obtain about $(0.220, 0.332, 0.210, 0.239)^\top$ (which does not sum up to exactly 1 because of rounding), which is very close to $\hat{\pi}$ as reported in Example 7.2.4. Furthermore, the serial dependence structure is represented very well now also for lag 2: $v(1) \approx 0.114$, $v(2) \approx 0.044$ and $\kappa(1) \approx 0.080$, $\kappa(2) \approx 0.025$, respectively. So if the additional number of parameters compared to the first-order model is acceptable, the second-order logit model appears to be preferable.

We conclude this section by pointing out that the regression approach can also be used for *ordinal* time series; see Kedem & Fokianos (2002) and Fokianos & Kedem (2003). To simplify the presentation, we first discuss a marginal regression model, but the model is easily extended to a conditional regression model.

Example 7.4.6 **(Ordinal regression model)** Let the states in S exhibit a natural ordering, $s_0 < s_1 < \ldots < s_d$. The idea is to assume that X_t or $\boldsymbol{Y}_t$, respectively, is generated from a latent real-valued random variable Q_t in the following way:

$$X_t = s_j \quad \text{iff} \quad Q_t - \gamma^\top Z_t \in [\eta_{j-1}; \eta_j), \tag{7.20}$$

where $-\infty = \eta_{-1} < \eta_0 < \dots < \eta_{d-1} < \eta_d = +\infty$ are *threshold parameters*. Here, Z_t is the covariate information at time t, and the Q_t are assumed to be i.i.d. If F_Q denotes the cdf of Q_t, then

$$\pi_{t,j} = P(X_t = s_j \mid Z_t = z_t) = F_Q(\eta_j + \gamma^\top z_t) - F_Q(\eta_{j-1} + \gamma^\top z_t),$$
$$P(X_t \le s_j \mid Z_t = z_t) = F_Q(\eta_j + \gamma^\top z_t) \quad \text{for } j = 0, \dots, d. \tag{7.21}$$

The model parameters to be estimated are $\eta_0, \dots, \eta_{d-1}$ as well as γ.

In the special case of F_Q being the cdf of the standard logistic distribution – that is, $F_Q(u) = 1/(1 + \exp(-u))$ – the resulting model is referred to as the *proportional odds model* or *ordered logit model*. Then (7.21) becomes

$$\ln\left(\frac{P(X_t \le s_j \mid Z_t = z_t)}{P(X_t > s_j \mid Z_t = z_t)} \right) = \eta_j + \gamma^\top z_t. \tag{7.22}$$

An example of the ordered logit model with $d + 1 = 6$ states $s_0 < s_1 < \dots < s_5$, and with threshold parameters $-\infty = \eta_{-1} < \dots < \eta_5 = +\infty$ according to (7.20), is shown in Figure 7.6a. There, the probability density function (pdf) $f_Q(u) = \exp(-u)/(1 + \exp(-u))^2$ of the standard logistic distribution is plotted together with the threshold values $\eta_0 = -2.2$, $\eta_1 = -1.3$, $\eta_2 = -0.4$, $\eta_3 = 0.7$, $\eta_4 = 1.9$ as well as with the resulting probability masses. As an example, the probability that the latent random variable Q_t takes a value between η_2 and η_3 (this corresponds to state 's_3') equals ≈ 0.267.

Remark 7.4.7 Note that the ordinal approach (7.21) can be embedded into the Definition 7.4.3 of categorical regression models. We have to set

$$\theta = \begin{pmatrix} \eta_0 \\ \vdots \\ \eta_{d-1} \\ \gamma \end{pmatrix}, \quad V_t = \begin{pmatrix} 1 & 0 & \cdots & 0 \\ 0 & \ddots & \ddots & \vdots \\ \vdots & \ddots & 1 & 0 \\ 0 & \cdots & 0 & 1 \\ Z_t & \cdots & Z_t & Z_t \end{pmatrix},$$

then $V_t^\top \theta = (\eta_0 + \gamma^\top Z_t, \dots, \eta_{d-1} + \gamma^\top Z_t)^\top$. Furthermore, we define the response function $h : \mathbb{R}^d \to \mathbb{S}_d^*$ by

$$h_1(u) = F_Q(u_1), \quad h_j(u) = F_Q(u_j) - F_Q(u_{j-1}) \quad \text{for } j = 2, \dots, d.$$

Then the model from Definition 7.4.3 becomes (7.21).

The ordinal regression approach of Example 7.4.6 is easily modified to obtain an autoregressive model. As for model (7.19), we skip one component of the

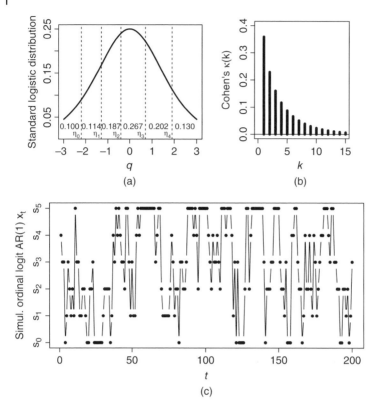

Figure 7.6 (a) Standard logistic distribution with threshold values; see Example 7.4.6. Ordinal logit AR(1) model from Example 7.4.8: (b) Cohen's κ, (c) simulated sample path.

binarization Y_t of X_t. But in view of the thresholds (7.20), this time, it is advantageous to define $Y_t^\star := (Y_{t,0}, \ldots, Y_{t,d-1})^\top$ following $\text{MULT}^*(1; \pi_{t,0}, \ldots, \pi_{t,d-1})$ (Example A.3.3). Combining approaches (7.19) and (7.22), we define an *ordinal autoregressive logit model* as (Fokianos & Kedem, 2003):

$$\ln\left(\frac{P(X_t \leq s_j \mid X_{t-1}, \ldots)}{P(X_t > s_j \mid X_{t-1}, \ldots)}\right) = \eta_j + \boldsymbol{\alpha}_1^\top \boldsymbol{Y}_{t-1}^\star + \ldots + \boldsymbol{\alpha}_p^\top \boldsymbol{Y}_{t-p}^\star. \tag{7.23}$$

Note that the d-dimensional parameter vectors $\boldsymbol{\alpha}_1, \ldots, \boldsymbol{\alpha}_p$ do not depend on j (certainly, this would also be possible), so that the model has only $d\,(p+1)$ parameters.

Example 7.4.8 (Ordinal logit autoregression) Let us pick up the ordinal logit approach with $d+1=6$ states $s_0 < s_1 < \ldots < s_5$ from Example 7.4.6, and let us construct an autoregressive model (7.23) of order $p = 1$. Note that this model is a (parsimoniously parametrized) ordinal Markov chain with transition probabilities

$$P(X_t = s_j \mid X_{t-1}) = F_Q(\eta_j + \alpha_1^\top Y_{t-1}^\star) - F_Q(\eta_{j-1} + \alpha_1^\top Y_{t-1}^\star) \qquad (= \pi_{t,j}),$$

see (7.21). In model (7.23), the event that X_{t-1} equals the largest state 's_5' is represented by the vector $Y_{t-1}^\star = (0, \dots, 0)^\top$, such that the expression for the log-odds (7.23) reduces to η_j. If we aim at having a model with positive dependence, then a large state should be followed by a large state with a high probability, which is not the case for the choice of the threshold values in Example 7.4.6; see Figure 7.6a. Therefore, for the present example, we shift these values by -3; a negative shift such that more probability mass is left above the largest threshold η_4. So we use $\eta_0 = -5.2$, $\eta_1 = -4.3$, $\eta_2 = -3.4$, $\eta_3 = -2.3$, $\eta_4 = -1.1$. As a result, the probabilities for falling between η_{j-1} and η_j (this is the conditional distribution for X_t given $X_{t-1} = s_5$) become approximately $0.005, 0.008, 0.019, 0.059, 0.159, 0.750$ for $j = 0, \dots, 5$. In particular, 's_5' is followed by 's_5' with about 75% probability.

The components of the autoregressive parameter vector α_1, in turn, are chosen as positive values. Since the smallest state corresponds to $Y_{t-1}^\star = (1, 0, 0, 0, 0)^\top$, the second smallest to $(0, 1, 0, 0, 0)^\top$ and so on, and since positive dependence requires that small values tend to be followed by small ones, we choose $\alpha_{1,0} \geq \alpha_{1,1} \geq \dots$: $\alpha_1 := (6, 4, 3.5, 3.5, 2)^\top$. So altogether, the model's transition matrix becomes

$$\mathbf{P} \approx \begin{pmatrix} 0.690 & 0.231 & 0.154 & 0.154 & 0.039 & 0.005 \\ 0.156 & 0.194 & 0.156 & 0.156 & 0.052 & 0.008 \\ 0.085 & 0.220 & 0.215 & 0.215 & 0.107 & 0.019 \\ 0.045 & 0.200 & 0.244 & 0.244 & 0.228 & 0.059 \\ 0.017 & 0.102 & 0.148 & 0.148 & 0.285 & 0.159 \\ 0.007 & 0.052 & 0.083 & 0.083 & 0.289 & 0.750 \end{pmatrix},$$

where not all columns sum to 1 because of rounding. As an example, the first column gives the conditional distribution for X_t given $X_{t-1} = s_0$, which implies that 's_0' is followed by 's_0' with about 69.0% probability, and by the next largest state 's_1' with about 15.6% probability. This "inertia" for the smallest and largest state, respectively, becomes visible in the simulated sample path shown in Figure 7.6c, and it causes Cohen's κ in Figure 7.6b to take positive values.

The stationary marginal distribution is computed according to the invariance equation (B.4) in the appendix as $(0.230, 0.106, 0.118, 0.141, 0.134, 0.271)^\top$, so the boundary states 's_0' and 's_5' are the most probable ones. Overall, this distribution is quite close to a uniform distribution, which explains the large values for the Gini index (6.1) and entropy (6.2), given by 0.973 and 0.964, respectively.

For further information on ordinal regression models, consult Kedem & Fokianos (2002) and Fokianos & Kedem (2003).

Part III

Monitoring Discrete-Valued Processes

In Parts I and II, we learned about models for discrete-valued processes and how to fit them to the given time series data. Once an appropriate model for the considered process has been found, it can be applied with respect to the further course of the process. The perhaps most common application of a given model is to forecast future realizations of the process, a topic that was discussed earlier in this book in the context of each of the respective models, especially in Section 2.6. In the following, we focus on another field for applying fitted models, namely the monitoring of the process. This is often called *statistical process control* (SPC). Instead of predicting outcomes conforming with the process model, the aim of SPC is to detect a possible deviation from the assumed (in-control) model as early as possible, but certainly without producing false alarms all the time. This might be useful in detecting the outbreak of disease epidemics, the deterioration of customer service, or a malfunction during production. An important tool for process monitoring is the control chart, or, more precisely in the context of a discrete-valued process, the *attributes control chart*. As a new observation arrives, a control statistic is computed and plotted on the chart; if the statistic violates the chart's control limits, an alarm is triggered to indicate that the process has possibly run out of control.

After providing a brief introduction to the discipline of SPC in general in Section 8.1, the remaining sections of Chapter 8 deal with types of control charts for count processes, where the tth control statistic is computed based on the counts available up to time t. Approaches like Shewhart, cumulative sum (CUSUM) and exponentially weighted moving-average (EWMA) control charts are discussed, together with the corresponding chart design and ways of analyzing their performance. Then in Chapter 9, analogous issues are considered for the case of categorical processes. Here, one strategy is to

An Introduction to Discrete-Valued Time Series, First Edition. Christian H. Weiss.
© 2018 John Wiley & Sons Ltd. Published 2018 by John Wiley & Sons Ltd.
Companion website: www.wiley.com/go/weiss/discrete-valuedtimeseries

take samples from the process and to determine the frequency distribution of the categories within each sample. This information is used to compute the statistic to be plotted on the chart. Continuous process monitoring is also possible, where each new observation is accompanied by a new control statistic.

8

Control Charts for Count Processes

8.1 Introduction to Statistical Process Control

Methods of *statistical process control* (SPC) help to monitor and improve processes in manufacturing and service industries, and they are also often used in fields such as public-health surveillance. For the given process, relevant quality characteristics are measured over time, thus leading to a (possibly multivariate) stochastic process $(X_t)_{\mathbb{N}}$ of continuous-valued or discrete-valued random variables (*variables data* or *attributes data*, respectively). Examples of such quality characteristics could be the diameter of a drill hole (variables data) or the number of non-conformities (attributes data) in a produced item, or the number of infections in a health-related example. One of the most important SPC tools is the *control chart*, which requires the relevant quality characteristics to be measured online. Control charts are applied to a process operating in a stable state (*in control*); that is, $(X_t)_{\mathbb{N}}$ is assumed to be stationary according to a specified model (the in-control model). As a new measurement arrives, it is used to compute a statistic (possibly also incorporating past values of the quality characteristic), which is then plotted on the control chart with its *control limits*. If the statistic violates the limits, then an alarm is triggered to signal that the process may not be stable anymore (*out of control*). So the process is interrupted, and it is checked if the alarm indeed results from an *assignable cause* (say, a shift or drift in the process mean); the time when the process left its in-control model is said to be the *change point* (more formal definitions are given below). In this case, corrective actions are required before continuing the process. If the process is still in its in-control state, the alarm is classified as a *false alarm*. An example of a control chart with limits 0 and 5 is shown in Figure 8.1, where the upper limit is violated at time $t = 224$. Note that the lower limit 0 can never be violated; that is, it is actually a one-sided (upper-sided) control chart. We shall discuss this control chart and the related application in much more detail in Example 8.2.2.3.

An Introduction to Discrete-Valued Time Series, First Edition. Christian H. Weiss.
© 2018 John Wiley & Sons Ltd. Published 2018 by John Wiley & Sons Ltd.
Companion website: www.wiley.com/go/weiss/discrete-valuedtimeseries

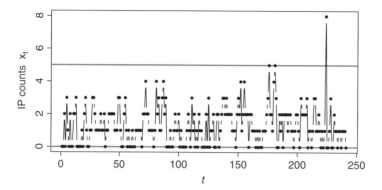

Figure 8.1 *c* chart of IP counts data with limits 0 and 5; see Example 8.2.2.3.

The use of control charts for prospective online monitoring, as described before, is commonly referred to as the *Phase-II* application. But control charts may also be applied in a retrospective manner to already available in-control data. This is called the *Phase-I* application of a control chart. During this iterative procedure, potential outliers are identified and removed from the data, and parameter estimates and the chart design are revised accordingly. A (successful) Phase-I analysis ends up with an estimated model characterizing the in-control properties of $(X_t)_{\mathbb{N}}$; this model is then used for designing the control charts to be used during Phase-II monitoring. More details about all these terms and concepts can be found, among others, in the textbook by Montgomery (2009) and in the survey paper by Woodall & Montgomery (2014).

In this book, we shall exclusively concentrate on attributes data processes, and we shall start with the monitoring of count processes. Typical examples from manufacturing industry are the number of non-conformities per produced item (range $\mathbb{N}_0$) or the number of defective items in a sample of size n (range $\{0, \ldots, n\}$). Non-manufacturing examples include counts of new cases of an infection (per time unit) in public-health surveillance, or counts of complaints by customers (per time unit) in a service industry. The majority of studies on the monitoring of such counts assumes the process to be i.i.d. in its in-control state (Woodall, 1997), but in this book, we shall attach more importance to the case of autocorrelated counts. In fact, there has been increasing research activity in this direction in recent years (Weiß, 2015b), and Alwan & Roberts (1995) have already shown that autocorrelation is indeed a common phenomenon in SPC-related count processes. Typical reasons are a high sampling frequency due to automated production environments in manufacturing industry, or varying service times (extending over more than one time unit) in service industry, or varying incubation times and infectivities of diseases in public-health surveillance.

In Section 8.2, we start with basic Shewhart charts for a count process, where the plotted statistic at time t is a function only of the most recent observation X_t (or of the most recent sample for sample-based monitoring). While the Shewhart charts themselves are rather simple, they offer an opportunity to introduce general design principles for control charts. These principles are applied in Section 8.3 when considering advanced control charts, such as the CUSUM and EWMA methods, where the plotted statistic at time t also uses past observations of the process and hence accumulates information about the process for a longer period of time. Later in Chapter 9, we shall move our focus towards the monitoring of categorical processes, but where methods for count data still might be useful for a sample-based monitoring approach.

Remark 8.1.1 (Change point methods) An approach related to the control chart are tests for a change point within a given time series. For the case of a count time series stemming from an INGARCH model, such change point tests were developed by Franke et al. (2012), Kang & Lee (2014) and Kang & Song (2015), while Torkamani et al. (2014) and Davoodi et al. (2015) considered an underlying INAR process; also see the references in Hudecová et al. (2015) and Kirch & Kamgaing (2016). Note that the main difference between change point tests and control charts is that the first are usually applied in an offline manner, to find the location of the change point in the available (and static) time series. Online versions of change point tests, where the in-control model is sequentially tested based on the available data at each time point, have been presented by Hudecová et al. (2015) and Kirch & Kamgaing (2015).

8.2 Shewhart Charts for Count Processes

The first control charts were proposed by Shewhart (1926, 1931). Because of this pioneering work, a number of standard control charts are referred to as *Shewhart control charts*; an extensive review of Shewhart control charts is given by Montgomery (2009). The characteristic feature of these charts is that the plotted statistic Z_t is a function only of the most recent observation X_t (or of the most recent sample for sample-based monitoring). Then Z_t is plotted on a chart against time t with time-invariant lower and upper *control limits* $l < u$ (as in Figure 8.1). An alarm is triggered at time t for the first time if

$$Z_1, \ldots, Z_{t-1} \in [l; u], \qquad \text{but } Z_t \notin [l; u]; \qquad (8.1)$$

then the process is interrupted to check for an assignable cause. The (random) *run length* of the control chart is defined as

$$L := \min\{t \in \mathbb{N} \mid Z_t \notin [l; u]\}; \qquad (8.2)$$

the corresponding run length distribution turns out to be of utmost importance when designing the control chart; that is, when choosing the control limits l and u.

8.2.1 Shewhart Charts for i.i.d. Counts

If monitoring a count process $(X_t)_{\mathbb{N}}$ with a Shewhart chart, the counts are commonly directly plotted on the chart as they arrive in time; that is, the plotted statistics are $Z_t = X_t$. If the range of the counts is unlimited, such a chart is referred to as a *c chart* (*c* for "count"). For the case of the finite range $\{0, \ldots, n\}$, a chart for $Z_t = X_t$ is said to be an *np chart*, while a *p chart* plots the relative quantities $Z_t = X_t/n$. This terminology is obviously motivated by the binomial distribution (Example A.2.1) and by the idea of a sample-based monitoring, with X_t or X_t/n expressing the absolute number or relative proportion, respectively, of "successes" in the sample being collected at time t. For simplicity, we shall always consider the case of an unlimited range (and hence *c* charts) in this section, but the presented concepts apply to *np* charts and *p* charts as well; see Section 9.1.1. A truly *two-sided c chart* has control limits $0 < l < u < \infty$ with $l, u \in \mathbb{N}$. One-sided charts are obtained by either setting $l = 0$ (*upper-sided c chart*) or $u = \infty$ (*lower-sided c chart*).

For the rest of this section, let us assume that $(X_t)_{\mathbb{N}}$ is serially independent. If the process is in-control, it is i.i.d. and has the in-control marginal distribution F_0. As an out-of-control scenario, we restrict to the case of a sudden shift; that is, at a certain time $\tau \in \mathbb{N}$ (called *change point*), the marginal distribution becomes F_1. This leads to the following (unconditional) *change point model* (Knoth, 2006):

> For $t < \tau$, the process is *in control*,
> while it is *out of control* for $t \geq \tau$ if $F_1 \neq F_0$.

For a change point $\tau = 1$, the process is out of control right from the beginning. If the control chart triggers an *alarm* at time L (rule (8.2)), we stop monitoring and conclude that the process might have run out of control. If indeed $\tau \leq L$ and $F_1 \neq F_0$, the alarm was correct; otherwise, it was a *false alarm*. In the first case, the difference $L - \tau + 1$ expresses the *delay* in detecting the change point. Here, the "+1" is used since even in the case of immediate detection $(L = \tau)$, we have one out-of-control observation (say, one defective item in a production process).

At this point, it is important to study the run length L of the control chart – see (8.2) – in more detail. If the process is in control, we wish the run length to be large (a *robust chart*), since then the run length expresses the time until the first false alarm. In contrast, it should be small for an out-of-control process, since the run length then goes along with the delay in detecting the process change. As for a significance test, the approach to designing the chart is to choose the control limits l, u in such a way that a certain degree of robustness against false alarms is guaranteed. For this purpose, one looks at properties of the in-control run length distribution. These could be quantiles, such as the median, but the main approach (although one that is sometimes criticized; see Kenett & Pollak (2012) as an example) is to consider the mean of the run length L; that is, the *average run length (ARL)*.[1] If there are several candidate designs leading to (roughly) the same in-control ARL, abbreviated as ARL_0, then one compares the out-of-control ARL performances of these charts to select the final chart design.

So the question is how to compute the ARL given a specific chart design l, u. If the process is in-control (that is, i.i.d. with marginal distribution F_0), then a signal is triggered at time t with probability

$$P_0(Z_t \notin [l; u]) = P_0(Z_t \le l - 1) + P_0(Z_t > u) = 1 - F_0(u) + F_0(l - 1).$$

Because of the *independence* of the plotted statistics, the distribution of L is a shifted geometric distribution (Example A.1.5), so it follows immediately that

$$ARL_0 = \frac{1}{1 - F_0(u) + F_0(l - 1)}. \tag{8.3}$$

Note that this formula also includes the one-sided cases by setting $F_0(-1) = 0$ and $F_0(\infty) = 1$.

If the distribution becomes out of control at time τ, then the delay in detecting this change is $L - \tau + 1$ (see above). Again, because of the independence of the plotted statistics (the *non-aging property*), this delay can still be described by a shifted geometric distribution, but using F_1 instead of F_0. Therefore, the out-of-control ARL for the considered i.i.d.-scenario is defined by setting $\tau = 1$ (since the true position of the change point does not affect the delay anyway); that is,

$$ARL_1 = \frac{1}{1 - F_1(u) + F_1(l - 1)}. \tag{8.4}$$

Note that ARLs should be interpreted with caution in practice, since the shifted geometric distribution is strongly skewed and has a large dispersion (the standard deviation nearly equals the mean; see Example A.1.5). This is illustrated by Figure 8.2, which shows the run length distribution corresponding to ARL = 250. Although an alarm is triggered *in the mean* after 250 plotted statistics, the median, for instance, equals only 173; that is, in 50% of all cases, the actual run length is not larger than 173. The quartiles range from 72 to 346, so again in 50% of all cases, the actual run length is outside even this region. For a further critical discussion, see Kenett & Pollak (2012).

---

1 Note that speaking of "the" ARL is only justified in the current section where the plotted statistics Z_t are serially independent. In Section 8.2.2, we shall recognize that there are several ways of defining an ARL if serial dependence is present.

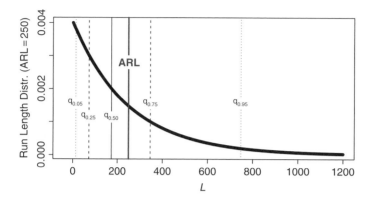

Figure 8.2 Distribution of run length L for ARL = 250, and some corresponding quantiles.

Example 8.2.1.1 (ARL performance of cchart) For low counts, it is often impossible to find a truly two-sided chart design satisfying a pre-specified requirement concerning the in-control ARL. Therefore, for illustration, let us consider the case of i.i.d. Poisson counts with the rather large in-control mean $\mu_0 = 6$. Furthermore, let us assume that the in-control ARL has to satisfy $\text{ARL}_0 \geq 250$. Then a possible c chart design is $l = 1$ and $u = 14$, leading to $\text{ARL}_0 \approx 257.8$ (note that for *discrete*-valued charts, it is usually not possible to meet a pre-specified ARL level exactly). A plot of the resulting ARL performance against varying values of the mean μ is shown in Figure 8.3 (black dots).

If the plotted function $\text{ARL}(\mu)$ would be maximal exactly for $\mu = \mu_0$, then the chart design would be said to be *ARL-unbiased*. In the present example, it is at least approximately ARL-unbiased: the plotted ARL function is roughly symmetric around $\mu = \mu_0$ such that the chart detects negative shifts in the

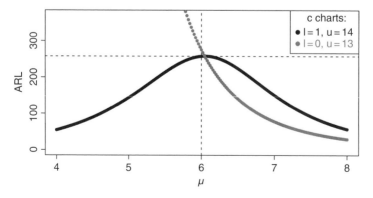

Figure 8.3 ARL performance of c chart against μ; see Example 8.2.1.1.

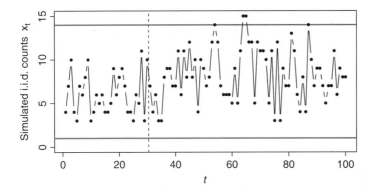

Figure 8.4 Two-sided c chart of simulated i.i.d. counts; Example 8.2.1.1 with change point $\tau = 31$.

mean just as well as positive shifts. As an example, ARL(8) $\approx$ 56.8; that is, if the mean is shifted from 6 to 8, we will require about 57 observations on average to detect such a change. The application of the two-sided c chart is illustrated in Figure 8.4, where the first 30 i.i.d. counts were simulated according to the in-control model Poi(6), and the remaining 70 i.i.d. counts (after the change point $\tau = 31$) according to the out-of-control model Poi(8). The first alarm is triggered at time $t = 64$; that is, with the delay 34.

For a positive shift such as considered above, an upper-sided c chart would certainly be more sensitive. With the chart design $l = 0$ and $u = 13$, we have ARL$_0 \approx 275.6$ and ARL(8) ≈ 29.3; see also the gray curve in Figure 8.3. Applied to the same simulated i.i.d. data as before, we obtain the chart shown in Figure 8.5, where the first alarm is now triggered at time $t = 54$ (delay 24).

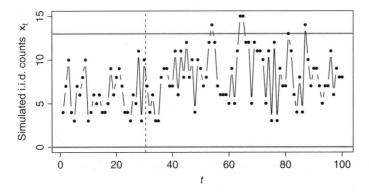

Figure 8.5 Upper-sided c chart of simulated i.i.d. counts; Example 8.2.1.1 with change point $\tau = 31$.

A possible way to achieve an ARL-unbiased c chart design that is close to any prespecified ARL_0-level was proposed by Paulino et al. (2016), and relies on a randomization of the emission of an alarm.

Usually, a control chart is designed as if the true in-control model is known precisely. In reality, however, the in-control model has to be estimated from given data (believed to stem from the presumed in-control model). Due to the uncertainty of parameter estimation, the true performance of the chart will usually deviate from the "believed" one, and this difference might be rather large. In view of (8.3) and the typically large values for ARL_0 (as in Example 8.2.1.1), the control limits correspond to rather extreme quantiles of the (estimated) in-control distribution. So already moderate misspecifications of the model parameters may lead to strong effects on the control limits and ARLs. Hence, for the data examples below, we should be aware that we always consider some kind of *conditional* ARL performance, conditioned on the fitted model.

A comprehensive literature review of the effect of estimated parameters on control chart performance is provided by Jensen et al. (2006). Probably the first such work in the attributes case is by Braun (1999), who considers the c and the p charts, while Testik (2007) investigates the effect of estimation on the CUSUM chart for i.i.d. Poisson counts; this chart is discussed in Section 8.3.1.

Example 8.2.1.2 (Effect of estimated parameters)   To get an idea of the effect of parameter estimation in the scenario considered in Example 8.2.1.1, we follow Testik (2007) and analyze the *conditional ARL performance* of the upper-sided c chart. Let $X_1, \ldots, X_n$ be the available in-control data, being i.i.d. according to $\text{Poi}(\mu_0)$. To estimate μ_0, we use the mean $\overline{X}_n$, where $n\,\overline{X}_n \sim \text{Poi}(n\,\mu_0)$ due to the additivity of the Poisson distribution (see Example A.1.1). Let $\overline{X}_n$ take the value 6, as in Example 8.2.1.1, such that we decide on limit $u = 13$ in view of obtaining $\text{ARL}_0 \geq 250$. If, in addition, the true μ_0 is 6, we would in fact have $\text{ARL}_0 \approx 275.6$. However, assuming that the true μ_0 is such that the observed value 6 equals the α-quantile of $\overline{X}_n$'s distribution, then we have the ARL_0 values shown in Table 8.1 instead. If 6 equals the lower quartile of $\overline{X}_n$, for example, then the true μ_0 is larger than 6, so we will observe false alarms more often (a situation that becomes worse with decreasing sample size n). Such a conditional performance analysis not only illustrates the possible deviations of the true ARL_0 from the intended one, it can also be used to revise the chart design. Picking up the idea of a "guaranteed conditional performance", as developed by Albers & Kallenberg (2004) and Gandy & Kvaløy (2013) for variables charts, one may assume that the observed value corresponds to, say, the 10%-quantile such that $\mu_0 \approx 6.678$ if $n = 25$. Then one uses this kind of "worst-case" value to derive the control limit, which equals 14 in the given example. For such a chart, we might feel 90% confident that the true ARL_0 is ≥ 250.

Table 8.1 Conditional ARL performance of upper-sided c chart for quantile levels α and sample sizes n; see Example 8.2.1.2.

	$\alpha = 0.25$		$\alpha = 0.50$		$\alpha = 0.75$	
n	μ_0	ARL_0	μ_0	ARL_0	μ_0	ARL_0
25	6.364	167.6	6.027	265.3	5.702	430.1
50	6.250	194.8	6.013	270.4	5.782	379.9
100	6.173	216.2	6.007	273.0	5.843	346.8
250	6.108	236.7	6.003	274.5	5.899	319.2

Table 8.2 Marginal ARLs of upper-sided c chart against sample size n.

n	25	50	100	250	500	1 000	5 000	10 000
$ARL_{0;\ marg}$	548.4	474.2	448.8	417.9	391.3	358.1	286.0	276.7

The effect of parameter estimation can also be illustrated by looking at the *marginal ARL*. For a given in-control value, say $\mu_0 = 6$, and for a given design rule, say taking u as the $(1 - 1/250)$-quantile of the fitted Poisson distribution, one computes the "expected ARL" as

$$ARL_{0;\ marg} := \sum_{s=1}^{\infty} ARL_0 \left(\overline{X}_n = s/n \right) \cdot P \left(\overline{X}_n = s/n \right),$$

where $ARL_0(\overline{X}_n = s/n)$ is computed by using the true μ_0 but varying designs, and where $P(\overline{X}_n = s/n) = p(s)$ with $p(s)$ being the pmf of $Poi(n\ \mu_0)$. We obtain the results shown in Table 8.2 that is, $ARL_{0;\ marg}$ converges only slowly to $ARL_0 \approx 275.6$ with increasing sample size n. In particular, although the marginal ARLs are larger than the intended ARL_0 level of 250, there is a non-negligible probability of ending up with $ARL_0 < 250$. If $n = 25$, for example, then this probability equals about 11.7%.

While (appropriately chosen) Shewhart charts are generally quite sensitive to very large shifts in the process (and they are also generally recommended for application in Phase I (Montgomery, 2009)), Example 8.2.1.1 has already demonstrated that these charts are not particularly well-suited to detecting small-to-moderate shifts. For this reason, in Section 8.3 we shall consider advanced control schemes that are more sensitive to small shifts, because these charts are designed to have an inherent memory.

8.2.2 Shewhart Charts for Markov-Dependent Counts

In this section, we skip the i.i.d.-assumption and allow the count process $(X_t)_{\mathbb{N}}$ to be a Markov chain, say, an INAR(1) process as in Section 2.1, a binomial AR(1) process as in Section 3.3, or an INARCH(1) process as in Example 4.1.6. But still, our aim is to plot the observed counts directly on the chart with limits $l < u$; that is, we choose again $Z_t = X_t$.

Because of the serial dependence of the plotted statistics, now the run length (8.2) no longer follows a simple geometric distribution. In addition, if we now look at the detection delay $L - \tau + 1$, the corresponding distribution generally depends on the position of the change point τ. Therefore, more refined ARL concepts have to be considered; a detailed survey of different ARL concepts is provided by Knoth (2006). In this book, the following ARL concepts are used:

- the *zero-state ARL*

$$\text{ARL} := E_1[L], \tag{8.5}$$

- the *conditional expected delay*

$$\text{ARL}^{(\tau)} := E_\tau[L - \tau + 1 \mid L \geq \tau], \tag{8.6}$$

- the *(conditional) steady-state ARL*

$$\text{ARL}^{(\infty)} := \lim_{\tau \to \infty} \text{ARL}^{(\tau)}, \tag{8.7}$$

where $E_\tau[\cdot]$ denotes the expectation related to the change point τ.

As before, we refer to the computed ARL value as the *in-control ARL* (*out-of-control ARL*) if $F_1 = F_0$ ($F_1 \neq F_0$), and the in-control ARL is signified by adding the index "0".

Obviously, the zero-state ARL is nothing other than $\text{ARL}^{(1)}$. For the case of serial independence, as considered in Section 8.2.1, we have $\text{ARL}^{(\tau)} = \text{ARL}^{(\infty)} = \text{ARL}$, but otherwise these ARLs may differ. The essential questions are:

- How can we compute these ARLs?
- Which one should be used in a specific application?

Let us start with the second question. When designing a chart, one first looks at the in-control behavior. In this context, it is reasonable to use the in-control zero-state ARL, $\text{ARL}^{(1)}$, as a measure of robustness against false alarms. Then, in a second step, one analyzes the out-of-control behavior. If there are reasons to expect, say, that the change will probably happen quite early, then it would be reasonable to evaluate the out-of-control performance of an $\text{ARL}^{(\tau)}$ with sufficiently small τ. In many applications, however, one will not have such information. However, we shall see below that $\text{ARL}^{(\tau)}$ often converges rather quickly to $\text{ARL}^{(\infty)}$. This implies that the steady-state ARL, $\text{ARL}^{(\infty)}$, might serve as a reasonable approximation for the true mean delay of detection after the unknown change point. Therefore, in this book, we shall evaluate the out-of-control performance in terms of $\text{ARL}^{(\infty)}$.

The first question is how to compute the different types of ARL. Certainly, it is always possible to approximate the ARLs through simulations; to simulate $\text{ARL}^{(\infty)}$, one will simulate $\text{ARL}^{(M)}$ with a large M as a substitute. But if considering a c chart applied to an underlying Markov chain, a numerically exact solution is also possible by using the *Markov chain* (MC) approach proposed by Brook & Evans (1972). Since this approach can also be used for the advanced control charts to be introduced below, we provide a rather general description in the sequel following Weiß (2011b).

In view of decision rule (8.1), we can assume a slightly simplified range for the plotted statistics Z_t: their range S is partitioned into the set Q of "no-alarm states" (because no alarm is triggered by the chart if Z_t takes a value in Q) and the set $\{a\}$ consisting of a single "alarm state" 'a'. This is justified since any kind of violation of the control limits will lead to the same action: stop the process and search for an assignable cause. Therefore, 'a' is an *absorbing state*; that is, it is no longer possible to leave this state. The set Q is equal to $\{l, \dots, u\}$ for the case of a two-sided c chart.

The MC approach now assumes a conditional change point model (Weiß, 2011b), as given in Definition 8.2.2.1; see also the survey about Markov chains in Appendix B.2.

Definition 8.2.2.1 (MC change point model) Let $(Z_t)_{\mathbb{N}}$ be a finite Markov chain with state space $S = Q \cup \{a\}$. We assume that

- $(Z_t)_{\mathbb{N}}$ is homogeneous before and after the change point $\tau \in \mathbb{N}$, and we denote the time-invariant transition probabilities

$$P(Z_t = i \mid Z_{t-1} = j) \quad \text{by} \quad \begin{cases} p_{i|j} & \text{for } t < \tau, \\ \tilde{p}_{i|j} & \text{for } t \geq \tau; \end{cases}$$

- the states in Q are *inessential* (Example B.2.1.1), while the absorbing state 'a' is essential; that is, for each $j \in Q$, there exists a lag $h \geq 1$ such that the h-step-ahead transition probability $p_{a|j}^{(h)} > 0$ (analogously with $\tilde{p}$);
- $(Z_t)_{t=1,\dots,\tau-1}$ is stationary.

If $\tilde{p}_{i|j} = p_{i|j}$ for all $i, j \in \{0, \dots, n\}$, then the whole process $(Z_t)_{\mathbb{N}}$ is stationary according to the in-control model. Furthermore, since 'a' is an absorbing state by definition, we have $p_{i|a} = \delta_{i,a} = \tilde{p}_{i|a}$ for all $i \in S$, where $\delta_{i,a}$ denotes the Kronecker delta. The requirement that Q consists of inessential states guarantees, among other things, that the probability of reaching 'a' in finite time equals 1.

Let us now describe the procedures for computing the different types of ARL; derivations and more details can be found in Brook & Evans (1972) and Weiß (2011b). For this purpose, define $\mathbf{Q}$ to be the transpose of the transition matrix for the states in Q; that is, $\mathbf{Q}^{\top} := (p_{i|j})_{i,j \in Q}$. Analogously, we set $\tilde{\mathbf{Q}}^{\top} := (\tilde{p}_{i|j})_{i,j \in Q}$. The requirement that Q consist of inessential states guarantees that the *fundamental matrices* $(\mathbf{I} - \mathbf{Q})^{-1}$ and $(\mathbf{I} - \tilde{\mathbf{Q}})^{-1}$ exist, where $\mathbf{I}$ denotes the identity matrix. Since 'a' is an absorbing state, the transition matrices of $(Z_t)_{\mathbb{N}}$ before and after the change point, respectively, are given by

$$P = \begin{pmatrix} Q^\top & 0 \\ 1^\top(I - Q^\top) & 1 \end{pmatrix} \quad \text{and} \quad \tilde{P} = \begin{pmatrix} \tilde{Q}^\top & 0 \\ 1^\top(I - \tilde{Q}^\top) & 1 \end{pmatrix}. \tag{8.8}$$

To compute the out-of-control *zero-state ARL* (for the in-control ARL, we just have to replace $\tilde{Q}$ by Q), we first compute the unique solution of the equation

$$(I - \tilde{Q})\mu = 1. \tag{8.9}$$

Here, the entries μ_j express the mean time to reach 'a' if $Z_1 = j$. After having specified the initial probabilities $P(Z_1 = j)$ for $j \in Q$ (see Remark 8.2.2.2), we collect these probabilities in the vector $\tilde{\pi}(1)$ (note that the change already happened at time 1). Then

$$\text{ARL} = \text{ARL}^{(1)} = 1 + \mu^\top \tilde{\pi}(1). \tag{8.10}$$

If $\tau \geq 2$, then there exist $\tau - 1$ in-control observations. So the entries μ_j of the solution to (8.9) express the mean delay to reach 'a' if $Z_{\tau-1} = j$. If the vector $\pi(1)$ consists of the probabilities $P(Z_1 = j)$ and if $\pi(\tau - 1) = (Q^\top)^{\tau-2}\pi(1)$ refers to the $P(Z_{\tau-1} = j)$ (Markov property, in control), then the *conditional expected delay* equals

$$\text{ARL}^{(\tau)} = \mu^\top \frac{\pi(\tau - 1)}{1^\top \pi(\tau - 1)} \quad \text{for } \tau \geq 2. \tag{8.11}$$

Finally, to compute the *steady-state ARL*, we need to take the limit $\lim_{\tau \to \infty} \pi(\tau - 1) = \lim_{\tau \to \infty}(Q^\top)^{\tau-2}\pi(1)$ according to (8.7). To be able to apply the *Perron–Frobenius theorem* – see Remark B.2.2.1 in Appendix B.2 for a summary – we have to assume that the non-negative matrix $Q^\top$ is primitive. For the corresponding Perron–Frobenius eigenvalue $\lambda_{\text{PF}} \leq 1$, there exists a strictly positive right eigenvector w; $w/(1^\top w)$ is the normed version of w. Then

$$\text{ARL}^{(\infty)} = \mu^\top \cdot \frac{w}{1^\top w}, \tag{8.12}$$

where the rate of convergence of $\text{ARL}^{(\tau)} \to \text{ARL}^{(\infty)}$ for $\tau \to \infty$ is determined by the second largest eigenvalue, which satisfies $|\lambda_2| < 1$.

Remark 8.2.2.2 (Initial probabilities) Before looking at a comprehensive example, just a few words concerning the probabilities $\tilde{\pi}(1)$ that are required for the zero-state ARL (8.10). The change point model in Definition 8.2.2.1 does not specify the distribution of Z_1 if already Z_1 is out of control. A simple approach would be to choose the stationary marginal distribution corresponding to the out-of-control transition probabilities $\tilde{p}_{i|j}$. But then, the process would be strictly stationary during its out-of-control state, which would differ from the behavior for $\tau \geq 2$. Another solution is to assume an invisible observation Z_0 stemming from the stationary in-control model; that is, the initial probability $P(Z_1 = i)$ is computed as $\sum_{j=0}^{\infty} p_j \cdot \tilde{p}_{i|j}$. We shall prefer this second solution in the sequel.

Example 8.2.2.3 (*c* **chart for IP counts data**)   Monitoring the activity of a web server can help detect intrusions into the server, or evaluate the attractiveness of a website. In the following, we consider the IP counts example presented by Weiß (2007), where each count represents the number of different IP addresses registered within periods of 2-min length. So the data give insights into the number of different users accessing the web server per time interval. We shall proceed in two steps: the available historical data is used for a Phase-I analysis to identify an appropriate in-control model. The control charts based on this model are then applied during Phase II to monitor "new" (simulated) data to detect a possible out-of-control situation.

For illustration, we assume the time series collected on 29 November 2005, between 10 a.m. and 6 p.m. as the available in-control data (length $T = 241$). They have mean $\bar{x} \approx 1.392$. As already argued in Weiß (2007), the data exhibit an AR(1)-like autocorrelation structure with $\hat{\rho}(1) \approx 0.219$ and a nearly equidispersed marginal distribution (dispersion index $\hat{I} \approx 1.054$, which is not significant; see (2.14) in Section 2.3). Therefore, it is reasonable to try to fit the Poisson INAR(1) model from Section 2.1.2 to the data. An INAR(1) model is also plausible in view of interpretation (2.4), since a user who is active at time t might also have been active at time $t - 1$. The full ML estimates for the Poisson INAR(1) model (see Section 2.2.2) are $\hat{\mu}_{\epsilon;\ \mathrm{ML}} \approx 0.997$ (std. err. 0.099) and $\hat{\alpha}_{\mathrm{ML}} \approx 0.243$ (std. err. 0.062).

This preliminary model is next used to find a *c* chart design, the zero-state in-control ARL (8.5) of which is required to be $> T$. Due to the low counts, only an upper-sided design is possible; that is, $l = 0$. Trying increasing values for the upper limit, $u = \ldots, 4, 5, 6, \ldots$, we get $\mathrm{ARL}_0 \approx \ldots, 94.9, 433.2, 2321.6, \ldots$, so we decide on $u = 5$ as the chart design. The resulting *c* chart was shown in Figure 8.1, and it triggers an alarm at time $t = 224$ (observation $x_{224} = 8$). Prompted by this alarm, Weiß (2007) did a further analysis of the full server log data and found out that all eight IP addresses at time $t = 224$ seem to be due to just one user, but who was routed into the internet through an area of changing IP addresses (instead of one unique IP address). Therefore, it is reasonable to correct this "outlier" by setting $x_{224} := 1$.

The above steps of the Phase-I analysis are repeated with the corrected data ($\bar{x} \approx 1.286$, $\hat{\rho}(1) \approx 0.292$, $\hat{I} \approx 0.933$), which leads to slightly different ML estimates – $\hat{\mu}_{\epsilon;\ \mathrm{ML}} \approx 0.905$ (std.err. 0.093) and $\hat{\alpha}_{\mathrm{ML}} \approx 0.298$ (std.err. 0.061) – but the same chart design, now with $\mathrm{ARL}_0 \approx 489.3$. Since no further points violate the upper limit $u = 5$, Phase-I analysis finishes with the aforementioned in-control model. Also further checks for model adequacy, see Section 2.4, confirm that the corrected data are well described by the fitted Poisson INAR(1) model.

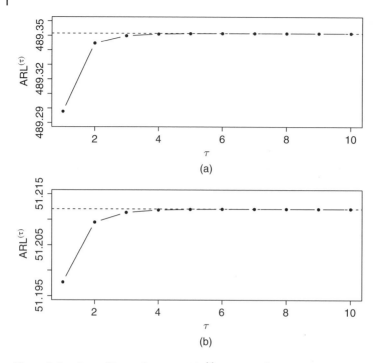

Figure 8.6 c chart of Example 8.2.2.3: $\text{ARL}^{(\tau)}$ against τ if (a) $\mu_\epsilon = \mu_{\epsilon,0}$, (b) $\mu_\epsilon = 1.5$.

The final in-control model is used for designing the control charts to be used during Phase II (without further considering the effect of estimation error according to the discussion before Example 8.2.1.2). For the moment, we leave it at the c chart with $u = 5$ and $\text{ARL}_0 \approx 489.3$. Before applying this chart for the prospective monitoring of new data, we first analyze the different types of conditional expected delay (8.6). The graphs in Figure 8.6 show the $\text{ARL}^{(\tau)}$ plotted against increasing τ, with the dashed line indicating the value of the steady-state ARL (8.7). Both in the in-control situation (ML-fitted model) and a particular out-of-control situation (innovations' mean $\mu_\epsilon = 1.5$), $\text{ARL}^{(\tau)}$ converges very quickly to $\text{ARL}^{(\infty)}$, such that $\text{ARL}^{(\infty)}$ becomes an excellent approximation for the true mean delay of detection; see the above discussion. This behavior is reasonable, since the second largest eigenvalue (with multiplicity 1) is rather small, taking a value of about 0.288, while the Perron–Frobenius eigenvalue $\lambda_{\text{PF}} \approx 0.998$. For the out-of-control zero-state ARL (8.10), the "invisible in-control X_0"-approach was used (Remark 8.2.2.2); if X_1 is assumed to follow the stationary out-of-control model, then $\text{ARL}^{(1)}$ would change from ≈ 51.2 to ≈ 50.6 for $\mu_\epsilon = 1.5$.

Generally, we see that the c chart is not particularly effective in detecting out-of-control situations; even in the case $\mu_\epsilon = 1.5$ (a positive shift by

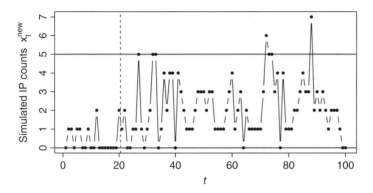

Figure 8.7 *c* chart of simulated IP counts; Example 8.2.2.3 and change point $\tau = 21$.

more than 50% compared to $\mu_{\epsilon,0} \approx 0.905$), it takes around 51 observations in the mean until an alarm is triggered. For illustration, we pick up this out-of-control scenario in the following way: 100 further observations from a Poisson INAR(1) process were simulated, initialized by the last observation $x_{241} = 0$ and following the in-control model for the first 20 observations, and then changing to $\mu_\epsilon = 1.5$ for the remaining 80 observations. So the true change point for $x_1^{\text{new}}, \ldots, x_{100}^{\text{new}}$, given by

$$0, 1, 1, 0, 1, 1, 0, 0, 1, 0, 0, 2, 0, 0, 0, 0, 0, 0, 0, 2, 1, 2, 0, 0, 1, 1, 5, 1, 0, 1, 2,$$
$$5, 5, 0, 1, 4, 2, 4, 4, 0, 4, 3, 2, 1, 1, 1, 2, 3, 3, 3, 2, 3, 3, 1, 1, 1, 1, 2, 3, 4, 1, 2,$$
$$3, 0, 2, 1, 1, 1, 1, 1, 3, 6, 5, 5, 3, 4, 0, 4, 1, 1, 1, 2, 2, 3, 4, 3, 3, 7, 2, 3, 2, 3, 2,$$
$$1, 2, 2, 2, 1, 0, 0,$$

equals $\tau = 21$. The corresponding c chart is shown in Figure 8.7. The first alarm is signaled at time $t = 72$; that is, with a delay of 52.

8.3 Advanced Control Charts for Count Processes

The basic c chart presented in Section 8.2 allows for continuous monitoring of a serially dependent count process. But the statistic plotted on the c chart at time t, which is simply the count observed at time t, does not include any information about the past observations of the process, or at least not explicitly, beyond the mere effect of autocorrelation. Therefore, the c chart (as any other Shewhart-type chart) is not particularly sensitive to small changes in the process. For this reason, several types of advanced control charts have been proposed, in which the plotted statistic at time t also uses past observations of the process and hence accumulates information about it for a longer period of time. In the sequel, we will discuss the most popular types of advanced control chart: CUSUM charts in Sections 8.3.1 and 8.3.2, and EWMA charts in Section 8.3.3. Further charts and references can be found in Woodall (1997) and Weiß (2015b).

8.3.1 CUSUM Charts for i.i.d. Counts

The traditional *cumulative sum* (CUSUM) control chart, being applied directly to the observations X_t of the process, is perhaps the most straightforward advanced candidate for monitoring processes of counts, because it preserves the discrete nature of the process by only using addition (but no multiplications). Initialized by a starting value $c_0^+ \geq 0$, the *upper-sided CUSUM* is defined by

$$C_0^+ = c_0^+, \qquad C_t^+ = \max\{0, X_t - k^+ + C_{t-1}^+\} \quad \text{for } t = 1, 2, \ldots, \qquad (8.13)$$

that is, by accumulating the deviations from the reference value $k^+ > 0$. Because of this accumulation, the plotted statistic at time t is not solely based on X_t but also incorporates the process in the past: $X_{t-1}, X_{t-2}, \ldots$ If the CUSUM statistic becomes negative, the $\max\{0, \cdot\}$ construction resets the CUSUM to zero.

The starting value is commonly chosen as $c_0^+ = 0$; a value $c_0^+ > 0$ is referred to as a *fast initial response* (FIR) feature, and it may help to detect an initial out-of-control state more quickly; see also the discussion below formulae (8.5)–(8.7). If k^+ and c_0^+ are taken as integer values, then also $(C_t^+)_{\mathbb{N}_0}$ is integer-valued. As another example, if $k^+, c_0^+ \in \{0, 1/2, 1, 3/2, \ldots\}$ then so is C_t^+, but in any case, we have a discrete range. In the sequel, we shall concentrate on integer-valued k^+, c_0^+. An alarm is triggered if C_t^+ violates the upper control limit $h^+ > 0$ (typically, $h^+ \geq k^+$).

While the upper-sided CUSUM is designed to detect increases in the process mean, the *lower-sided CUSUM*, defined by

$$C_0^- = c_0^-, \qquad C_t^- = \max\{0, k^- - X_t + C_{t-1}^-\} \quad \text{for } t = 1, 2, \ldots, \qquad (8.14)$$

aims at uncovering decreases in the mean. If (C_t^+, C_t^-) are monitored simultaneously, then this chart combination is referred to as a *two-sided CUSUM chart*. A book with a lot of background information about CUSUM charts is the one by Hawkins & Olwell (1998).

In this section, we assume the monitored count process $(X_t)_{\mathbb{N}}$ to be i.i.d. in its in-control state, a situation that was also considered in the article by Brook & Evans (1972). Because of the accumulation according to (8.13), however, the statistics $(C_t^+)_{\mathbb{N}_0}$ are no longer i.i.d., but constitute a Markov chain (analogous arguments apply to the lower-sided CUSUM (8.14)) with transition probabilities

$$P(C_t^+ = a \mid C_{t-1}^+ = b) = \begin{cases} P(X_t = k^+ - b + a) & \text{if } a > 0, \\ P(X_t \leq k^+ - b) & \text{if } a = 0, \end{cases}$$

and the initial statistic satisfies $P(C_1^+ = a) = P(C_1^+ = a \mid C_0^+ = c_0^+)$. Therefore, the MC approach as described in Section 8.2.2 is applicable, with $Q = \{0, \dots, h^+\}$. In fact, Brook & Evans (1972) introduced their MC approach for exactly this type of control chart and considered the application to i.i.d. Poisson counts.

Example 8.3.1.1 (ARL performance of CUSUM chart) Let us pick up the situation of Example 8.2.1.1, where we assumed i.i.d. Poisson counts with in-control mean $\mu_0 = 6$. Among others, we applied an upper-sided c chart with limits $l = 0$ and $u = 13$ to this process, leading to $\mathrm{ARL}_0 \approx 275.6$.

For an upper-sided CUSUM chart with $c_0^+ = 0$, we choose $k^+ = 7$ close to but larger than μ_0. Trying different values for h^+, we finally choose $h^+ = 11$ for the upper limit such that the zero-state in-control ARL is close to the above one: $\mathrm{ARL}_0 \approx 252.6$. While all types of conditional expected delay (8.6) coincide for the i.i.d. Shewhart charts, they lead to different values for the CUSUM chart. This is illustrated by Figure 8.8, where $\mathrm{ARL}^{(\tau)}$ is plotted against τ (dashed line: $\mathrm{ARL}^{(\infty)}$ from (8.7)): once for the in-control scenario, then again for the out-of-control scenario $\mu = 8$. Furthermore, a slightly modified CUSUM design, with an additional FIR feature ($c_0^+ = 3$), is also considered.

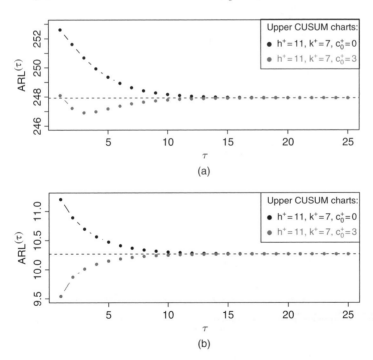

Figure 8.8 CUSUM charts of Example 8.3.1.1: $\mathrm{ARL}^{(\tau)}$ against τ if (a) $\mu = \mu_0$, (b) $\mu = 8$.

It becomes clear that ARL$^{(\tau)}$ again converges quickly to ARL$^{(\infty)}$, similar to Example 8.2.2.3, with the rate of convergence being determined by the second largest eigenvalue ≈ 0.695 (multiplicity 1). The FIR feature leads to decreased ARL$^{(\tau)}$ for small τ (note the non-monotonic behavior in (a)), but the effect vanishes with increasing τ. Hence, a FIR feature just influences the detection of very early process changes. Although it was not possible to calibrate the c chart and CUSUM charts perfectly, it becomes clear that the CUSUM chart is more sensitive to $\mu = 8$, with steady-state ARL ≈ 10.3 instead of ≈ 29.3. A more detailed comparison is provided in Example 8.3.3.1. For illustration, the CUSUM chart (without FIR feature) is applied to the simulated i.i.d. data from Example 8.2.1.1 (change point $\tau = 31$, μ shifts from 6 to 8), see Figure 8.9. The first alarm is triggered at time $t = 43$; that is, with the delay 13.

Finally, while the zero-state ARL is generally rather artificial when evaluating the out-of-control performance, it has a nice interpretation for the CUSUM chart without a FIR feature: it expresses the mean delay of detecting the change point if the CUSUM statistics are in the least favorable state (namely 0) at the time of the change point ("worst-case scenario").

We conclude this section by pointing out the relationship between the CUSUM scheme (8.13) and the *sequential probability ratio test* (SPRT); see Sections 6.1–6.2 in Hawkins & Olwell (1998) for more details. The likelihood function (see Remark B.2.1.2) for i.i.d. counts is given by

$$L(\theta) := P(X_T = x_T, \ldots, X_1 = x_1 \mid \theta) = \prod_{t=1}^{T} p_{x_t}(\theta),$$

so we obtain the likelihood ratio (LR) as

$$LR(\theta_0, \theta_1) := \frac{L(\theta_1)}{L(\theta_0)} = \prod_{t=1}^{T} \frac{p_{x_t}(\theta_1)}{p_{x_t}(\theta_0)}.$$

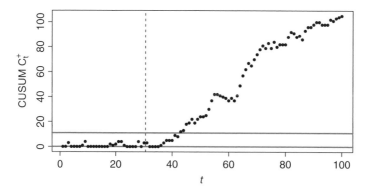

Figure 8.9 CUSUM chart of simulated i.i.d. counts in Example 8.3.1.1; change point $\tau = 31$.

The SPRT now monitors the logarithmic likelihood ratio (log-LR)

$$\ell R(\theta_0, \theta_1) := \ln LR(\theta_0, \theta_1) = \sum_{t=1}^{T} \underbrace{(\ln p_{x_t}(\theta_1) - \ln p_{x_t}(\theta_0))}_{=: \ \ell R_t} \tag{8.15}$$

for increasing T. This procedure can be rewritten recursively by accumulating the contributions ℓR_t to the log-LR at times $t \geq 1$, thus leading to a type of one-sided CUSUM scheme:

$$\tilde{C}_0 = 0, \qquad \tilde{C}_t = \tilde{C}_{t-1} + \ell R_t.$$

Note the relation to the random walk in Example B.1.6. Comparing this type of CUSUM recursion with the one given in (8.13), we see that the $\max\{0, \ \cdot\}$ construction is missing, so this CUSUM is not reset to zero if the CUSUM statistic becomes negative. As pointed out by Lorden (1971), the CUSUM (8.13) is equivalent to monitoring a slight modification of (8.15):

$$\max_{1 \leq j \leq t} \sum_{s=j}^{t} \ell R_s \qquad \text{for } t = 1, 2, \dots \tag{8.16}$$

If this statistic was not positive at time $t - 1$ but $\ell R_t > 0$, then the statistic at time t just equals ℓR_t, which corresponds to the above resetting feature.

Example 8.3.1.2 (Log-LR CUSUM for Poisson counts) Let us look at the example of i.i.d. Poisson counts with parameter values μ_0 or μ_1, where $\mu_1 > \mu_0$. Then we obtain

$$L(\mu) = e^{-T\mu} \prod_{t=1}^{T} \mu^{x_t} \ \bigg/ \ \left(\prod_{t=1}^{T} x_t! \right),$$

$$LR(\mu_0, \mu_1) = e^{-T(\mu_1 - \mu_0)} \prod_{t=1}^{T} \left(\frac{\mu_1}{\mu_0} \right)^{x_t},$$

$$\ell R_t = -(\mu_1 - \mu_0) + x_t \, (\ln \mu_1 - \ln \mu_0).$$

Accumulating the re-scaled contributions $\ell R_t / (\ln \mu_1 - \ln \mu_0)$, we arrive at the upper-sided CUSUM

$$C_0^+ = 0, \quad C_t^+ = \max \left\{ 0, \ X_t - \underbrace{\frac{\mu_1 - \mu_0}{\ln \mu_1 - \ln \mu_0}}_{=: \ k^+} + C_{t-1}^+ \right\} \qquad \text{for } t = 1, 2, \dots$$

Therefore, if we have a reasonable out-of-control scenario that is to be detected as early as possible, the optimal choice for k^+ is $\frac{\mu_1 - \mu_0}{\ln \mu_1 - \ln \mu_0}$. As an example, if we compute this value for $\mu_1 = 8$ and $\mu_0 = 6$, as in Example 8.3.1.1, then we obtain about 6.952. So our choice $k^+ = 7$ above was indeed reasonable from this perspective.

Remark 8.3.1.3 (Log-LR CUSUM) Let us look back again to the log-LR CUSUM scheme (8.16), which, if omitting the logarithm, equals

$$\max_{1\le j\le t} \prod_{s=j}^{t} \frac{p_{x_s}(\theta_1)}{p_{x_s}(\theta_0)} = \max_{1\le j\le t} \frac{p_{x_j}(\theta_1)\cdots p_{x_t}(\theta_1)}{p_{x_j}(\theta_0)\cdots p_{x_t}(\theta_0)}.$$

As argued by Kenett & Pollak (1996), we can rewrite this as

$$\max_{1\le j\le t} \frac{p_{x_1}(\theta_0)\cdots p_{x_{j-1}}(\theta_0)\cdot p_{x_j}(\theta_1)\cdots p_{x_t}(\theta_1)}{p_{x_1}(\theta_0)\cdots p_{x_t}(\theta_0)},$$

so j corresponds to a possible position of the change point. Because of the max-imization in j, the statistic indeed selects the most probable position of the change point in view of the available data (maximum likelihood principle).

This illustrates another nice feature of the CUSUM approach (8.13), see also Section 1.9 in Hawkins & Olwell (1998): if the CUSUM chart $(C_t^+)_{t=1,2,\ldots}$ triggers an alarm (for the first time) at $t = n$, and if the CUSUM statistic was equal to zero for the last time at $t = m < n$, then m is the estimated position of the change point, and $X_{m+1}, \ldots, X_n$ might be used to estimate the out-of-control value θ_1 of the process parameters. In the data example discussed in Example 8.3.1.1 and shown in Figure 8.9, the last zero before the alarm ($n = 43$) is at time $m = 35$, so the position of the change point is estimated at 35 (while the true position is $\tau = 31$). The estimate of the out-of-control mean equals

$$k^+ + \frac{c_n^+ - c_m^+}{n - m} = 8.5 = \frac{1}{n - m} \sum_{t=m+1}^{n} x_t.$$

To summarize, the CUSUM chart not only allows us to detect that there was a change in the process, it also allows us to estimate the position of the change point and the extent of the change.

8.3.2 CUSUM Charts for Markov-dependent Counts

Now, let us turn back to the case of a Markov-dependent count process $(X_t)_{\mathbb{N}}$, as in Section 8.2.2. If we apply the upper-sided CUSUM scheme (8.13) to such a process, then the statistics $(C_t^+)_{\mathbb{N}_0}$ no longer constitute a Markov chain, so the MC approach of Brook & Evans (1972) is not directly applicable. But, as shown in Weiß & Testik (2009) and Weiß (2011b), ARL computations are possible by considering the *bivariate* process $(X_1, C_1^+), (X_2, C_2^+), \ldots$, which is a bivariate

Markov chain with transition probabilities

$$p_{x,c|y,d} := P(X_t = x, C_t^+ = c \mid X_{t-1} = y, C_{t-1}^+ = d)$$
$$= \delta_{c,\max\{0,\,x-k^++d\}} \cdot p_{x|y},$$

$$\pi_{x,c}(1) := P(X_1 = x, C_1^+ = c \mid C_0^+ = c_0^+)$$
$$= \delta_{c,\max\{0,\,x-k^++c_0^+\}} \cdot p_x.$$

(8.17)

In view of the CUSUM decision rule, it is clear that the set Q of "no-alarm states" is contained in $\mathbb{N}_0 \times \{0, \dots, h^+\}$. However, since values of X_t larger than $k^+ + h^+$ will always push C_t^+ beyond h^+, the set Q is indeed finite. Excluding impossible transitions (say, from an alarm state back to a no-alarm state), Weiß & Testik (2009) showed that

$$Q := \{(x,c) \in \mathbb{N}_0^2 \mid c \leq h^+,\ c + k^+ - h^+ \leq x \leq c + k^+\},$$

(8.18)

which is of size $s := \frac{1}{2}(h^+ - k^+ + 1)(h^+ + k^+ + 2) + (h^+ + 1)k^+$. So the matrices $\mathbf{Q}, \tilde{\mathbf{Q}}$ required for the MC approach (8.8) are of dimension s^2, which will often be a rather large number. It should be noted, however, that many entries of $\mathbf{Q}, \tilde{\mathbf{Q}}$ will be equal to 0 according to (8.17); that is, $\mathbf{Q}, \tilde{\mathbf{Q}}$ are sparse matrices. Therefore, the MC approach for ARL computation can be implemented efficiently using sparse matrix techniques; see Section 3 in Weiß (2011b) for possible software solutions.

Example 8.3.2.1 (CUSUM chart for IP counts data) Let us continue Example 8.2.2.3, where we designed a c chart with $u = 5$ and $\text{ARL}_0 \approx 489.3$ for the fitted Poisson INAR(1) model. The corresponding marginal mean is about 1.289, so for the upper CUSUM chart, it is reasonable to try $k^+ = 2, 3$. In view of obtaining a zero-state in-control ARL close to the above value of 489.3, we ultimately arrive at the following designs: $(h^+, k^+, c_0^+) = (8, 2, 7)$ (with FIR feature) and $(3, 3, 0)$ (without FIR feature), leading to $\text{ARL}_0 \approx 489.0$ and $\text{ARL}_0 \approx 475.6$, respectively. Before further analyzing these designs, a few technical details. For design 1 – that is, $(h^+, k^+) = (8, 2)$ – we have matrix dimension 60×60 (see also the formula for s below (8.18)), but among the 3600 entries, only 371 are non-zero. For design 2 – that is, $(h^+, k^+) = (3, 3)$ – we have $16^2 = 256$ entries, 88 of which are non-zero. The Perron–Frobenius eigenvalues are both around 0.998, but, more interesting for us, the second largest eigenvalues are about 0.732 and 0.294, respectively, each with multiplicity 1. So we expect the convergence $\text{ARL}^{(\tau)} \to \text{ARL}^{(\infty)}$ to be much more slow for design 1 than for design 2.

Looking at the graphs in Figure 8.10, this difference in convergence speed becomes obvious. Furthermore, the first values of $\text{ARL}^{(\tau)}$ differ markedly from $\text{ARL}^{(\infty)}$ for design 1, because of the FIR feature $c_0^+ = 7$. The effect of this FIR feature essentially vanishes if $\tau > 10$; that is, if the change point does not happen during the first ten observations after the start of monitoring. Then, on average,

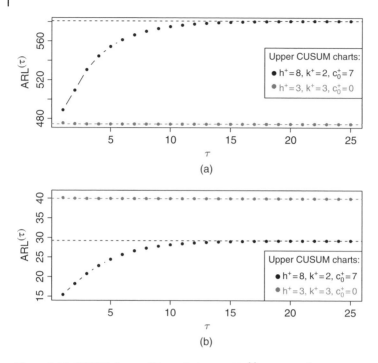

Figure 8.10 CUSUM charts of Example 8.3.2.1: $\mathrm{ARL}^{(\tau)}$ against τ if (a) $\mu_{\epsilon} = \mu_{\epsilon,0}$, (b) $\mu_{\epsilon} = 1.5$.

design 1 will be much more robust against false alarms than design 2. But if we have a change in μ_{ϵ} from $\mu_{\epsilon,0} \approx 0.905$ to a value of 1.5, then design 1 is indeed more sensitive on average (see also below).

Now we apply both CUSUM charts to the simulated data from Example 8.2.2.3; see Figure 8.11. While the c chart triggers its first alarm with a delay of 52, design 1 signals at time $t = 39$ (delay 19), and design 2 at $t = 33$ (delay 13). So both are much faster than the c chart in this example, and design 2 is faster than design 1, although on average, it will be the other way round.

To conclude this example, let us compare the steady-state out-of-control performance between the three considered charts against increasing values of μ_{ϵ}. Figure 8.12 shows that the CUSUM design 1, although being most robust against false alarms in the in-control case, quickly leads to the smallest $\mathrm{ARL}^{(\infty)}$ values. Only for very large shifts do the $\mathrm{ARL}^{(\infty)}$ performances equate again.

The idea of applying the MC approach to the bivariate process of observed counts and CUSUM statistics also essentially applies to the lower-sided CUSUM scheme (8.14), but the set Q then becomes infinite; that is, ARLs can only be computed approximately (see Yontay et al. (2013) for details).

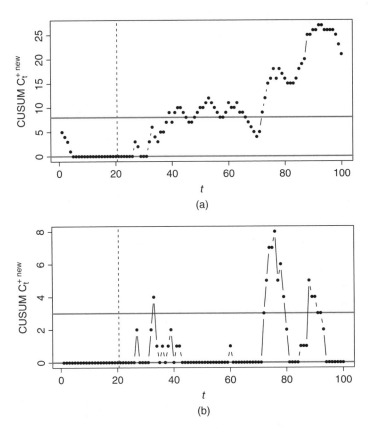

Figure 8.11 CUSUM charts of simulated IP counts for Example 8.3.2.1 with change point $\tau = 21$: (a) $(h^+, k^+, c_0^+) = (8, 2, 7)$, (b) $(h^+, k^+, c_0^+) = (3, 3, 0)$.

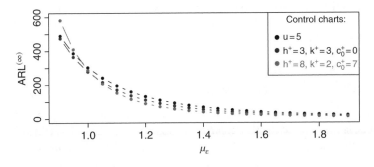

Figure 8.12 c chart and CUSUM charts for Example 8.3.2.1: $\text{ARL}^{(\infty)}$ against μ_ε.

If using a two-sided scheme, then the MC approach has to be applied to the trivariate Markov chain $(X_t, C_t^+, C_t^-)_{\mathbb{N}}$. Although here, the set Q is finite again, computations become very slow because of the immense matrix dimensions; more details and feasible approximations are presented by Yontay et al. (2013).

Remark 8.3.2.2   (Log-LR CUSUM charts)   Besides applying the standard CUSUM schemes (8.13) and (8.14), one may also look at the log-likelihood ratio (log-LR) related to the Markov model for $(X_t)_{\mathbb{N}}$, as in (8.15) and Example 8.3.1.2. From the Markov property, it follows that the contribution to the log-LR of the tth observation equals

$$\ell R_1 = \ln\left(\frac{P_{\mu_1}(X_1)}{P_{\mu_0}(X_1)}\right), \qquad \ell R_t = \ln\left(\frac{P_{\mu_1}(X_t|X_{t-1})}{P_{\mu_0}(X_t|X_{t-1})}\right) \quad \text{for } t \geq 2.$$

If the log-LR approach is applied to an INAR(1) process, then the required transition probabilities are given by (2.5), by (3.21) for a binomial AR(1) model, by (4.6) for a (Poisson) INARCH(1) model, or by (4.11) for a binomial INARCH(1) model. As long as the transition probabilities for $(X_t)_{\mathbb{N}}$ are of a feasible form, this approach will lead to a useful LR-CUSUM scheme. This was exemplified by Weiß & Testik (2012) for the case of the INARCH(1) model from Example 4.1.6, where (4.6) leads to

$$\begin{aligned}
\ell R_t &= \ln\left(\frac{P_{\beta_1,\alpha_1}(X_t|X_{t-1})}{P_{\beta_0,\alpha_0}(X_t|X_{t-1})}\right) \\
&= -(\beta_1 - \beta_0) - (\alpha_1 - \alpha_0)\, X_{t-1} + X_t\, \ln\left(\frac{\beta_1 + \alpha_1\, X_{t-1}}{\beta_0 + \alpha_0\, X_{t-1}}\right)
\end{aligned}$$

for $t \geq 2$.

Although we exemplified the log-LR approach for Markov count processes here, it can also be used for completely different types of count process. As an example, Höhle & Paul (2008) derived such a log-LR CUSUM chart for counts stemming from the seasonal log-linear model (5.6), which proved to be useful for the surveillance of epidemic counts. A related study is the one by Sparks et al. (2010).

8.3.3   EWMA Charts for Count Processes

Another advanced approach for process monitoring, which is also very popular in applications, is the *exponentially weighted moving-average* (EWMA) control chart, which dates back to Roberts (1959). The standard EWMA recursion is defined by

$$Z_0 = z_0, \quad Z_t = \lambda \cdot X_t + (1 - \lambda) \cdot Z_{t-1} \quad \text{for } t = 1, 2, \dots \tag{8.19}$$

with $\lambda \in (0; 1]$; that is, it is a weighted mean of all available observations, where the weights decrease exponentially with increasing time lag j:

$$Z_t = \lambda \sum_{j=0}^{t-1} (1 - \lambda)^j X_{t-j} + (1 - \lambda)^t z_0.$$

An application of (8.19) to the case of Poisson counts was presented by Borror et al. (1998). The EWMA recursion (8.19), however, has an important drawback compared to the CUSUM approach of Sections 8.3.1 and 8.3.2 if applied to count processes: it does not preserve the discrete range, except the boundary case $\lambda = 1$, which just corresponds to a c chart. On the contrary, the range of possible values of Z_t changes in time, which rules out, among other things, the possibility of an exact ARL computation by the MC approach. As a simple numerical example, assume that $z_0 = 1$ and $\lambda = \frac{1}{2}$; then Z_1 takes a value in $\{\frac{1}{2}, 1, \frac{3}{2}, \ldots\}$ and Z_2 in $\{\frac{1}{4}, \frac{1}{2}, \frac{3}{4}, \ldots\}$, and so on.

Therefore, Gan (1990a) suggests plotting rounded values of the statistic (8.19):

$$Q_t = \text{round}(\lambda \cdot X_t + (1 - \lambda) \cdot Q_{t-1}) \quad \text{for } t = 1, 2, \ldots \tag{8.20}$$

with $\lambda \in (0; 1]$, which are initialized by $Q_0 := q_0 \in \mathbb{N}_0$. Note that the statistics Q_t can take only integer values from $\mathbb{N}_0$, and $\lambda = 1$ again leads to a c chart. q_0 might be chosen as the rounded value of the in-control mean. An alarm is triggered if Q_t violates one of the control limits $0 \le l < u$.

In the i.i.d. case, as considered by Gan (1990a), the statistics $(Q_t)_{\mathbb{N}}$ constitute a Markov chain with transition probabilities

$$P(Q_t = a \mid Q_{t-1} = b) = P\left(\lambda X_t + (1 - \lambda) b \in \left[a - \tfrac{1}{2}; a + \tfrac{1}{2}\right)\right)$$
$$= P\left(X_t \in \left[\frac{a - (1 - \lambda) b - 1/2}{\lambda}; \frac{a - (1 - \lambda) b + 1/2}{\lambda}\right)\right),$$

and the initial probabilities are obtained by replacing b by q_0. So the MC approach of Brook & Evans (1972) is applicable, analogous to the CUSUM case discussed in Section 8.3.1, but now with $Q = \{l, \ldots, u\}$. Note that the lower limit can only be violated if $\text{round}(\lambda \cdot 0 + (1 - \lambda) \cdot l) < l$ holds; that is, if $l > 1/(2\lambda)$. Other choices of l lead to a purely upper-sided EWMA chart.

Example 8.3.3.1 (ARL performance of EWMA chart) Let us pick up the situation of Examples 8.2.1.1 and 8.3.1.1 again; see also Table 5 in Gan (1990a). Setting $q_0 = \mu_0 = 6$, and trying different values for (u, λ), we find two upper-sided designs, with ARL$_0$ being close to the previous values: $(u, \lambda) = (11, 0.7)$ and $(u, \lambda) = (9, 0.4)$, leading to ARL$_0$ at about 264.3 and 256.2, respectively. Such an upper-sided design is always obtained by setting $l = 0$,

Table 8.3 ARL$^{(\tau)}$ against increasing τ for two EWMA charts.

(l, u, λ)	μ	ARL$^{(1)}$	ARL$^{(2)}$	ARL$^{(3)}$	...	ARL$^{(\infty)}$
$(0, 11, 0.7)$	6	264.32	264.27	264.18	...	264.18
	8	20.02	20.03	19.99	...	20.00
$(1, 9, 0.4)$	6	256.21	255.57	255.37	...	255.32
	8	13.73	13.64	13.62	...	13.62

but as aforementioned, any other value $l \le 1/(2\lambda)$ would lead to an identical chart. In view of keeping Q (and hence the dimension of the involved matrices) minimal, we choose $l = 0 \le 1/(2 \cdot 0.7) \approx 0.71$ for the first design, but $l = 1 \le 1/(2 \cdot 0.4) = 1.25$ for the second one. Plots of ARL$^{(\tau)}$ against τ are omitted this time, since they do not provide new insight compared to previous graphs. In both cases, ARL$^{(\tau)}$ converges quickly to ARL$^{(\infty)}$, since the second-largest eigenvalues are not particularly large, at about 0.304 and 0.584, respectively (see Table 8.3).

These few lines already illustrate that the out-of-control performance usually becomes better for decreasing λ. Since λ controls the memory, with $\lambda = 1$ corresponding to the memory-less c chart, this behavior is reasonable.

We finish this example with a comparison of the steady-state out-of-control performances (against increasing μ) between the two EWMA charts considered here, and the c and CUSUM charts of Example 8.3.1.1. Figure 8.13 shows that both EWMA designs outperform the c chart (better with lower λ), but they do not reach the sensitivity of the CUSUM chart. The latter is particularly well-suited in detecting small-to-moderate shifts, while for large shifts, all charts perform reasonably well.

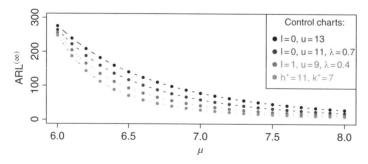

Figure 8.13 c, CUSUM and EWMA charts of Example 8.3.3.1: ARL$^{(\infty)}$ against μ.

If the underlying count process $(X_t)_\mathbb{N}$ is itself a Markov chain, then we proceed by analogy to Section 8.3.2 and consider the bivariate process $(X_t, Q_t)_\mathbb{N}$ (Weiß, 2009e). $(X_t, Q_t)_\mathbb{N}$ constitutes a bivariate Markov chain with range $\mathbb{N}_0^2$ and with transition probabilities

$$
\begin{aligned}
p_{x,a|y,b} &:= P(X_t = x, Q_t = a \mid X_{t-1} = y, Q_{t-1} = b) \\
&= \mathbb{1}_{[a-1/2;a+1/2)}(\lambda\, x + (1 - \lambda)\, b) \cdot p_{x|y},
\end{aligned}
$$

$$
\begin{aligned}
\pi_{x,a}(1) &:= P(X_1 = x, Q_1 = a \mid Q_0 = q_0) \\
&= \mathbb{1}_{[a-1/2;a+1/2)}(\lambda\, x + (1 - \lambda)\, q_0) \cdot p_x,
\end{aligned}
\tag{8.21}
$$

where $\mathbb{1}_A(z) := \mathbb{1}(z \in A)$ denotes the indicator function. So ARLs can be computed again by adapting the MC approach; see Weiß (2009e) for details. Here, the set Q of "no-alarm states" is derived as

$$
Q := \Big\{ (x, q) \in \mathbb{N}_0^2 \mid l \le q \le u, \\
\Big\lceil \tfrac{1}{\lambda}\Big(q - (1 - \lambda)u - \tfrac{1}{2}\Big) \Big\rceil \le x \le \Big\lceil \tfrac{1}{\lambda}\Big(q - (1 - \lambda)l + \tfrac{1}{2}\Big) \Big\rceil - 1 \Big\},
\tag{8.22}
$$

and the resulting matrices $\mathbf{Q}, \tilde{\mathbf{Q}}$ are again sparse matrices.

Example 8.3.3.2 (EWMA chart for IP counts data) Let us continue Examples 8.2.2.3 and 8.3.2.1. We find two upper-sided EWMA designs with a comparable in-control performance (always $q_0 := \text{round}(\mu_0) = 1$):

- $(l, u, \lambda) = (0, 4, 0.65)$ with $\text{ARL}_0 \approx 441.1$ (a matrix with $14^2 = 196$ entries, 85 of which are non-zero and second-largest eigenvalue about 0.290),
- $(l, u, \lambda) = (1, 3, 0.35)$ with $\text{ARL}_0 \approx 567.9$ (a matrix with $16^2 = 256$ entries, 104 of which are non-zero and second-largest eigenvalue about 0.627).

In addition, the last design is truly upper-sided, despite having $l > 0$; see the discussion before Example 8.3.3.1.

Although the second design is much more robust against false alarms, the out-of-control steady-state ARLs for $\mu = 1.5$ are nearly the same, at 41.7 and 40.8, respectively. A more detailed analysis of the steady-state ARL as a function of μ (plots are omitted this time) shows that both EWMA designs are generally more sensitive than the c chart, but do not reach the sensitivity of the two CUSUM charts from Example 8.3.2.1.

Applied to the simulated data from Example 8.2.2.3, we obtain the EWMA charts shown in Figure 8.14. Surprisingly, the second chart is rather late in detecting the out-of-control situation: it triggers its first alarm at $t = 73$; that is, with delay 53. Therefore, in the given example, the second EWMA design performs worst of all the considered charts (c chart: delay 52; CUSUM 1: delay 19; CUSUM 2 and EWMA 1: delay 13). This can be explained by the combination of small λ and additional rounding in (8.20), which leads to a strong smoothing effect, as visible in Figure 8.14b. Even if Q_{t-1} already

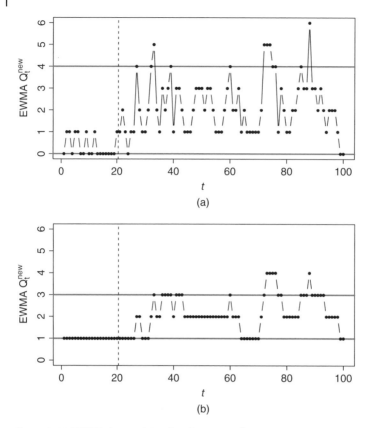

Figure 8.14 EWMA charts of simulated IP counts for Example 8.3.3.2 with change point $\tau = 21$: (a) $(l, u, \lambda) = (0, 4, 0.65)$, (b) $(l, u, \lambda) = (1, 3, 0.35)$.

reached the upper limit u, X_t is required to be larger than $u + 1/(2\lambda)$ to lift Q_t beyond u.

A possible disadvantage of the rounded EWMA approach (8.20) became clear from the second design in Example 8.3.3.2: for small values of λ, which are generally recommended if small mean shifts are to be detected, one may observe some kind of "oversmoothing"; that is, Q_t becomes piecewise constant in time t and rather insensitive to process changes. Therefore, Weiß (2011c) proposed a modification of (8.20), where a refined rounding operation is used: for $s \in \mathbb{N}$, the operation s-round maps x onto the nearest fraction with denominator s. For $s = 1$, we obtain the standard rounding operation, while 2-round rounds onto values in $\{0, 1/2, 1, 3/2, \dots\}$, for example. The resulting s-EWMA chart follows the recursion

$$Q_t^{(s)} = s\text{-round}\left(\lambda \cdot X_t + (1 - \lambda) \cdot Q_{t-1}^{(s)}\right) \quad \text{for } t = 1, 2, \dots \tag{8.23}$$

with $\lambda \in (0; 1]$. If $(X_t)_{\mathbb{N}}$ is a Markov chain, then $(X_t, Q_t^{(s)})_{\mathbb{N}}$ again is a discrete Markov chain, now with range $\mathbb{N}_0 \times \mathbb{Q}_{0,s}^+$, where $\mathbb{Q}_{0,s}^+ := \{\frac{r}{s} \mid r \in \mathbb{N}_0\}$ is the set of all non-negative rationals with denominator s. So again, it is possible to adapt the MC approach of Brook & Evans (1972) for ARL computation; see Weiß (2011c) for details.

9

Control Charts for Categorical Processes

Having discussed control charts for count processes in the Sections 8.2 and 8.3, we shall now turn to another type of attributes data process $(X_t)_{\mathbb{N}}$, namely categorical processes, as introduced in Part II. Although now the range (state space) $S = \{s_0, s_1, \ldots, s_d\}$ of X_t consists of a finite number $d + 1$ of (unordered) categories with $d \in \mathbb{N}$, count values will still play an important role since the most obvious way of evaluating categorical data is by counting the occurrences of categories; see also Chapter 6.

For quality-related applications, X_t often describes the result of an inspection of an item, which either leads to the classification $X_t = s_i$ for an $i = 1, \ldots, d$ iff the tth item was non-conforming of type s_i, or $X_t = s_0$ for a conforming item. A typical example is the one described by Mukhopadhyay (2008), in which a non-conforming ceiling fan cover is classified according to the most predominant type of paint defect, say "poor covering" or "bubbles". Another field of application is the monitoring of network traffic data with different types of audit events; see Ye et al. (2002) for details. For monitoring such categorical processes, we shall consider two general strategies: if the process evolves too fast to be monitored continuously, then segments are taken from the process at selected times. For each of the resulting samples, a statistic is computed and plotted on a control chart. Here, it is important to carefully consider the serial dependence *within* the sample; see Section 9.1 for further details. In other cases, it is possible to continuously monitor the process, but then the serial dependence has to be taken into account *between* the plotted statistics. Control charts for this scenario are presented in Section 9.2. In both cases, we shall first concentrate on the special case of a binary process (that is, $d = 1$) and then extend our discussion to the general categorical case.

An Introduction to Discrete-Valued Time Series, First Edition. Christian H. Weiss.
© 2018 John Wiley & Sons Ltd. Published 2018 by John Wiley & Sons Ltd.
Companion website: www.wiley.com/go/weiss/discrete-valuedtimeseries

9.1 Sample-based Monitoring of Categorical Processes

In this section, we assume that the categorical process $(X_t)_\mathbb{N}$ cannot be monitored continuously. Instead, samples are taken as non-overlapping segments[1] from the process at times $t_1, t_2, \ldots$, each being of a certain length $n > 1$. Note that we restrict ourselves to a constant segment length n for simplicity, but at least the Shewhart-type charts could be directly adapted to varying length n_k by using varying control limits (Montgomery, 2009, Section 7.3.2). The time distance $t_k - t_{k-1} > n$ is assumed to be sufficiently large such that we do not need to worry about the serial dependence *between* the samples; just the serial dependence *within* the samples. After having collected the segment, a certain type of sample statistic is computed and then plotted on an appropriately designed control chart.

9.1.1 Sample-based Monitoring: Binary Case

In view of its practical importance, let us first focus on the special case of a binary process $(X_t)_\mathbb{N}$, with the range coded as $\{0, 1\}$ as in Example 6.3.2. Having available the binary segment $X_{t_k}, \ldots, X_{t_k+n-1}$, one commonly determines either the sample sum $N_k^{(n)} = X_{t_k} + \ldots + X_{t_k+n-1}$ (say, counts of non-conforming items) or the corresponding sample fraction of '1's. Since the sample fraction differs from the count just by a factor $1/n$, we shall always consider the resulting count process $(N_k^{(n)})_\mathbb{N}$ in the sequel. The original binary process $(X_t)_\mathbb{N}$ is now monitored by monitoring this derived count process $(N_k^{(n)})_\mathbb{N}$. At this point, the fundamental premise of this section should be remembered: although $(X_t)_\mathbb{N}$ might exhibit serial dependence, due to taking sufficiently distant segments, we shall assume $(N_k^{(n)})_\mathbb{N}$ to be serially independent, and hence i.i.d. in its in-control state.

For monitoring $(N_k^{(n)})_\mathbb{N}$ (being i.i.d. in its in-control state), any of the concepts discussed in Chapter 8 can be used, it just has to be adapted to the finite range $\{0, \ldots, n\}$ of $N_k^{(n)}$. This difference sometimes manifests itself in the name of the resulting control charts. If the counts are plotted directly on a Shewhart-type chart, for instance, it is no longer referred to as a *c* chart, but as an *np chart*; see also the discussion in Section 8.2.1, as well as Montgomery (2009). If the sample fractions are plotted, it is called a *p chart*. Despite this different terminology, the *np* chart still has two control limits l, u satisfying $0 \le l < u \le n$, which includes the one-sided charts as boundary cases (upper-sided if $0 = l < u < n$). Also ARLs are computed as before; see (8.3) and (8.4).

Remark 9.1.1.1 **(ARL vs. ATS)** The ARL values related to the *np* chart applied to $(N_k^{(n)})_\mathbb{N}$ have to be treated with some caution. To illustrate this, let

---

1 Also see the "rational subgroup" concepts discussed in Section 5.3.4 in Montgomery (2009).

us assume that the underlying process $(X_t)_{\mathbb{N}}$ refers to the quality of manufactured items, and that we have fixed sampling intervals $t_k - t_{k-1} = K > n$; say, $t_k := k \cdot K - n + 1$. If the chart triggers its first alarm after plotting the rth sample statistic $N_r^{(n)}$ (that is, the run length equals r), then the number of manufactured items until this alarm is much larger, given by $r \cdot K$. Therefore, in such a situation, it is sometimes preferred to look at the *average time to signal* (ATS), where "time" refers to the original process $(X_t)_{\mathbb{N}}$, not to the number of plotted statistics. In the given example, we have ATS $= K \cdot$ ARL. Note that the ATS is sometimes referred to as the *average number of events* (ANE) or the *average number of observations to signal* (ANOS) instead.

Concerning the distribution of the sample counts, the serial dependence structure of the underlying binary process $(X_t)_{\mathbb{N}}$ is of importance. If $(X_t)_{\mathbb{N}}$ is i.i.d. with $P(X_t = 1) = \pi \in (0; 1)$ (say, the probability of a non-conforming item), then each sample sum $N_k^{(n)} = X_{t_k} + \dots + X_{t_k + n - 1}$ is binomially distributed according to Bin(n, π) (Example A.2.1). So the statistics $(N_k^{(n)})_{\mathbb{N}}$ constitute themselves as an i.i.d. process of binomial counts. But if $(X_t)_{\mathbb{N}}$ exhibits serial dependence, in contrast, the distribution of $N_k^{(n)}$ will deviate from a binomial one.

In Deligonul & Mergen (1987), Bhat & Lal (1990) and Weiß (2009f), the case of $(X_t)_{\mathbb{N}}$ being a binary Markov chain with success probability $\pi \in (0; 1)$ and autocorrelation parameter $\rho \in \left(\max \left\{ \frac{-\pi}{1-\pi}, \frac{1-\pi}{-\pi} \right\}; 1 \right)$ was considered (Example 7.1.3); that is, with the transition matrix given by (7.6). In this case, $N_k^{(n)} = X_{t_k} + \dots + X_{t_k + n - 1}$ follows the so-called *Markov binomial distribution* MB(n, π, ρ) (which coincides with Bin(n, π) iff $\rho = 0$). While the mean of $N_k^{(n)}$ is not affected by the serial dependence, the variance in particular changes (*extra-binomial variation* if $\rho > 0$, see the discussion in the context of Equation 2.3):

$$E[N_k^{(n)}] = n\pi, \quad V[N_k^{(n)}] = n\pi(1 - \pi)\frac{1+\rho}{1-\rho}\underbrace{\left(1 - \frac{2\rho(1 - \rho^n)}{n(1 - \rho^2)}\right)}_{\approx 1 \text{ for large } n}. \tag{9.1}$$

The pmf is given by (Kedem, 1980, Corollary 1.1)

$$P(N_k^{(n)} = 0) = (1 - \pi)p_{0|0}^{n-1}, \qquad P(N_k^{(n)} = n) = \pi p_{1|1}^{n-1},$$

and by $P(N_k^{(n)} = j) = \pi p_{1|1}^{j-1} \, p_{0|1}^2 \, p_{0|0}^{n-j-2}$

$$\sum_{a=0}^{2} \binom{2}{a} \left(\frac{p_{0|0}}{p_{0|1}}\right)^a \sum_{r=0}^{j-1} \binom{n-j-1}{r+1-a} \binom{j-1}{r} \left(\frac{p_{1|0} \, p_{0|1}}{p_{1|1} \, p_{0|0}}\right)^r \tag{9.2}$$

for $0 < j < n$ (*zero inflation* if $\rho > 0$; see the discussion in Appendix A.2). If the time distance $t_k - t_{k-1}$ between successive segments from $(X_t)_{\mathbb{N}}$ is sufficiently large, the resulting process of counts $(N_k^{(n)})_{\mathbb{N}}$ can still be assumed to be approximately i.i.d. (note that the correlation between X_t and X_s decays exponentially with $\rho^{|t-s|}$), but with a marginal distribution different from a binomial one. This difference in the distribution of $N_k^{(n)}$ certainly has to be considered when designing a corresponding control chart (Weiß, 2009f).

In addition, advanced control schemes, such as the EWMA or CUSUM charts discussed in Section 8.3, can be used for monitoring $(N_k^{(n)})_{\mathbb{N}}$. Assuming the counts $N_k^{(n)}$ to be binomially distributed in their in-control state (that is, $(X_t)_{\mathbb{N}}$ is assumed to be i.i.d.), Gan (1993) applied the CUSUM scheme described in Section 8.3.1 for process monitoring, while Gan (1990b) used the modified EWMA chart from Section 8.3.3 (with rounding operation as in (8.20)) for this purpose. The application of such an EWMA chart to the case of $(X_t)_{\mathbb{N}}$ being a binary Markov chain – that is, with $N_k^{(n)}$ following the Markov binomial distribution – was considered by Weiß (2009f). The computation of ARLs is done in the same way as described in Sections 8.3.1 and 8.3.3, by just using the pmf of the (Markov) binomial distribution. A completely different approach for a sample-based monitoring of an underlying binary Markov chain was recently proposed by Adnaik et al. (2015), who did not compute the sample sums $N_k^{(n)}$ as the charting statistics, but instead used some kind of likelihood ratio statistic for each of the successive segments. Finally, Höhle (2010) proposed a log-LR CUSUM chart for monitoring the $N_k^{(n)}$ under the assumption that these counts follow a marginal (beta-)binomial logit regression model; see Section 7.4.

Example 9.1.1.2 **(Medical diagnoses)** In Weiß & Atzmüller (2010), several binary time series concerning the diagnostic behavior of different examiners were discussed (abdominal ultrasound examinations; time passes according to the arrival of patients). Each of these time series refers to a particular diagnosis and a particular examiner, where the value '1' states that the considered diagnosis cannot be excluded for the current patient; that is, suitable countermeasures are required. As explained in Weiß & Atzmüller (2010), the time series are expected to stem from a serially independent process, since the patients usually arrive independently of each other at the examiner, but the marginal distribution may change over time due to a change in the diagnostic behavior of the examiner (due to, say, learning). The aim is to monitor the diagnostic behavior of the considered examiners in view of detecting such changes.

Obviously, the above processes evolve sufficiently slowly in time that continuous process monitoring is feasible; we shall consider such approaches in Example 9.2.1.2. Here, just as an exercise, we shall take segments from one of these processes to illustrate the application of some of the control charts discussed above. We consider the time series concerning the diagnosis "M165" (fatty liver) of examiner "Mf542a5" (Weiß & Atzmüller, 2010, Example 3.1), and we focus on the data collected between times 251 and 1178, since these were justified to be reasonable Phase-I data.

So, altogether, a binary time series of length $T = 928$ (= $29 \cdot 32$) is available, which is assumed to stem from an i.i.d. model. We decide on a sampling interval $K = 29$ (Remark 9.1.1.1) and sample size $n = 10$, so altogether 32 counts $n_k^{(10)}$ are obtained, which should follow a unique binomial distribution, Bin$(10, \pi)$. In fact, an inspection of a time series plot as well as of an autocorrelation plot give no reason to doubt an i.i.d.-assumption for $n_1^{(10)}, \ldots, n_{32}^{(10)}$; the resulting moment estimate for π equals about 0.241.

For this fitted binomial model, we now develop designs for the upper np chart and the upper CUSUM chart, to detect a possible future increase in the number of "fatty liver" diagnoses. For the np chart, ARLs are computed according to (8.3) and (8.4), and the design $u = 5$ (and $l = 0$) leads to ATS$_0 = K \cdot$ ARL$_0 \approx$ 1777 (Remark 9.1.1.1). Applying this chart in Phase I to the sample counts, no alarm is triggered, which confirms the in-control assumption.

For the CUSUM chart, ARLs are computed by the MC approach, as described in Section 8.3.1, and comparable zero-state in-control ATS values are obtained for the designs $(h^+, k^+) = (3, 3)$ and $(h^+, k^+) = (1, 4)$, with $c_0^+ = 0$ (no FIR) in both cases: ATS$_0 \approx 1507$ and ATS$_0 \approx 1511$, respectively. The steady-state out-of-control performance of the charts is analyzed in Figure 9.1. While the np chart is difficult to compare with the CUSUM charts, since it is visibly more robust in its in-control state, it is easy to decide between the CUSUMs: the design $(h^+, k^+) = (3, 3)$ leads to clearly better ATS$^{(\infty)}$ performance.

Note that the worse CUSUM, $(h^+, k^+) = (1, 4)$, results in a very similar decision rule to the np chart: values for $N_k^{(n)} > 5$ always lead to an alarm, and additionally, any pattern of the form "5, 4, 4, ..., 4, 5" (with arbitrary number of '4'). So this particular CUSUM chart just extends the considered np chart by some kind of runs rule. This is illustrated by the simulated Phase-II sample ($\pi_0 \approx 0.241$ shifts to $\pi = 0.35$ at $\tau = 26$) in Figure 9.2. While the CUSUM design $(h^+, k^+) = (3, 3)$ clearly performs best in this particular example (not shown; first alarm at sample $k = 29$), it is interesting to compare the two

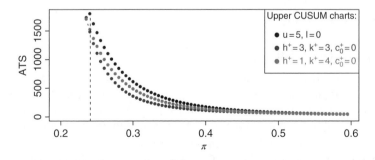

Figure 9.1 np chart and CUSUM charts of Example 9.1.1.2: ATS$^{(\infty)}$ against π.

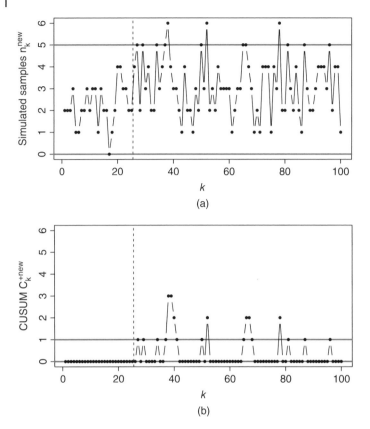

Figure 9.2 Simulated sample counts for Example 9.1.1.2: (a) upper-sided *np* chart with $u = 5$ and (b) CUSUM chart $(h^+, k^+, c_0^+) = (1, 4, 0)$.

remaining charts: both trigger alarms at samples $k = 38, 52, 78$ caused by sample counts of 6, but the CUSUM in (b) has an additional alarm at $k = 66$. The latter is caused by the sample counts $n_{65}^{(10)} = n_{66}^{(10)} = 5$, which corresponds to the above pattern with zero '4's.

9.1.2 Sample-based Monitoring: Categorical Case

Let us return to the truly categorical case; that is, where the range of $(X_t)_{\mathbb{N}}$ consists of more than two states, $S = \{s_0, \ldots, s_d\}$ with $d > 1$. As in Section 6.2, we denote the time-invariant marginal probabilities by $\boldsymbol{\pi} = (\pi_{s_0}, \ldots, \pi_{s_d})^\top$. If the number of different states, $d + 1$, is small, it would be feasible to monitor the process with d simultaneous binary charts; say, by using the *p-tree method* described in Duran & Albin (2009). However, here we shall concentrate on charting procedures in which the information about the process is comprised in a univariate statistic: After having taken a segment from the process, we first compute the resulting frequency distribution as a summary, which then serves as the base for deriving the statistic to be plotted on the control chart. To keep it consistent with the binary case from Section 9.1.1, we concentrate on absolute frequencies: $N_k^{(n)} = (N_{k;0}^{(n)}, \ldots, N_{k;d}^{(n)})^\top$ with $N_{k;j}^{(n)}$ being the absolute frequency of the state $s_j \in S$ in the sample $X_{t_k}, \ldots, X_{t_k+n-1}$, such that $\sum_{j=0}^{d} N_{k;j}^{(n)} = n$. With Y_t denoting the binarization of X_t, we may express $N_k^{(n)} = Y_{t_k}^{(n)} + \cdots + Y_{t_k+n-1}^{(n)}$.

If the underlying categorical process $(X_t)_{\mathbb{N}}$ is even serially independent (so altogether i.i.d.), then the distribution of each $N_k^{(n)}$ is a multinomial one; see Example A.3.3. This case was considered by Marcucci (1985) and Mukhopadhyay (2008), among others, who proposed plotting *Pearson's χ^2-statistic* on a control chart,

$$S_k^{(n)} = \sum_{j=0}^{d} \frac{(N_{k;j} - n\pi_{0;s_j})^2}{n\pi_{0;s_j}}, \tag{9.3}$$

where $\pi_0 := (\pi_{0;s_0}, \ldots, \pi_{0;s_d})^\top$ refers to the in-control values of the categorical probabilities. So in the in-control case, the process $(S_k^{(n)})_{\mathbb{N}}$ is i.i.d. with a marginal distribution that might be approximated by a χ_d^2-distribution (see Horn (1977) concerning the goodness of this approximation). As an alternative, Weiß (2012) proposed using a control statistic that measures the relative change of categorical dispersion. As the underlying categorical dispersion measure, the *Gini index* (6.1) might be used. If $(X_t)_{\mathbb{N}}$ is i.i.d., following the in-control model, then

$$G_k^{(n)} = \frac{1 - n^{-2} \sum_{j=0}^{d} N_{k;j}^2}{1 - \sum_{j \in S} \pi_{0;j}^2} \tag{9.4}$$

is approximately normally distributed, with a mean of $1 - 1/n$ and variance $\frac{4}{n} \left(\sum_{j \in S} \pi_{0;j}^3 - \left(\sum_{j \in S} \pi_{0;j}^2 \right)^2 \right) / \left(1 - \sum_{j \in S} \pi_{0;j}^2 \right)^2$; see Section 6.2. These approximate distributions for $S_k^{(n)}$ or $G_k^{(n)}$ may be used during chart design. But since the quality of these approximations is often rather bad (note that n is often quite small and that the control limit is usually chosen as an extreme quantile), the final design and evaluation of the ARL performance requires simulations in practice.

Remark 9.1.2.1 (ARL simulation) The standard approach for ARL simulation is to generate the process for a certain number of replications (usually between 10,000 and 1 million), and to apply the considered chart to each of these process replications. The ith process is stopped when the first alarm occurs, and the time until this alarm (the actual run length l_i) is stored. The mean $\bar{l}$ of all these run lengths is the approximation to the true ARL.

The above charts $S_k^{(n)}$ or $G_k^{(n)}$, however, are Shewhart charts for i.i.d. statistics. So we just need to know the marginal distribution of $S_k^{(n)}$ or $G_k^{(n)}$, then ARLs are computed according to (8.3) and (8.4). Therefore, here, we can also simulate just one very long time series and then compute the control limit as an appropriate quantile from the simulated values. The ARL, in turn, is computed from the relative frequency of observations violating the control limits.

Example 9.1.2.2 (Paint defects) We pick up the application scenario described by Mukhopadhyay (2008), where manufactured ceiling fan covers were checked for possible paint defects. If the fan cover has no paint defect, then it is conforming and categorized as 'ok'. Otherwise, if non-conforming, it is classified according to the most predominant defect, with the $d = 6$ defect categories:

- poor covering ('pc')
- overflow ('of')
- patty defect ('pt')
- bubbles ('bb')
- paint defect ('pd')
- buffing defect ('bd').

In Mukhopadhyay (2008), an underlying i.i.d. process is assumed, such that the $N_k^{(n)}$ are multinomially distributed. The reported data example shows varying sample sizes (varying around ≈ 150), and the overall observed marginal frequencies are

$$\approx (0.769, 0.081, 0.059, 0.021, 0.023, 0.022, 0.025) =: \pi_0.$$

For illustration, we assume this distribution as the in-control distribution, and we take segments of constant length $n = 150$. Since we do not specify a certain sampling interval, we just look at the ARLs in this example.

In quality-related applications such as this, the probability $\pi_{0;\text{ok}}$ of a conforming unit is typically large, while the defect probabilities are much smaller. So in the sense of Section 6.2, the in-control distribution π_0 is expected to exhibit low dispersion. The aim is to detect a deterioration in the quality, which goes along with a decrease of $\pi_{0;\text{ok}}$ and hence (often) with an increase in dispersion. Therefore, we use an upper-sided version of the $G_k^{(150)}$ chart, besides the $S_k^{(150)}$ chart, which is upper-sided anyway. For chart design, we first try to use the asymptotic distributions of $S_k^{(150)}$ and $G_k^{(150)}$, respectively, and invert (8.3) to find the control limit for a specified value of ARL_0. To check the quality of this asymptotic approximation, we then determine the actual in-control ARL by simulations (10,000 replications; see Remark 9.1.2.1). The results are shown in Table 9.1.

Table 9.1 ARL$_0$ performance of $S_k^{(150)}$ and $G_k^{(150)}$ chart with approximate chart design.

Chart	$S_k^{(150)}$ chart			$G_k^{(150)}$ chart		
Design-ARL$_0$	50	100	200	50	100	200
SimulatedARL$_0$	39.9	69.5	117.1	56.1	129.4	268.5

Obviously, none of the approximations is satisfactory, since we get only roughly the intended in-control behavior. So a further simulation-based fine-tuning of the control limit is necessary for both charts.

For illustration purposes, let us further investigate the designs $u = 18.55$ for the $S_k^{(150)}$ chart, and $u = 1.28$ for the $G_k^{(150)}$ chart, both leading to ARL$_0 \approx 115$. Two out-of-control scenarios are considered:

- the probability of poor covering is increased by a factor $1 + \delta$; that is, $\pi_{1;\mathrm{pc}} = (1 + \delta)\pi_{0;\mathrm{pc}}$, while all other probabilities are uniformly decreased
- the probability of there being no paint defect is decreased by a factor $1 - \delta$; that is, $\pi_{1;\mathrm{ok}} = (1 - \delta)\pi_{0;\mathrm{ok}}$, while all other probabilities are uniformly increased

The resulting simulated ARL performance is shown in Figure 9.3; the non-smoothness is caused by simulation error. In both cases, the Gini chart is much more sensitive to small shifts δ, while the $S_k^{(150)}$ chart performs equally well (scenario 2) or even better (scenario 1) for larger shifts δ. This behavior can be explained to some extent by looking at the "true" values of the monitored statistics; that is, at $\sum_{j \in S} \frac{(n\pi_{1;j} - n\pi_{0;j})^2}{n\pi_{0;j}} = n\left(\sum_{j \in S} \frac{\pi_{1;j}^2}{\pi_{0;j}} - 1\right)$ and $\frac{1 - \sum_{j \in S}\pi_{1;j}^2}{1 - \sum_{j \in S}\pi_{0;j}^2}$, respectively, as functions of δ. Both statistics increase with increasing deviation, but the Pearson in a convex way and the Gini in a concave one.

Finally, Figure 9.4 shows the $S_k^{(150)}$ chart and $G_k^{(150)}$ charts obtained for a simulated sample of length 100, where the first 20 samples stem from the in-control model, and the remaining 80 samples stem from the out-of-control Scenario 1 with $\delta = 0.3$ (that is, $\pi_{1;\mathrm{pc}} \approx 0.105$). Both charts trigger several alarms after the change point, with the first alarm for the $G_k^{(150)}$ chart at sample 34. Note that the "true" Gini value shifts from 1 to about 1.067 after the change point; that is, the true (Gini) dispersion is increased by about 6.7%.

A sample-based approach is also possible if $(X_t)_{\mathbb{N}}$ is serially dependent. But then, certainly, the distributions of $N_k^{(n)}$ and hence of $S_k^{(n)}$ and $G_k^{(n)}$ will deviate from those given above for the i.i.d. case. If, for instance, $(X_t)_{\mathbb{N}}$ is an NDARMA process (Section 7.2), then the effect on the distribution can be quantified in terms of the constant c from (7.12). Considering the complete vector $N_k^{(n)}$, the

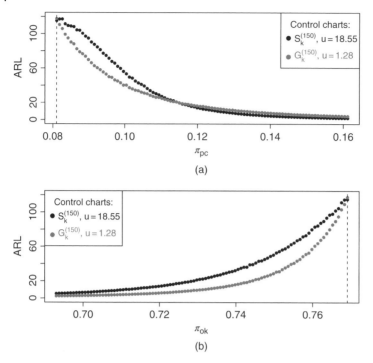

Figure 9.3 Simulated ARL performances for (a) Scenario 1 and (b) Scenario 2; see Example 9.1.2.2.

covariance matrix Σ from Example A.3.3 is asymptotically inflated by the factor c (Weiß, 2013b). For the Gini statistic $G_k^{(n)}$, variance and mean change (approximately) according to (7.13) and (7.14), respectively, while Weiß (2013b) showed that $S_k^{(n)}/c$ is approximately χ_d^2-distributed.

Example 9.1.2.3 **(Markov multinomial distribution)** If $(X_t)_{\mathbb{N}}$ is the DAR(1) process of Example 7.2.2, then $c = (1 + \phi_1)/(1 - \phi_1)$ (in-control value). Furthermore, analogous to the Markov binomial distribution from (9.2) above, the distribution of $N_k^{(n)}$ is called the *Markov multinomial distribution* by Wang & Yang (1995), who also provide a closed-form formula for the joint pgf of $N_k^{(n)}$. Note that the jth component $N_{k;j}^{(n)}$ follows the $MB(n, \pi_j, \phi_1)$ distribution, since for this particular type of Markov chain, also each component of the binarization $(Y_t)_{\mathbb{N}}$ is itself a binary Markov chain (see Remark 7.2.3).

Höhle (2010) proposed a log-LR CUSUM chart if the $N_k^{(n)}$ stem from a marginal multinomial logit regression model; see Section 7.4.

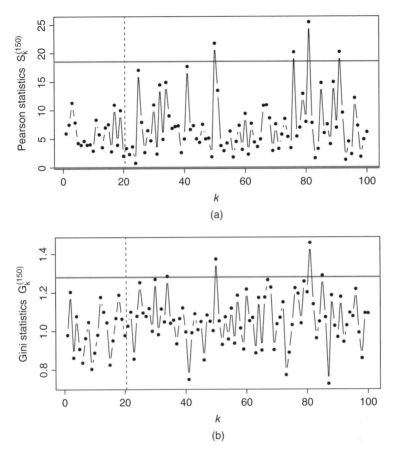

Figure 9.4 (a) $S_k^{(150)}$ chart and (b) $G_k^{(150)}$ chart applied to simulated sample; see Example 9.1.2.2.

9.2 Continuously Monitoring Categorical Processes

If the process evolves sufficiently slowly, then it is possible to implement a continuous monitoring approach of the categorical process $(X_t)_\mathbb{N}$. So as a new categorical observation X_t arrives, the next control statistic is computed and plotted on the control chart.

9.2.1 Continuous Monitoring: Binary Case

As in Section 9.1.1, let us first focus on the special case of a binary process $(X_t)_\mathbb{N}$. Perhaps the best-known approach for (quasi) continuously monitoring a binary process is by plotting run lengths $Q_1, Q_2, \ldots$ on an appropriately designed chart:

$$Q_1 := \text{number of observations until first occurrence of '1'}$$
$$Q_n := \text{number of observations after } (n-1)\text{th occurrence of '1'} \qquad (9.5)$$
$$\text{until } n\text{th occurrence of '1', for } n \geq 2.$$

As an example,

$$(x_t) = \Big(\underbrace{0,0,1,}_{q_1=3} \underbrace{0,0,0,0,1,}_{q_2=5} \underbrace{1,}_{q_3=1} \underbrace{0,0,0,1,}_{q_4=4} \underbrace{0,0,1,}_{q_5=3} \dots\Big).$$

The monitoring of such runs is a reasonable approach, especially for high-quality processes where $\pi = P(X_t = 1)$ is very small. Small π implies that long runs are observed, but if π increases (deterioration of quality), the runs become shorter (and vice versa). So the detection of a decrease in the run lengths is often particularly relevant. Having fixed a truly two-sided design $1 < l < u$, we stop monitoring with the nth run if either $Q_n < l$ for the first time, or if already u zeros have been observed since the last run (because then, Q_n will necessarily become larger than u, but we do not need to wait until the run is finished).

Concerning performance evaluation, Remark 9.1.1.1 should be remembered. The ARL – that is, the average number of plotted runs until the first alarm – would be quite misleading, since a single run might comprise a rather large number of original observations. Therefore, the ATS is clearly preferable as a measure of chart performance.

If $(X_t)_{\mathbb{N}}$ is i.i.d. (Bourke, 1991; Xie et al., 2000; Weiß, 2013c), then $(Q_n)_{\mathbb{N}}$ is also i.i.d. according to the shifted geometric distribution (Example A.1.5). The ATS can be computed according to (Weiß, 2013c)

$$\text{ARL} = (1 - (1-\pi)^{l-1} + (1-\pi)^u)^{-1},$$
$$\text{ATS} = \text{ARL} \cdot \frac{1}{\pi}(1 - (1-\pi)^u). \qquad (9.6)$$

Example 9.2.1.1   (ATS performance of runs chart)   Let us apply formula (9.6) to evaluate the performance of some runs chart designs. Trying to find (nearly) *ATS-unbiased* designs (see Example 8.2.1.1 and also Xie et al. (2000)) with an ATS_0 around 600–700 for the in-control values $\pi_0 = 0.01, 0.02, 0.03$ (high-quality processes), we ultimately obtain the designs $l = 9, u = 259$ ($\pi_0 = 0.01$, $\text{ATS}_0 \approx 612$), $l = 3, u = 178$ ($\pi_0 = 0.02$, $\text{ATS}_0 \approx 725$), and $l = 2, u = 131$ ($\pi_0 = 0.03$, $\text{ATS}_0 \approx 675$). The plot of the respective $\text{ATS}(\pi)$ functions against π in Figure 9.5 confirms the designs to be quasi unbiased. It should be noted, however, that these designs do not lead to unbiased ARL performances. However, the practical meaning of ARLs for a runs chart (with values ranging between 0 and 25 in the present examples) is questionable anyway.

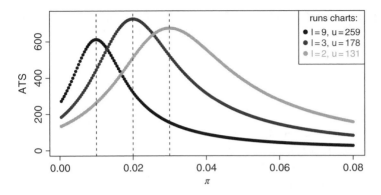

Figure 9.5 ATS performance of runs chart against π; see Example 9.2.1.1.

For illustration, a binary sequence and the respective runs were simulated, where the simulation stopped after having completed the 100th run (which happened after 3454 binary observations). The data leading to the first 20 runs stem from the in-control model; that is, they are i.i.d. with $\pi = \pi_0 := 0.02$. Afterwards, the defect probability was shifted to $\pi = 0.04$ (deterioration of quality). The sequence of runs is plotted on a control chart with design $l = 3, u = 178$, the in-control performance of which is described by $ATS_0 \approx 725$ and $ARL_0 \approx 14.9$, and the out-of-control performance by $ATS \approx 315$ and $ARL \approx 12.6$ (note that runs are shorter on average for $\pi = 0.04$ than for $\pi = 0.02$, so the ATS is more strongly affected by such a change). The resulting runs chart is plotted in Figure 9.6, using a log scale to improve the readability.

The first alarm is triggered for the 6th run; that is, it is a false alarm (up to this event, altogether 467 binary observations had been generated). The alarm is caused by violating the upper limit ($q_6 = 206 > u$); that is, it would indicate an

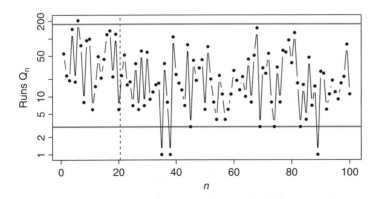

Figure 9.6 Runs chart applied to simulated data; see Example 9.2.1.1.

improvement of the quality. With $\text{ARL}_0 \approx 14.9$, it is not surprising to observe such a false alarm within the first 20 runs (this in-control period altogether comprises 1281 binary observations).

The next alarm is a true alarm, which is caused by the 35th run $q_{35} = 1 < l$ (that is, with the 1621th binary observations), and indicates a deterioration of quality. So the delay in detecting the out-of-control situation is 15 runs ($\text{ARL} \approx 12.6$) or 340 binary observations ($\text{ATS} \approx 315$), respectively.

The monitoring of runs is straightforwardly extended to the Markov case; see Blatterman & Champ (1992) and Lai et al. (2000). The runs $Q_1, Q_2, \ldots$ from a binary Markov chain $(X_t)_{\mathbb{N}}$ according to Example 7.1.3 are still serially independent, but their distribution is no longer shifted geometric. While Q_1 has to be treated separately, for Q_n with $n \geq 2$, we obviously have

$$P(Q_n = 1) = p_{1|1} \overset{(7.6)}{=} \pi(1-p) + p, \quad \text{and for } q \geq 2:$$
$$P(Q_n = q) = p_{1|0}\, p_{0|0}^{q-2}\, p_{0|1} \overset{(7.6)}{=} \pi(1-\pi)(1-p)^2\,(1 - \pi(1-p))^{q-2}.$$

If '1's are observed more frequently, the runs become quite short on average. In such a case, the CUSUM procedure proposed by Bourke (1991) to monitor the run length in $(X_t)_{\mathbb{N}}$ is more appropriate. This *geometric CUSUM* control chart is essentially equivalent to the *Bernoulli CUSUM* control chart, which was proposed by Reynolds & Stoumbos (1999) for an i.i.d. binary process $(X_t)_{\mathbb{N}}$, and which was extended to the case of a binary Markov chain, as in Example 7.1.3, by Mousavi & Reynolds (2009). These charts are constructed in an analogous way to (8.15): the CUSUM chart is defined by accumulating the contributions to the log-likelihood ratio (log-LR) at times $t \geq 1$. The contribution by the tth observation equals

$$\ell R_t = \ln\left(\frac{P_{\pi_1}(X_t)}{P_{\pi_0}(X_t)}\right) = \ln\left(\frac{\pi_1^{X_t}(1-\pi_1)^{1-X_t}}{\pi_0^{X_t}(1-\pi_0)^{1-X_t}}\right) \qquad \text{(i.i.d. case)}$$
$$= X_t \ln\frac{\pi_1(1-\pi_0)}{\pi_0(1-\pi_1)} + \ln\frac{1-\pi_1}{1-\pi_0},$$

where π_1 refers to the relevant out-of-control parameter value of π, while π_0 represents the in-control value. In the Markov case (Example 7.1.3; see also Remark 8.3.2.2), ℓR_1 is computed as before, while

$$\ell R_t = \ln\left(\frac{P_{\pi_1}(X_t|X_{t-1})}{P_{\pi_0}(X_t|X_{t-1})}\right)$$
$$= \ln\left(\frac{(1-p)\,\pi_1^{X_t}(1-\pi_1)^{1-X_t} + p\,\delta_{X_t,X_{t-1}}}{(1-p)\,\pi_0^{X_t}(1-\pi_0)^{1-X_t} + p\,\delta_{X_t,X_{t-1}}}\right) \qquad \text{for } t \geq 2.$$

An upper-sided CUSUM chart (we restrict to this case, since usually increases in π are to be detected) can now be constructed analogously to (8.13) by defining

$$\tilde{C}_0^+ = 0, \quad \tilde{C}_t^+ = \max\{0, \ell R_t + \tilde{C}_{t-1}^+\} \quad \text{for } t = 1, 2, \ldots \tag{9.7}$$

In the i.i.d. case, the CUSUM (9.7) might be rewritten in the form

$$C_t^+ = \max\left\{0, X_t - \underbrace{\ln \frac{1 - \pi_0}{1 - \pi_1} \Big/ \ln \frac{\pi_1(1 - \pi_0)}{\pi_0(1 - \pi_1)}}_{=: k^+} + C_{t-1}^+\right\}. \tag{9.8}$$

Note that the plotted statistics of these CUSUM charts go along with the observations, so ARL = ATS. To allow for an exact ARL computation with the MC approach (Section 8.2.2), k^+ can be required to take the form $1/m$ with an $m \in \mathbb{N}$ (Reynolds & Stoumbos, 1999); a similar strategy is proposed by Mousavi & Reynolds (2009) for the case of $(X_t)_\mathbb{N}$ being a binary Markov chain. In this case (with h^+ being a multiple of $1/m$), the resulting transition matrices $\mathbf{Q}, \tilde{\mathbf{Q}}$ for the MC approach are sparse matrices (see Section 8.3.2), since only a few combinations $(a, b) \in \{0, 1/m, \ldots, h^+\}^2$ are possible at all for $P(C_t^+ = a \mid C_{t-1}^+ = b)$. We have

$$P(C_t^+ = a \mid C_{t-1}^+ = b) = \begin{cases} 1 - \pi & \text{if } a = 0, \, b \leq \dfrac{1}{m} \text{ or } a > 0, \, b = a + \dfrac{1}{m}, \\ \pi & \text{if } a > 0, \, b = a + \dfrac{1}{m} - 1. \end{cases}$$

Note that a lower-sided CUSUM chart can be constructed in an analogous way to (9.8). For a log-LR CUSUM with respect to an underlying logit regression model (Section 7.4), see Höhle (2010).

Example 9.2.1.2 (CUSUM chart for diagnosis data) Let us return to the binary time series $x_1, \ldots, x_{928}$ from Example 9.1.1.2, for the diagnosis of "fatty liver" by examiner "Mf542a5" (Weiß & Atzmüller, 2010). Now investigating the full time series, the SACF plot still indicates serial independence, and the overall mean leads to the estimate $\hat{\pi} \approx 0.241$ for the probability of diagnosing a fatty liver. With a corresponding mean run length of only about 4.14, a runs chart is not appropriate. So we shall concentrate on the CUSUM chart (9.8) for the fitted i.i.d. model.

To get an idea of reasonable values for k^+, we compute $\ln \frac{1-\pi_0}{1-\pi_1} \Big/ \ln \frac{\pi_1(1-\pi_0)}{\pi_0(1-\pi_1)}$ for $\pi_0 := \hat{\pi}$ and $\pi_1 = 0.3, 0.4, 0.5$, which leads to about $0.270, 0.317, 0.364$, respectively. So we choose $k^+ = 1/4$ (which should be better for small shifts) and $k^+ = 1/3$ (which should be better for somewhat larger shifts). Trying to obtain in-control zero-state ARLs close to the ATS$_0$ values from Example 9.1.1.2 (note that for the considered CUSUM, we have ARL = ATS), we ultimately end up with the designs $h^+ = 53/4, k^+ = 1/4$ (ARL$_0 \approx 1642$) and $h^+ = 15/3, k^+ = 1/3$

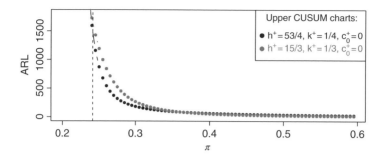

Figure 9.7 CUSUM charts of Example 9.2.1.2: $\text{ARL}^{(\infty)}$ against π.

$(\text{ARL}_0 \approx 1661)$. The corresponding transition matrices for the MC approach have only 105 and 30 non-zero entries, respectively. The steady-state ARL performance of these designs is illustrated by Figure 9.7, with the first design indeed being more sensitive to small shifts.

As an illustration, we now apply these two CUSUMs in Phase II. For this purpose, we use the remaining 449 observations from Example 3.1 in Weiß & Atzmüller (2010), which were collected after an eight-month break. Both CUSUM designs do not lead to an alarm (design $(h^+, k^+) = (53/4, 1/4)$ is shown in Figure 9.8), so the examiner seems to remain in the in-control state that had previously been identified.

We conclude by pointing out another approach for continuously monitoring a binary process: the EWMA chart discussed in Section 8.3.3, which was applied to binary processes by Yeh et al. (2008) and Weiß & Atzmüller (2010), among others.

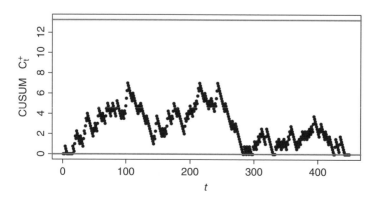

Figure 9.8 CUSUM chart $(h^+, k^+) = (53/4, 1/4)$ for fatty liver data; see Example 9.2.1.2.

9.2.2 Continuous Monitoring: Categorical Case

Generally, while it is quite natural to check for runs in a binary process, it is more difficult to define a run for the truly categorical case in a reasonable way. One possible solution was discussed in Weiß (2012), where one waits for a segment $(s, \ldots, s)$ of length k, where s is taken from a specified subset $S^* \subset S$. But as pointed out in Weiß (2012), waiting times for completely different types of patterns might also be relevant, depending on the actual application scenario. Because of this ambiguity, we shall not further consider the monitoring of runs in a categorical process here.

Instead, we follow the path of Section 9.2.1 and consider CUSUM charts derived from the log-likelihood ratio approach. So let π_0 denote again the in-control value of the marginal distribution π, and let π_1 be the relevant out-of-control value. A CUSUM scheme for the case of an underlying i.i.d. process was proposed by Ryan et al. (2011). In this case, the contribution to the log-LR at time $t \geq 1$, $\ell R_t := \ln(P_{\pi_1}(X_t)/P_{\pi_0}(X_t))$, can be expressed as either

$$\ell R_t = \ln\left(\frac{\pi_{1;X_t}}{\pi_{0;X_t}}\right) \quad \text{or} \quad \ell R_t = \sum_{j=0}^{d} Y_{t,j} \ln\left(\frac{\pi_{1;s_j}}{\pi_{0;s_j}}\right), \tag{9.9}$$

where the latter version uses the binarization Y_t of X_t. The log-LR approach also applies to serially dependent categorical processes. As an example, in analogy to the work by Mousavi & Reynolds (2009) concerning a binary Markov chain (see Section 9.2.1), we consider a categorical Markov chain. Then, ℓR_1 is computed as before, where

$$\ell R_t = \ln\left(\frac{P_{\pi_1}(X_t|X_{t-1})}{P_{\pi_0}(X_t|X_{t-1})}\right) \quad \text{for } t \geq 2.$$

If we are concerned with the particular case of a DAR(1) process as in Examples 7.2.2 and 9.1.2.3, it then follows that

$$\ell R_t = \ln\left(\frac{(1 - \phi_1)\,\pi_{1;X_t} + \delta_{X_t,X_{t-1}}\,\phi_1}{(1 - \phi_1)\,\pi_{0;X_t} + \delta_{X_t,X_{t-1}}\,\phi_1}\right) \quad \text{for } t \geq 2. \tag{9.10}$$

The log-LR-based CUSUM statistic at time t is defined as before, by the recursion

$$\tilde{C}_t^+ = \max\{0,\ \ell R_t + \tilde{C}_{t-1}^+\}, \quad \text{where} \quad \tilde{C}_0^+ := 0. \tag{9.11}$$

An alarm is triggered once $\tilde{C}_t$ violates the upper control limit $h^+ > 0$ for the first time. For a log-LR CUSUM with respect to an underlying logit regression model (Section 7.4), see Höhle (2010).

Example 9.2.2.1 (**CUSUM chart for paint defects**) Let us continue Example 9.1.2.2, where we monitored the quality of manufactured ceiling fan covers as described by Mukhopadhyay (2008). The possible states s_j are either 'ok' ($j = 0$) or one of the $d = 6$ defect categories: 'pc', 'of', 'pt', 'bb', 'pd', 'bd' ($j = 1, \ldots, 6$). The in-control distribution of the resulting i.i.d. process $(X_t)_{\mathbb{N}}$ equals

$$\boldsymbol{\pi}_0 = (0.769, 0.081, 0.059, 0.021, 0.023, 0.022, 0.025).$$

Two out-of-control scenarios were considered in Example 9.1.2.2. Scenario 1 assumes the probability of poor covering ($j = 1$) to be increased by the factor $1 + \delta$; that is, $\pi_{1;\mathrm{pc}} = (1 + \delta)\,\pi_{0;\mathrm{pc}}$, while all other probabilities are decreased according to $\pi_{1;s_j} = \frac{1-\pi_{1;\mathrm{pc}}}{1-\pi_{0;\mathrm{pc}}}\,\pi_{0;s_j}$. Hence, (9.9) becomes

$$\begin{aligned}
\ell R_t &= Y_{t,1}\,\ln(\pi_{1;\mathrm{pc}}/\pi_{0;\mathrm{pc}}) + \sum_{j\neq 1} Y_{t,j}\,\ln(\pi_{1;s_j}/\pi_{0;s_j}) \\
&= Y_{t,1}\,\ln(1+\delta) + \ln\left(\frac{1-(1+\delta)\,\pi_{0;\mathrm{pc}}}{1-\pi_{0;\mathrm{pc}}}\right)(1 - Y_{t,1}) \\
&= Y_{t,1}\,\ln\left(\frac{(1+\delta)(1-\pi_{0;\mathrm{pc}})}{1-(1+\delta)\,\pi_{0;\mathrm{pc}}}\right) - \ln\left(\frac{1-\pi_{0;\mathrm{pc}}}{1-(1+\delta)\,\pi_{0;\mathrm{pc}}}\right),
\end{aligned}$$

and a slightly modified version of (9.11) (see (9.8)) is given by

$$C_t^+ = \max\left\{0,\ \underbrace{Y_{t,1} - \ln\left(\frac{1-\pi_{0;\mathrm{pc}}}{1-(1+\delta)\,\pi_{0;\mathrm{pc}}}\right)\Big/\ln\left(\frac{(1+\delta)(1-\pi_{0;\mathrm{pc}})}{1-(1+\delta)\,\pi_{0;\mathrm{pc}}}\right)}_{=:k^+} + C_{t-1}^+\right\}.$$

So Scenario 1 just leads to a type of Bernoulli CUSUM chart (Scenario 2 from Example 9.1.2.2 could be treated in an analogous way by using $\pi_{1;\mathrm{ok}} = (1 - \delta)\pi_{0;\mathrm{ok}}$ and $\pi_{1;s_j} = \frac{1-\pi_{1;\mathrm{ok}}}{1-\pi_{0;\mathrm{ok}}}\pi_{0;s_j}$ otherwise). If we want to compute the ARLs exactly by the MC approach (Section 8.2.2), we can again choose k^+ in the form $1/m$ with an $m \in \mathbb{N}$ (Reynolds & Stoumbos, 1999), otherwise, simulations are required (Remark 9.1.2.1). For more complex out-of-control scenarios $\boldsymbol{\pi}_1$, the categorical CUSUM chart (9.11) differs from a simple Bernoulli CUSUM, but following the approach described in Appendix A of Ryan et al. (2011), an adaption of the MC approach is still possible.

For illustration, we continue with Scenario 1. As in Example 9.2.1.2, we compute the value for k^+ for some reasonable shift sizes – say $\delta = 0.3, 0.5$ – which imply choosing $k^+ = 1/11$ or $k^+ = 1/10$. Aiming at an in-control zero-state ARL around 1000, we find the designs $h^+ = 69/11, k^+ = 1/11$ (ARL$_0 \approx 1055$) and $h^+ = 52/10, k^+ = 1/10$ (ARL$_0 \approx 1051$). Both charts show nearly the same steady-state ARL performance with respect to changes according to Scenario 1 (see Figure 9.9), with slightly lower ARLs for the first design. For instance, if $\delta = 0.3$, we have ARL$^{(\infty)} \approx 238$ vs. ARL$^{(\infty)} \approx 254$. Applied to a simulated sample

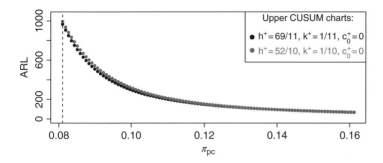

Figure 9.9 CUSUM charts of Example 9.2.2.1: $ARL^{(\infty)}$ against π_{pc}.

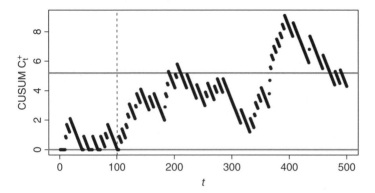

Figure 9.10 CUSUM chart $(h^+, k^+) = (52/10, 1/10)$ for simulated sample; see Example 9.2.2.1.

path, where the first 100 observations follow π_0 and the remaining 400 follow π_1 with $\delta = 0.3$, both charts look very similar, with chart 2 (shown in Figure 9.10) being slightly faster in detecting the process change (the first alarm is triggered at time $t = 190$ vs. $t = 205$).

While the designed CUSUM chart does well for a change in the anticipated direction, we may get bad performance in a misspecified shift scenario; see the discussion in Ryan et al. (2011). Imagine, for instance, a modification of Scenario 1, where the positive shift does not occur for poor covering ('pc') but for bubbles ('bb'). In this case, the value of $\pi_{1;\,pc}$ would be decreased compared to $\pi_{0;\,pc}$; that is, the designed CUSUM would even become more robust. So if flexibility concerning the out-of-control scenario is required, one has to apply different (Bernoulli) CUSUM charts simultaneously.

An EWMA control chart for the monitoring of a categorical process is proposed by Ye et al. (2002).

Part IV

Appendices

To make this book more self-contained, Appendix A summarizes some basic facts about common types of count distributions, although these facts are also available in many textbooks; some references are provided in Appendix A. Appendix B has been added to this book for the same reason, and is about stochastic processes, including both general properties and brief summaries for specific types, like ARMA and GARCH processes. Appendix B also gives recommendations for textbooks containing more detailed treatments of these topics.

Appendix C, in contrast, is about the models, methods and examples discussed in the main part of this book. It provides an overview of the datasets and the software implementations used for creating this book, the latter written in the language R (R Core Team, 2016). Some general remarks are also given, to help readers interested in translating the code into other programming languages.

An Introduction to Discrete-Valued Time Series, First Edition. Christian H. Weiss.
© 2018 John Wiley & Sons Ltd. Published 2018 by John Wiley & Sons Ltd.
Companion website: www.wiley.com/go/weiss/discrete-valuedtimeseries

A

Examples of Count Distributions

This appendix provides a brief survey of common models for count data and their stochastic properties. In particular, we discuss dispersion properties and types of generating functions, the latter being useful tools when analyzing count models. The presented models are divided into univariate and multivariate models, and into models for an infinite range like $\mathbb{N}_0$ or a finite range like $\{0, \ldots, n\}$.

A.1 Count Models for an Infinite Range

The most famous model for a count random variable having the range $\mathbb{N}_0$ (full set of non-negative integers) is the Poisson distribution.

Example A.1.1 (Poisson distribution) The *Poisson distribution* Poi(λ) with parameter $\lambda > 0$ is defined by the pmf

$$P(X = k) = e^{-\lambda} \frac{\lambda^k}{k!} \qquad \text{for } k \in \mathbb{N}_0.$$

From its generating functions (also see Table 2.2),

$$\text{pgf}(z) = \exp\left(\lambda(z-1)\right) \qquad \text{and} \qquad \text{cgf}(z) = \lambda(e^z - 1),$$

the characteristic equi-cumulant property follows,

$$\kappa_1 = \kappa_2 = \ldots = \lambda.$$

In particular, mean and variance are equal to each other (equal to λ), a property referred to as *equidispersion*. It is also useful to know that the factorial moments are given by $\mu_{(r)} = \lambda^r$ for $r \in \mathbb{N}$. Another important property is the additivity of the Poisson distribution: if $X_1 \sim \text{Poi}(\mu_1)$, $X_2 \sim \text{Poi}(\mu_2)$ and both are independent, then $X_1 + X_2 \sim \text{Poi}(\mu_1 + \mu_2)$. These and further properties are provided in Johnson et al. (2005, Chapter 4).

An Introduction to Discrete-Valued Time Series, First Edition. Christian H. Weiss.
© 2018 John Wiley & Sons Ltd. Published 2018 by John Wiley & Sons Ltd.
Companion website: www.wiley.com/go/weiss/discrete-valuedtimeseries

The equidispersion property of the Poisson distribution can be stated equivalently in terms of the index of dispersion $I = \sigma^2/\mu$ according to (2.1), by saying that the Poisson distribution always satisfies $I = 1$ (a related approach focussing on the probability for observing a zero, based on the zero index (2.2), is considered below). Values for I deviating from 1, in turn, express a violation of the Poisson model.

Example A.1.2 **(Compound Poisson distribution)** We adapt the notation and definitions from Chapter XII in Feller (1968). Let $Y_1, Y_2, \ldots$ be independent and identically distributed (i.i.d.) random variables with the range being contained in $\mathbb{N}$; let v denote the upper limit of the range (we allow the case $v = \infty$). Denote the pgf of the Y_i (*compounding distribution*) by $H(z)$; in the case of a finite range with upper limit $v < \infty$ and $h_v > 0$, H becomes the polynomial $H(z) = h_1 z + \ldots + h_v z^v$.

Let $N \sim \text{Poi}(\lambda)$ be Poisson distributed, independently of $Y_1, Y_2, \ldots$ Then $X := Y_1 + \ldots + Y_N$ is said to be *compound Poisson distributed*; that is, $X \sim \text{CP}_v(\lambda, H)$. The generating functions of X are given by

$$\text{pgf}(z) = \exp\left(\lambda(H(z) - 1)\right) \quad \text{and} \quad \text{cgf}(z) = \lambda(H(e^z) - 1),$$

while the pmf can be computed recursively as (Kemp, 1967, "Panjer recursion")

$$P(X = 0) = e^{-\lambda}, \quad P(X = k) = \frac{\lambda}{k} \cdot \sum_{j=1}^{\min\{k,v\}} j\, h_j\, P(X = k - j) \quad \text{for } k \geq 1.$$

If the rth moment $\mu_{Y,r}$ of the Y_i exists, then the rth cumulant of X is given by

$$\kappa_{X,r} = \lambda \cdot \mu_{Y,r}.$$

This implies that the variance-mean ratio I of X equals $\mu_{Y,2}/\mu_{Y,1}$; that is, the CP_v-distribution is equidispersed iff $v = 1$ (note that the CP_1-distribution coincides with the Poi-distribution). Otherwise, we have $I > 1$; that is, any CP_v-distribution with $v > 1$ is overdispersed.

The compound Poisson distributions are uniquely characterized by the property of being *infinitely divisible* (Feller, 1968); that is, their pgf satisfies that $\text{pgf}(z)^{1/n}$ is itself a pgf for all $n \in \mathbb{N}$. The CP distribution is also additive; see Lemma 2.1.3.2 for more details.

Note that in the literature, there are several other names for the compound Poisson distribution: for example, it is sometimes referred to as the *Poisson-stopped sum distribution* because the summation $X = Y_1 + \ldots + Y_N$ is stopped by the Poisson random variable N.

The family of CP_v-distributions includes a number of important special cases. Besides $\text{CP}_1 = \text{Poi}$, the distributions of Examples A.1.3–A.1.6 also belong to the CP-family.

Example A.1.3 **(Poisson distribution of order** v**)** If $v < \infty$ (finite compounding structure), the resulting CP distribution is also referred to as a *Hermite distribution* of order v (Puig & Valero, 2007). If, in addition, the compounding distribution is the uniform distribution on $\{1, \ldots, v\}$ – that is, if $h_x = 1/v$ for all $x = 1, \ldots, v$ – then this distribution is also known as the *Poisson distribution of order* v (Johnson et al., 2005). It is abbreviated hereafter as $\mathrm{Poi}_v(\lambda)$, where $\mathrm{Poi}_1 = \mathrm{Poi}$. The first four moments are

$$\mu = \frac{\lambda(v+1)}{2}, \qquad \sigma^2 = \frac{\lambda(v+1)(2v+1)}{6}, \qquad \bar{\mu}_3 = \frac{\lambda v(v+1)^2}{4},$$

$$\bar{\mu}_4 = \frac{\lambda(v+1)(2v+1)(3v^2+3v-1)}{30} + 3\sigma^4.$$

In particular, $I = (2v+1)/3$.

The negative binomial distribution belongs to the CP_∞-family.

Example A.1.4 **(Negative binomial distribution)** The *negative binomial distribution* $\mathrm{NB}(n, \pi)$ with $n \in (0; \infty)$ and $\pi \in (0; 1)$ has the pmf

$$P(X = k) = \binom{n+k-1}{k} \cdot (1-\pi)^k \cdot \pi^n \qquad \text{for } k \in \mathbb{N}_0,$$

where the binomial coefficient is defined as

$$\binom{n+k-1}{k} = \frac{\Gamma(n+k)}{\Gamma(n)\, k!} = \frac{(n+k-1)_{(k)}}{k!},$$

with $\Gamma(\cdot)$ denoting the gamma function. The pgf equals

$$\mathrm{pgf}(z) = \left(\frac{\pi}{1 - (1-\pi)\, z}\right)^n.$$

The case $n = 1$ is referred to as the *geometric distribution* $\mathrm{Geom}(\pi)$; see also Example A.1.5. The NB-distribution is compound Poisson with

$$\lambda := -n \ln \pi \quad \text{and} \quad H(z) = \frac{\ln(1 - (1-\pi)\, z)}{\ln \pi} = \sum_{k=1}^{\infty} \frac{(1-\pi)^k}{-k\, \ln \pi}\, z^k.$$

So the compounding distribution is just the *logarithmic series distribution* $\mathrm{LSD}(\pi)$. The first four moments of $\mathrm{NB}(n, \pi)$ are

$$\mu = n\,\frac{1-\pi}{\pi}, \qquad \sigma^2 = n\,\frac{1-\pi}{\pi^2}, \qquad \bar{\mu}_3 = \frac{n(1-\pi)(2-\pi)}{\pi^3},$$

$$\bar{\mu}_4 = \frac{3n^2(1-\pi)^2 + n(1-\pi)(\pi^2 - 6\pi + 6)}{\pi^4}.$$

In particular, $I = 1/\pi$. For $n \to \infty$ but μ fixed, it approaches the $\mathrm{Poi}(\mu)$ distribution. Finally, the NB distribution is also additive, in the sense that for $X_1 \sim \text{NB}(n_1, \pi)$, $X_2 \sim \text{NB}(n_2, \pi)$ and both being independent, their sum satisfies $X_1 + X_2 \sim \text{NB}(n_1 + n_2, \pi)$. These and further properties are provided in Johnson et al. (2005, Chapter 5).

Example A.1.5 (Geometric distribution) As mentioned in Example A.1.4, the special case $\text{NB}(1, \pi)$ of a negative binomial distribution with $n = 1$ is commonly referred to as a *geometric distribution*, abbreviated as $\text{Geom}(\pi)$. It can be interpreted as the number of failures until observing the first success in a sequence of i.i.d. replications of a Bernoulli experiment (see also Example A.2.1), that is, an experiment with only two outcomes ("failure" or "success") and with success probability π. Properties of the geometric distribution follow immediately from Example A.1.4.

In some applications, not the number of failures but the number of trials (waiting time) until the first success is relevant, which corresponds to shifting X to $Y := X + 1$. We refer to Y as following a *shifted geometric distribution*. Among others, we have

$$P(Y = k) = (1 - \pi)^{k-1} \cdot \pi \quad \text{for } k \in \mathbb{N}, \qquad \mu = \frac{1}{\pi}, \qquad \sigma^2 = \frac{1 - \pi}{\pi^2}.$$

Example A.1.6 (Generalized Poisson distribution) Another popular member of the CP_∞-family is Consul's *generalized Poisson distribution*, $\text{GP}(\lambda, \theta)$, with $\lambda > 0$ and $0 \le \theta < 1$. Its pmf equals

$$P(X = x) = \frac{\lambda}{x!} (\lambda + \theta x)^{x-1} \exp\left(-(\lambda + \theta x)\right) \qquad \text{for } x \in \mathbb{N}_0,$$

so $\theta = 0$ leads to the $\text{Poi}(\lambda)$-distribution. Mean and variance equal

$$E[X] = \frac{\lambda}{1 - \theta}, \qquad V[X] = \frac{\lambda}{(1 - \theta)^3}, \qquad \text{that is,} \qquad I = \frac{1}{(1 - \theta)^2}.$$

The GP distribution is also additive, in the sense that for $X_1 \sim \text{GP}(\lambda_1, \theta)$, $X_2 \sim \text{GP}(\lambda_2, \theta)$, both being independent, their sum satisfies $X_1 + X_2 \sim \text{GP}(\lambda_1 + \lambda_2, \theta)$. These and further properties are provided in Johnson et al. (2005, Section 7.2.6).

All distributions in Examples A.1.2–A.1.6 are overdispersed. The opposite phenomenon, $I < 1$ – that is, a variance smaller than the mean – is referred to as *underdispersion*. The $\text{GP}(\lambda, \theta)$-distribution of Example A.1.6 might be extended to also cover underdispersion, by allowing θ to be negative. But then it is necessary to truncate the range of the GP-distribution, where the degree of truncation depends on the actual values of the model parameters. The truncation of the range also causes the problem that the probabilities according to the above pmf no longer sum to 1, and otherwise exact properties become only approximate (Johnson et al., 2005, p. 336). An analogous criticism – that is, that essential properties of the distribution are known only approximately – also applies to the families of Efron's *double Poisson (DP) distributions* (Johnson et al., 2005, Section 11.1.8) and of *Conway–Maxwell (COM) Poisson distributions* (Shmueli et al., 2005).

Therefore, in Weiß (2013a), two other distribution families are recommended for underdispersed counts: the Good distribution and the power-law weighted Poisson (PL) distribution.

Example A.1.7 **(Good distribution)** Using the polylogarithm

$$\text{Li}_v(z) = \sum_{x=1}^{\infty} z^x \cdot x^{-v} \qquad \text{for } |z| < 1,$$

the *Good distribution* is defined by the pmf

$$P(X = x) = \frac{q^{x+1} \cdot (x+1)^{-v}}{\text{Li}_v(q)} \qquad \text{with } 0 < q < 1 \text{ and } v \in \mathbb{R}.$$

Its pgf equals

$$\text{pgf}(z) = \frac{1}{z} \frac{\text{Li}_v(qz)}{\text{Li}_v(q)},$$

and moments are obtained from the relation (Kulasekera & Tonkyn, 1992):

$$E[(X+1)^k] = \frac{\text{Li}_{v-k}(q)}{\text{Li}_v(q)}.$$

In particular, mean and variance are given by

$$E[X] = \frac{\text{Li}_{v-1}(q)}{\text{Li}_v(q)} - 1, \qquad V[X] = \frac{\text{Li}_{v-2}(q)}{\text{Li}_v(q)} - \frac{\text{Li}_{v-1}^2(q)}{\text{Li}_v^2(q)},$$

and we have underdispersion ($I < 1$) for values of v smaller than about $-\ln 2/\ln(4/3) \approx -2.41$.

Example A.1.8 **(PL distribution)** The *power-law weighted Poisson distribution* (PL distribution) as introduced by del Castillo & Pérez-Casany (1998) is based on the weight function $w(x) = (x+a)^v$ with $a > 0$ and $v \in \mathbb{R}$, which is applied to $Y \sim \text{Poi}(\lambda)$ via

$$C_v(\lambda, a) = e^\lambda \cdot E_\lambda[w(Y)] = \sum_{y=0}^{\infty} \frac{\lambda^y (y+a)^v}{y!}.$$

If even $v \in \mathbb{N}$, then

$$C_v(\lambda, a) = e^\lambda \cdot \sum_{k=0}^{v} \binom{v}{k} a^{v-k} E_\lambda[Y^k],$$

that is, $C_v(\lambda, a)$ simply equals e^λ times a polynomial in λ. As an example, $C_1(\lambda, a) = e^\lambda(\lambda + a)$ and $C_2(\lambda, a) = e^\lambda((\lambda + a)^2 + \lambda)$.

The pmf of the PL distribution is given by

$$P(X = x) = \frac{\lambda^x \cdot (x + a)^v}{C_v(\lambda, a) \cdot x!}.$$

So the pgf equals

$$\text{pgf}(z) = \sum_{x=0}^{\infty} \frac{(\lambda z)^x \cdot (x + a)^v}{C_v(\lambda, a) \cdot x!} = \frac{C_v(\lambda z, a)}{C_v(\lambda, a)},$$

and the factorial moments are given by

$$\mu_{(k)} = \lambda^k \cdot \frac{C_v(\lambda, a + k)}{C_v(\lambda, a)}.$$

del Castillo & Pérez-Casany (1998) proved that X shows underdispersion iff $v > 0$. Weiß (2013a) recommends using the distributions $\text{PL}_v(\lambda, a)$ with given $v \in \mathbb{N}$ for underdispersed counts, where closed-form expressions for the pmf, pgf, mean and variance are easily obtained from the above formulae.

Another characteristic property of the Poisson distribution (Example A.1.1) is the probability of observing a zero: the zero index $I_{zero} = 1 + \ln(p_0)/\mu$ according to (2.2) equals 0 for the Poisson distribution, but may differ otherwise. For a compound Poisson distribution $\text{CP}_v(\lambda, H)$, for instance, we obtain (Example A.1.2)

$$I_{zero} = 1 + \frac{\ln e^{-\lambda}}{\lambda \, H'(1)} = 1 - \frac{1}{H'(1)}, \tag{A.1}$$

which equals 0 iff $v = 1$, and which satisfies $I_{zero} \in (0; 1)$ otherwise (zero inflation; note that $H'(1) \geq 1$ is the mean of the compounding distribution). A more flexible approach towards zero modification is summarized in the following example.

Example A.1.9 (Zero-modified distributions) If $\tilde{p}_0, \tilde{p}_1, \ldots$ abbreviate the probability masses of the parent distribution, then the distribution defined by the pmf

$$P(X = x) = \delta_{x,0} \cdot \omega + (1 - \omega) \, \tilde{p}_x \qquad \text{with } \frac{-\tilde{p}_0}{1 - \tilde{p}_0} \leq \omega \leq 1$$

is referred to as a *zero-modified distribution* (Johnson et al., 2005, Section 8.2.3). Here, $\delta_{k,l}$ is the Kronecker delta, which equals 1 iff $k = l$ and 0 otherwise. Obviously, $\omega > 0$ corresponds to zero inflation, and $\omega < 0$ to zero deflation (both with respect to the parent distribution). It immediately follows that

$$\text{pgf}(z) = \omega + (1 - \omega) \cdot \widetilde{\text{pgf}}(z), \qquad \text{mgf}(z) = \omega + (1 - \omega) \cdot \widetilde{\text{mgf}}(z),$$

where $\widetilde{\mathrm{pgf}}(z)$ and $\widetilde{\mathrm{mgf}}(z)$ denote the parent's generating functions. Hence,

$$\mu_n = (1 - \omega)\, \tilde{\mu}_n, \qquad \text{especially } I = \tilde{I} + \omega\, \tilde{\mu}.$$

So zero inflation accompanies overdispersion and vice versa.

If the parent distribution in Example A.1.9 is the Poisson distribution $\mathrm{Poi}(\lambda)$, then the distribution obtained for $\omega > 0$ is said to be the *zero-inflated Poisson distribution*, $\mathrm{ZIP}(\lambda, \omega)$.

We conclude this section with a brief look at a distribution for a slightly different type of infinite and integer-valued range: the full range of integers $\mathbb{Z}$.

Example A.1.10 **(Skellam distribution)** Let $X_1 \sim \mathrm{Poi}(\mu_1)$ and $X_2 \sim \mathrm{Poi}(\mu_2)$ be two independent and Poisson-distributed random variables (Example A.1.1). Define $Y = X_1 - X_2$ as the difference between them. Then the distribution of Y is referred to as the *Skellam distribution*, with parameters $\mu_1, \mu_2 > 0$, and the range of Y is $\mathbb{Z}$. If

$$I_n(z) := \sum_{k=0}^{\infty} \frac{\left(\frac{z}{2}\right)^k \cdot \left(\frac{z}{2}\right)^{k+n}}{k! \cdot (k + n)!} \quad \text{for } n \in \mathbb{N}_0, \qquad I_{-n}(z) := I_n(z),$$

denotes the *modified Bessel function of the first kind*, then the pmf of Y is given by

$$P(Y = y) = e^{-\mu_1 - \mu_2}\, (\mu_1/\mu_2)^{y/2}\, I_y(2\sqrt{\mu_1 \mu_2}) \qquad \text{for } y \in \mathbb{Z}.$$

All odd cumulants of Y (including the mean of Y) are equal to $\mu_1 - \mu_2$ (hence, the distribution is skewed if $\mu_1 \neq \mu_2$), and all even cumulants (including the variance of Y) are equal to $\mu_1 + \mu_2$. These and further properties are provided in Johnson et al. (2005, Section 4.12.3).

A.2 Count Models for a Finite Range

The distributions surveyed up until now all have unlimited ranges. In some applications, however, it is known in advance that an upper bound $n \in \mathbb{N}$ exists; this can never be exceeded. Then the distributions discussed in the sequel become relevant. Thse have a finite range of the form $\{0, \ldots, n\}$.

Example A.2.1 **(Binomial distribution)** A random variable X with range $\{0, \ldots, n\}$ for an $n \in \mathbb{N}$ is *binomially distributed* according to $\mathrm{Bin}(n, \pi)$ with $\pi \in (0; 1)$ if its distribution is given by

$$P(X = k) = \binom{n}{k} \cdot \pi^k \cdot (1 - \pi)^{n-k} \qquad \text{for } 0 \leq k \leq n.$$

The case $n = 1$ is referred to as the *Bernoulli distribution* (see also Example A.1.5). The mean and variance of the $\text{Bin}(n, \pi)$-distribution are given by $E[X] = n\pi$ and $V[X] = n\pi(1 - \pi)$; that is, we have $I = 1 - \pi < 1$. The pgf equals

$$\text{pgf}(z) = (1 - \pi + \pi z)^n,$$

so factorial moments are computed as

$$\mu_{(r)} = n_{(r)} \, \pi^r = n \cdots (n - r + 1) \cdot \pi^r \qquad \text{for } 1 \leq r \leq n.$$

The binomial distribution $\text{Bin}(n, \pi)$ can be regarded as the distribution of the sum of n independent Bernoulli trials with success probability π; that is, of the number of successes among n replications of the Bernoulli experiment. This implies the additivity of the binomial distribution: if $X_1 \sim \text{Bin}(n, \pi)$, $X_2 \sim \text{Bin}(m, \pi)$ and both are independent, then $X_1 + X_2 \sim \text{Bin}(n + m, \pi)$. Note that a similar property holds for the Poisson distribution (Example A.1.1). In fact, both distributions are related to each other by the *Poisson limit theorem*: for $n \to \infty$ but keeping $\mu = n\pi$ fixed, the binomial distribution $\text{Bin}(n, \pi)$ approaches $\text{Poi}(\mu)$. These and further properties are provided in Johnson et al. (2005, Chapter 3).

We have seen that in a Poisson sense, the binomial distribution is underdispersed. However, since we are concerned with a different type of random phenomenon anyway – one with a finite range – it is more appropriate to evaluate the dispersion behavior in terms of the binomial index of dispersion $I_{\text{Bin}} = \sigma^2/(\mu \, (1 - \mu/n))$ according to (2.3), with the binomial distribution satisfying $I_{\text{Bin}} = 1$.

A distribution with $I_{\text{Bin}} > 1$ shows "overdispersion with respect to a binomial model", a phenomenon that we refer to as *extra-binomial variation*.

Example A.2.2   **(Beta-binomial distribution)**   A popular approach for generalizing the binomial distribution $\text{Bin}(n, \pi)$ is to assume that π is random itself, distributed according to a beta distribution, say $\text{BETA}(\frac{1-\phi}{\phi} \, v, \frac{1-\phi}{\phi} (1 - v))$ with $v, \phi \in (0; 1)$, having mean v and variance $\phi \, v(1 - v)$. So conditioned on the value of π, we just have the $\text{Bin}(n, \pi)$ distribution for the sample sum X. The unconditional distribution of X is referred to as the *beta-binomial distribution* $\text{BB}(n; v, \phi)$. Its pmf equals

$$P(X = k) = \binom{n}{k} \cdot \frac{B\left(k + \frac{1-\phi}{\phi} \, v, n - k + \frac{1-\phi}{\phi} (1 - v)\right)}{B\left(\frac{1-\phi}{\phi} \, v, \frac{1-\phi}{\phi} (1 - v)\right)},$$

where $B(\cdot)$ denotes the beta function. Mean and variance are given by

$$E[X] = n \, v, \qquad V[X] = n \, v \, (1 - v) \cdot (1 + \phi \, (n - 1)).$$

So $I_{\text{Bin}} = 1 + \phi\,(n-1) \in (1; n)$ determines the degree of extra-binomial variation, which increases for increasing ϕ. These and further properties are provided in Johnson et al. (2005, Section 6.2.2).

Another distribution with extra-binomial variation is the *Markov binomial distribution*, which is briefly discussed in the context of Equations 9.1 and 9.2.

The probability of observing a zero equals $(1 - \pi)^n$ in the binomial case. Increasing this probability according to the approach in Example A.1.9, with $\omega > 0$, leads to the *zero-inflated binomial distribution*, ZIB(n, π, ω). However, the BB$(n; \pi, \phi)$ distribution according to Example A.2.2 is also easily seen to be zero-inflated with respect to a binomial distribution, since the zero probability is obtained as

$$P(X = 0) = \frac{1-\pi}{1} \cdot \frac{1 - \pi + \frac{\phi}{1-\phi}}{1 + \frac{\phi}{1-\phi}} \cdots \frac{1 - \pi + (n-1)\frac{\phi}{1-\phi}}{1 + (n-1)\frac{\phi}{1-\phi}} > (1-\pi)^n.$$

A.3 Multivariate Count Models

Although this book is primarily concerned with univariate count data, a few models for multivariate count data are sketched here. Much more information can be found in the book by Johnson et al. (1997). For the particular case of bivariate distributions, the book by Kocherlakota & Kocherlakota (1992) is also recommended.

Example A.3.1 (Multivariate Poisson distribution) The simplest way of defining a multivariate version of the Poisson distribution of Example A.1.1 is by assuming $d + 1$ independent random variables $\epsilon_0, \dots, \epsilon_d$ with $\epsilon_i \sim \text{Poi}(\lambda_i)$ for $i = 0, \dots, d$. Then

$$X := (\epsilon_1 + \epsilon_0, \dots, \epsilon_d + \epsilon_0)^\top \tag{A.2}$$

is said to be *multivariately Poisson distributed* according to MPoi $(\lambda_0; \lambda_1, \dots, \lambda_d)$. Due to the additivity of the Poisson distribution, each component is itself univariately Poisson distributed, namely $X_i \sim \text{Poi}(\lambda_i + \lambda_0)$, but the common summand ϵ_0 causes the components to be cross-correlated:

$$\text{Cov}[X_i, X_j] = \lambda_0, \qquad \text{Corr}[X_i, X_j] = \lambda_0 \cdot ((\lambda_i + \lambda_0)\,(\lambda_j + \lambda_0))^{-1/2}.$$

Because of the mutual independence of $\epsilon_0, \dots, \epsilon_d$, we have $E[z_1^{X_1} \cdots z_d^{X_d}] = E[z_1^{\epsilon_1}] \cdots E[z_d^{\epsilon_d}] \cdot E[(z_1 \cdots z_d)^{\epsilon_0}]$, so the joint pgf is given by

$$\text{pgf}(z_1, \dots, z_d) = \exp\left(\sum_{i=1}^{d} \lambda_i z_i + \lambda_0 \cdot z_1 \cdots z_d - \lambda_\bullet \right)$$

with $\lambda_\bullet := \sum_{i=0}^{d} \lambda_i$. Sometimes, one more generally defines a random variable X with range $\mathbb{N}_0^d$ to be multivariately Poisson distributed if its pgf is of the form

$$\text{pgf}(z_1, \dots, z_d) = \exp\left(\sum_{i=1}^{d} a_i \cdot z_i + \sum_{1 \leq i < j \leq d} a_{ij} \cdot z_i z_j \right.$$
$$\left. + \dots + a_{1\dots d} \cdot z_1 \cdots z_d - a_\bullet \right),$$

where $a_\bullet = \sum_{i=1}^{d} a_i + \sum_{1 \leq i < j \leq d} a_{ij} + \dots + a_{1\dots d}$. In the special case $d = 2$ – that is, in the case of the *bivariate Poisson distribution* – the two approaches are equivalent. The corresponding pmf is given by

$$P(X = x) = e^{-\lambda_0 - \lambda_1 - \lambda_2} \frac{\lambda_1^{x_1}}{x_1!} \frac{\lambda_2^{x_2}}{x_2!} \sum_{i=0}^{\min\{x_1, x_2\}} \binom{x_1}{i} \binom{x_2}{i} i! \left(\frac{\lambda_0}{\lambda_1 \lambda_2}\right)^i.$$

For the general d-variate case as well as for recursive schemes for pmf computation, see Tsiamyrtzis & Karlis (2004). These and further properties are provided in Johnson et al. (1997, Chapter 37)

The approach of Example A.3.1 for constructing a d-variate distribution could be applied to other univariate additive count models as well.

Example A.3.2 (Multivariate negative binomial distribution)   There are a number of ways of defining a *multivariate* extension to the *negative binomial distribution* from Example A.1.4. Using an analogous parametrization to the one in Example A.1.4, the pmf of the MNB($n, \pi_1, \dots, \pi_d$) distribution with $\pi_0 := 1 - \pi_1 - \dots - \pi_d$ and $\pi_0, \pi_1, \dots, \pi_d \in (0; 1)$ is defined by

$$P(X = x) = \frac{\Gamma\left(n + \sum_{i=1}^{d} x_i\right)}{\Gamma(n) \prod_{i=1}^{d} x_i!} \pi_0^n \prod_{i=1}^{d} \pi_i^{x_i},$$

and its pgf follows as

$$\text{pgf}(z_1, \dots, z_d) = \left(\frac{\pi_0}{1 - \sum_{i=1}^{d} \pi_i z_i}\right)^n.$$

The ith component X_i is univariately NB-distributed according to NB($n, \pi_0/(\pi_0 + \pi_i)$). The cross-correlation between X_i and X_j is given by

$$Corr[X_i, X_j] = \sqrt{\frac{\pi_i \pi_j}{(\pi_0 + \pi_i)(\pi_0 + \pi_j)}}.$$

For these and further properties, see Chapter 36 in Johnson et al. (1997).

Extending the binomial distribution (Example A.2.1), with its finite range, is more demanding. The most famous multivariate binomial distribution is the multinomial distribution.

Example A.3.3 **(Multinomial distribution)** Instead of summing n independent Bernoulli trials, the *multinomial distribution* $\text{MULT}(n; \pi_0, \ldots, \pi_d)$, with $n \in \mathbb{N}$ and probabilities $\pi_0, \ldots, \pi_d > 0$ such that $\pi_0 + \ldots + \pi_d = 1$, is defined by summing n independent copies of a binary random vector Y, where exactly one of the components takes the value 1 and all others are equal to 0. So the possible range of Y consists of the unit vectors $e_0, \ldots, e_d \in \{0, 1\}^{d+1}$, where $e_j = (e_{j,0}, \ldots, e_{j,d})^\top$ is defined by $e_{j,i} = \delta_{j,i}$ (Kronecker delta; e_j has a one in its jth component) for $j = 0, \ldots, d$. The probabilities $P(Y = e_j) = \pi_j$ are assumed.

As a result, $N \sim \text{MULT}(n; \pi_0, \ldots, \pi_d)$ has the range $\{n \in \{0, \ldots, n\}^{d+1} \mid n_0 + \ldots + n_d = n\}$, and its pmf is given by

$$P(N = n) = \binom{n}{n_0, \ldots, n_d} \cdot \pi_0^{n_0} \cdots \pi_d^{n_d}.$$

The covariance matrix equals

$$n \cdot \Sigma, \quad \text{where } \Sigma = (\sigma_{ij}) \quad \text{is given by } \sigma_{ij} = \begin{cases} \pi_i(1 - \pi_i) & \text{if } i = j, \\ -\pi_i \pi_j & \text{if } i \neq j. \end{cases}$$

Each component N_j of N is binomially distributed according to $\text{Bin}(n, \pi_j)$. These and further properties are provided in Johnson et al. (1997, Chapter 35).

Note that it sometimes simplifies the notation if one omits one of the $d + 1$ components; if we write $(N_1, \ldots, N_d)^\top \sim \text{MULT}^*(n; \pi_1, \ldots, \pi_d)$, where $N_1 + \ldots + N_d \leq n$ and $\pi_1 + \ldots + \pi_d < 1$, then we assume the extended vector $(n - N_1 - \ldots - N_d, N_1, \ldots, N_d)^\top$ to be distributed according to $\text{MULT}(n; 1 - \pi_1 - \ldots - \pi_d, \pi_1, \ldots, \pi_d)$.

The use of the multinomial distribution as a multivariate extension of the binomial distribution is limited by the fact that each binomial component N_j has the same population size n, and that the sum of the components has to be equal to this value n. The importance of the multinomial distribution is in the fact that the binary random vectors Y can be understood as a *binarization* of a categorical random variable X with range $\{s_0, \ldots, s_d\}$ by defining $Y = e_j$ if $X = s_j$. Then N represents the realized absolute frequencies of n independent replications of X, and N/n gives the corresponding relative frequencies (proportions).

Remark A.3.4 **(Compositional data)** If the number n of replications becomes infinitely large, $n \to \infty$, then the vector of random proportions becomes a continuous random vector with the $(d + 1)$-*part unit simplex* as its range,

$$\mathbb{S}_{d+1} := \{u \in (0; 1)^{d+1} \mid u_0 + \ldots + u_d = 1\}.$$

The corresponding data, which express the "proportions of some whole" (Aitchison, 1986, p. 1), are referred to as *compositional data* (CoDa); books on this topic are the ones by Aitchison (1986) and Pawlowsky-Glahn & Buccianti (2011). We shall not further discuss CoDa in this book since we are concentrating on the discrete-valued case, but the relation between the multinomial setup and CoDa could be useful, for example for approximating the multinomial distribution for large n by an appropriate CoDa model, or for generalizing the multinomial distribution by analogy to the beta-binomial approach from Example A.2.2: the vector $(\pi_0, \ldots, \pi_d)$ of multinomial probabilities also belongs to the simplex $\mathbb{S}_{d+1}$ and might be assumed to be random itself. If, for instance, the most basic CoDa distribution, the *Dirichlet distribution*, is used for this purpose (see Chapter 49 in Kotz et al. (2000) and Chapter 2 in Ng et al. (2011) for a comprehensive survey), then one obtains the *Dirichlet-multinomial distribution* as a counterpart to the beta-binomial distribution; see Section 35.13 in Johnson et al. (1997) for further details.

Another type of generalized multinomial distribution is the *Markov multinomial distribution*, which is briefly discussed in Example 9.1.2.3.

Example A.3.5 **(Bivariate binomial distributions)** A more flexible generalization, but restricted to the bivariate case, is the bivariate binomial distribution of type II (Kocherlakota & Kocherlakota, 1992). It starts by defining the *bivariate Bernoulli distribution* with parameters $0 < \alpha_1, \alpha_2 < 1$ and $0 < \alpha < \min\{\alpha_1, \alpha_2\}$ through the bivariate random vectors Y with range $\{0, 1\}^2$, the joint probabilities $p_{ij} := P(Y = (i, j)^\top)$ of which have to satisfy

$$p_{11} = \alpha, \quad p_{11} + p_{10} = \alpha_1, \quad p_{11} + p_{01} = \alpha_2.$$

Instead of α, one may consider

$$\phi_\alpha := Corr[Y_1, Y_2] = \frac{\alpha - \alpha_1 \alpha_2}{\sqrt{\alpha_1 \alpha_2 (1 - \alpha_1)(1 - \alpha_2)}}$$

as a third parameter besides α_1, α_2, which has to satisfy the restriction

$$\max \left\{ -\sqrt{\frac{\alpha_1 \alpha_2}{(1 - \alpha_1)(1 - \alpha_2)}}, \; -\sqrt{\frac{(1 - \alpha_1)(1 - \alpha_2)}{\alpha_1 \alpha_2}} \right\}$$
$$< \phi_\alpha < \min \left\{ \sqrt{\frac{\alpha_1 (1 - \alpha_2)}{(1 - \alpha_1)\alpha_2}}, \; \sqrt{\frac{(1 - \alpha_1)\alpha_2}{\alpha_1 (1 - \alpha_2)}} \right\}.$$

If $Y_1, \ldots, Y_k$ are $k \geq 1$ i.i.d. bivariately Bernoulli-distributed random variables, then the sample sum $W := Y_1 + \cdots + Y_k$ is said to follow a *bivariate binomial distribution of Type I*, abbreviated as $\mathrm{BVB}_1(k; \alpha_1, \alpha_2, \phi_\alpha)$. As a result, the components W_i are univariately binomially distributed according to $\mathrm{Bin}(k, \alpha_i)$, but still have the unique sample size k for each component. To overcome this limitation and to allow for varying sample sizes $n_1, n_2 > 0$ with $0 \leq k \leq \min\{n_1, n_2\}$, one defines two further random variables U, V such that W, U, V are independent, and where $U \sim \text{Bin}(n_1 - k, \alpha_1)$ and $V \sim \text{Bin}(n_2 - k, \alpha_2)$. Then $X :=$ $(W_1 + U, W_2 + V)^\top$ is said to follow a *bivariate binomial distribution of Type II*, abbreviated to $\text{BVB}_{\text{II}}(n_1, n_2, k; \alpha_1, \alpha_2, \phi_\alpha)$. Its pmf is given by

$$
P(X = x) = \sum_{j_1=0}^{\min\{x_1, n_1-k\}} \binom{n_1 - k}{j_1} \alpha_1^{j_1} (1 - \alpha_1)^{n_1 - k - j_1}
$$

$$
\cdot \sum_{j_2=0}^{\min\{x_2, n_2-k\}} \binom{n_2 - k}{j_2} \alpha_2^{j_2} (1 - \alpha_2)^{n_2 - k - j_2}
$$

$$
\cdot \sum_{i=\max\{0, x_1 - j_1 + x_2 - j_2 - k\}}^{\min\{x_1 - j_1, x_2 - j_2\}} \binom{k}{i, \ x_1 - j_1 - i, \ x_2 - j_2 - i, \ k + i + j_1 + j_2 - x_1 - x_2} \alpha^i
$$

$$
(\alpha_1 - \alpha)^{x_1 - j_1 - i} (\alpha_2 - \alpha)^{x_2 - j_2 - i} (1 + \alpha - \alpha_1 - \alpha_2)^{k + i + j_1 + j_2 - x_1 - x_2}.
$$

Because of the additivity of the binomial distribution, the ith component satisfies $X_i \sim \text{Bin}(n_i, \alpha_i)$ for $i = 1, 2$, and the cross-correlation between X_1 and X_2 is given by $Corr[X_1, X_2] = \phi_\alpha \, k / \sqrt{n_1 n_2}$, which might take positive or negative values.

More details on these and other multivariate discrete distributions are provided by Johnson et al. (1997) and Kocherlakota & Kocherlakota (1992).

B

Basics about Stochastic Processes and Time Series

This appendix aims to provide a brief survey of some relevant terms and concepts from a basic course about stochastic processes and (continuous-valued) time series analysis. Many more details can be found in introductory textbooks such as those by Box et al. (2015), Brockwell & Davis (2016), Cryer & Chan (2008), Falk et al. (2012), Shumway & Stoffer (2011) and Wei (2006). For Section B.4.2, Lütkepohl (2005) is also recommended as a comprehensive textbook about multivariate time series. For the Markov chains discussed in Section B.2, the book by Seneta (1983) offers a lot of further detail.

B.1 Stochastic Processes: Basic Terms and Concepts

A (discrete-time) *stochastic process* (or simply *process*) is a sequence $(X_t)_{\mathcal{T}}$ of random variables $X_t : \Omega \to S$ defined on a probability space $(\Omega, \mathfrak{A}, P)$ and with range S (the *state space* of the process), where $\mathcal{T}$ is a discrete and linearly ordered set. For simplicity, we usually choose $\mathcal{T} = \mathbb{Z} := \{\ldots, -1, 0, 1, \ldots\}$ or $\mathcal{T} = \mathbb{N}_0 := \{0, 1, \ldots\}$. We distinguish between continuous-valued processes and discrete-valued processes, depending on whether the range S is a continuous or discrete set, respectively. If the range S of the random variables is equal to the set $\mathbb{R}$ of real numbers (or to a connected subset thereof), we refer to $(X_t)_{\mathcal{T}}$ as a *real-valued process*, while a *count process* even requires $S \subseteq \mathbb{N}_0$.

If the event $\omega \in \Omega$ is realized, this leads to a sequence $(X_t(\omega))_{\mathcal{T}}$, a *realization* (sample path) of the process. A *time series* $(x_t)_{\mathcal{T}_0}$, where $\mathcal{T}_0 \subseteq \mathcal{T}$ has to be a finite set in practice, is then understood as an observable part of such a realization,

$$(x_t)_{\mathcal{T}_0} \subseteq (X_t(\omega))_{\mathcal{T}} \qquad \text{for a fixed } \omega \in \Omega.$$

If we now speak about a *model for the time series*, we indeed refer to a model for the underlying process. In the sense of Kolmogorov's extension theorem, a stochastic process is uniquely characterized through its finite-dimensional distributions; that is, the joint distributions of any finite selection of random

An Introduction to Discrete-Valued Time Series, First Edition. Christian H. Weiss.
© 2018 John Wiley & Sons Ltd. Published 2018 by John Wiley & Sons Ltd.
Companion website: www.wiley.com/go/weiss/discrete-valuedtimeseries

variables $X_{t_1}, \ldots, X_{t_n}$ with $t_1 < \ldots < t_n \in \mathcal{T}$. Besides these complete distributions, certain moment properties are especially relevant for practice.

Definition B.1.1 (**Moments**) If the process $(X_t)_{\mathbb{Z}}$ consists of *square-integrable* real-valued random variables, then

- *mean* $\mu(t) := E[X_t]$,
- *variance* $\sigma^2(t) := V[X_t]$,
- *autocovariance* (at lag k) $\gamma_t(k) := Cov[X_t, X_{t-k}]$,
- *autocorrelation* (at lag k) $\rho_t(k) := Corr\,[X_t, X_{t-k}] = \frac{Cov\,[X_t, X_{t-k}]}{\sigma(t)\,\sigma(t-k)}$

are (generally) functions of the time t, with time lag $k \in \mathbb{Z}$.

We abbreviate the *autocorrelation function* as ACF. Note that $\sigma^2(t) = \gamma_t(0)$.

Example B.1.2 (**White noise**) The most basic type of process is so-called *white noise*. $(\epsilon_t)_{\mathbb{Z}}$ is said to be (strong) white noise if the ϵ_t are independent and identically distributed (i.i.d.) random variables. In the following, we shall usually consider real-valued and *square-integrable white noise*; because of the i.i.d.-assumption, the above first and second-order moments (Definition B.1.1) are constant in time; see also the stationarity concepts in Definition B.1.3 below.

In some textbooks, $(\epsilon_t)_{\mathbb{Z}}$ is only required to consist of square-integrable and uncorrelated random variables, with time-invariant first- and second-order moments; such kind of process is referred to as *weak white noise* hereafter.

Requiring for the time-invariance of certain stochastic properties of the considered process leads to concept of *stationarity*.

Definition B.1.3 (**Stationarity**) A process $(X_t)_{\mathbb{Z}}$ is said to be *(strictly) stationary* if the joint distributions of any finite selection of random variables $X_{t_1}, \ldots, X_{t_n}$ with $t_1 < \ldots < t_n \in \mathbb{Z}$ are time-invariant; that is, if the joint distribution functions of $(X_{t_1}, \ldots, X_{t_n})$ and $(X_{t_1+k}, \ldots, X_{t_n+k})$ are identical for any $k \in \mathbb{Z}$.

A real-valued process $(X_t)_{\mathbb{Z}}$ is said to be *weakly stationary* if it is square-integrable, and if mean and autocovariance are time-invariant; that is, if $\mu(t) = \mu$ and $\gamma_t(k) = \gamma(k)$ for all $t \in \mathbb{Z}$ and for any lag $k \in \mathbb{Z}$, which includes the variance for $k = 0$.

So the white noise from Example B.1.2 is stationary, and if it is also real-valued and square-integrable, then it is also weakly stationary. A weak white noise is weakly stationary by definition.

Since, for a weakly stationary process, mean and autocovariances are constant in time, they can be estimated from a given time series of the process by computing the respective empirical counterparts.

Definition B.1.4 **(Sample moments)** Let $x_1, \ldots, x_T$ be a time series from a weakly stationary real-valued process $(X_t)_{\mathbb{Z}}$. Then we define the

- *sample mean* $\bar{x} = \frac{1}{T} \sum_{t=1}^{T} x_t$,
- *sample variance* $s^2 = \frac{1}{T} \sum_{t=1}^{T} (x_t - \bar{x})^2$,

and for lags $k = 0, \ldots, T-1$, we define the

- *sample autocovariance* $\hat{\gamma}(k) = \frac{1}{T} \sum_{t=k+1}^{T} (x_t - \bar{x})(x_{t-k} - \bar{x})$,
- *sample autocorrelation* $\hat{\rho}(k) = \dfrac{\hat{\gamma}(k)}{\hat{\gamma}(0)} = \dfrac{\sum_{t=k+1}^{T} (x_t - \bar{x})(x_{t-k} - \bar{x})}{\sum_{t=1}^{T} (x_t - \bar{x})^2}$.

A plot of the sample autocorrelation function (SACF) $\hat{\rho}(k)$ against the time lag k is referred to as a *correlogram*.

Note that $\hat{\gamma}(0) = s^2$. The definitions are extended to negative lags by setting $\hat{\gamma}(-k) := \hat{\gamma}(k)$ and $\hat{\rho}(-k) := \hat{\rho}(k)$.

As an example, let us consider the sample mean $\bar{X}$ as an estimator of μ. If the underlying process is i.i.d. (white noise), then weak moment conditions suffice to guarantee that $\sqrt{T}\,(\bar{X} - \mu)$ is asymptotically normally distributed; see, for example, the Lindeberg–Lévy central limit theorem (CLT). To obtain a similar assertion for serially dependent (but still stationary) processes, the dependence structure of the process has to be analyzed in more detail. The traditional approach is to express the extent of serial dependence in terms of *mixing properties*, but more recently, *weak dependence conditions* have also been proposed as an alternative. The following definition summarizes two important mixing conditions for which appropriate CLTs are available; a survey about further mixing properties and their applications is provided by Bradley (2005). The relationship between weak dependence and mixing in the case of discrete-valued processes is discussed by Doukhan et al. (2012).

Definition B.1.5 **(ϕ-mixing, α-mixing)** Let $(X_t)_{\mathbb{Z}}$ be a stationary process on the probability space $(\Omega, \mathfrak{A}, P)$. It is said to be *$\phi$-mixing* if there exists a non-negative sequence $(\phi_n)_{\mathbb{N}}$ of weights with $\phi_n \to 0$ for $n \to \infty$ such that for each $t \in \mathbb{Z}$, $k \in \mathbb{N}$ and all events $E_1 \in \mathfrak{A}(X_t, X_{t-1}, \ldots)$, $E_2 \in \mathfrak{A}(X_{t+k}, X_{t+k+1}, \ldots)$, the following inequality holds:

$$|P(E_1 \cap E_2) - P(E_1)\,P(E_2)| \leq \phi_k\,P(E_1).$$

It is called α-*mixing* instead (also *strongly mixing*), with sequence of weights $(\alpha_n)_{\mathbb{N}}$, if the weaker requirement

$$|P(E_1 \cap E_2) - P(E_1)\, P(E_2)| \le \alpha_k$$

holds.

Both types of mixing process can be understood as a process in which the distant future is approximately independent of the past and present. A useful CLT for ϕ-mixing processes is given, among others, on p. 200 in Billingsley (1999), and a CLT for α-mixing processes is presented in Theorem 1.7 of Ibragimov (1962). Both theorems make additional requirements on the speed of convergence of the weights towards 0; these requirements are satisfied if the weights can be shown to decrease geometrically quickly.

Examples of non-stationary processes include processes exhibiting trend or seasonality. A particularly simple example of a non-stationary process is the random walk.

Example B.1.6 **(Random walk)** If $(\epsilon_t)_{\mathbb{N}}$ is real-valued white noise, then the process $(X_t)_{\mathbb{N}}$ defined by

$$X_1 = \epsilon_1, \qquad X_t = X_{t-1} + \epsilon_t \quad \text{for } t \ge 2,$$

is referred to as a *random walk*. Explicitly, it holds that

$$X_t = \sum_{s=1}^{t} \epsilon_s.$$

Hence, if the white noise is even square-integrable with mean μ_ϵ and variance σ_ϵ^2, we obtain that

$$\mu(t) = \mu_\epsilon\, t, \qquad \sigma^2(t) = \sigma_\epsilon^2\, t,$$

$$\gamma_t(k) = \sigma_\epsilon^2\, (t-k), \qquad \rho_t(k) = \sqrt{1 - \frac{k}{t}} \qquad \text{for } k \ge 0.$$

If $\mu_\epsilon \ne 0$, then the mean is not constant in time; in such a case, one also speaks of a *random walk with drift*.

By construction, the random walk at future time $t+1$ remembers its present value X_t and then adds a random disturbance ϵ_{t+1}; further past values $X_{t-1}, X_{t-2}, \ldots$ are without influence on X_{t+1} if the value of X_t is given. A process with such a "memory of length 1" is called a Markov chain.

Definition B.1.7 **(Markov process)** A process $(X_t)_{\mathbb{Z}}$ with state space S is said to be a *pth-order Markov process* with $\mathrm{p} \in \mathbb{N}$ if for all $t \in \mathbb{Z}$ and for each measurable set $A \subseteq S$,

$$P(X_t \in A \mid X_{t-1} = x_{t-1}, \ldots) = P(X_t \in A \mid X_{t-1} = x_{t-1}, \ldots, X_{t-p} = x_{t-p})$$

holds for all $x_i \in S$. In the case $\mathrm{p} = 1$, $(X_t)_{\mathbb{Z}}$ is commonly called a *Markov chain*.

In this book, only the case of a discrete state space is of relevance. Then, the Markov property of Definition B.1.7 simplifies to

$$
\begin{aligned}
P(X_t = x_t \mid X_{t-1} &= x_{t-1}, \ldots) \\
&= P(X_t = x_t \mid X_{t-1} = x_{t-1}, \ldots, X_{t-p} = x_{t-p}),
\end{aligned}
\tag{B.1}
$$

which has to hold for all $x_t \in S$. In fact, for theoretical analysis, it suffices to concentrate on the first-order case, since any pth-order Markov process with state space S can be transformed into a first-order Markov process with state space S^p by considering the vector-valued process $(\boldsymbol{X}_t)_{\mathbb{Z}}$ with $\boldsymbol{X}_t := (X_t, \ldots, X_{t-p+1})^{\top}$. The next section summarizes important properties of such discrete-valued Markov chains.

B.2 Discrete-Valued Markov Chains

In this section, the process $(X_t)_{\mathbb{N}_0}$ is assumed to be a (first-order) Markov chain with state space S. S is either finite or countably infinite; if S is not even ordinal, then it is assumed that its values are at least arranged in a certain lexicographical order to simplify notations: $S = \{s_0, s_1, \ldots\}$. Here, we summarize some basic facts about such discrete-valued Markov chains, but much more information is provided in, for example, the book by Seneta (1983) or in Chapter XV of Feller (1968).

B.2.1 Basic Terms and Concepts

From now on, we shall always assume that the Markov chain $(X_t)_{\mathbb{N}_0}$ is even *(time-)homogeneous*; that is, for any $i, j \in S$, the *transition probabilities* do not vary with time:

$$
P(X_t = i \mid X_{t-1} = j) = p_{i|j} \qquad \text{for all } t \in \mathbb{N}.
\tag{B.2}
$$

The corresponding *transition matrix* $\mathbf{P} = (p_{i|j})_{i,j=s_0, s_1, \ldots}$ might be of infinite dimension, depending on the cardinality of S. Similarly, all marginal probabilities at a given time are summarized as a (possibly time-dependent) vector:

$$
\boldsymbol{p}_t := (P(X_t = s_0),\ P(X_t = s_1), \ldots)^{\top}.
$$

Knowing the marginal distribution at time 0 (*initial distribution*), the transition matrix allows us to compute all further probabilities recursively via

$$
\boldsymbol{p}_t = \mathbf{P}\,\boldsymbol{p}_{t-1} = \ldots = \mathbf{P}^t\,\boldsymbol{p}_0.
\tag{B.3}
$$

In particular, the h-step-ahead transition probabilities $p_{i|j}^{(h)} := P(X_t = i \mid X_{t-h} = j)$ are the entries of the matrix $\mathbf{P}^h$.

Example B.2.1.1 (Classification of states) With the help of the h-step-ahead transition probabilities $p_{i|j}^{(h)}$, the Markov chain's states S can be classified. If, for given $i, j \in S$, there exists a time lag h such that $p_{i|j}^{(h)} > 0$, we say that j *leads* to i. If both j leads to i and i leads to j, then i and j are said to be *communicating* states.

If for the state $i \in S$, we find a state j such that i leads to j, but j does not lead to i, then i is an *inessential* state. If an *essential* state i, in turn, leads to a state j, then it necessarily communicates with j.

In view of (B.3), it is obvious that the homogeneous Markov chain $(X_t)_{\mathbb{N}_0}$ becomes stationary (see Definition B.1.3) if the marginal distributions remain fixed in time; that is, if $\boldsymbol{p}_t = \ldots = \boldsymbol{p}_0 =: \boldsymbol{p}$. So in view of (B.3), we need to look at so-called *invariant probability vectors*; that is, at probability vectors $\boldsymbol{\pi}$ satisfying the invariance equation

$$\mathbf{P}\,\boldsymbol{\pi} = \boldsymbol{\pi}. \tag{B.4}$$

Then $(X_t)_{\mathbb{N}_0}$ is stationary iff the initial distribution is an invariant vector; that is, the invariant distributions are the possible stationary marginal distributions of $(X_t)_{\mathbb{N}_0}$. The essential question is as follows:

Which conditions must be satisfied by the transition matrix $\mathbf{P}$ such that an invariant distribution (uniquely?) exists at all? Possible answers to this question are presented in Section B.2.2 below. Before this, let us briefly discuss the maximum likelihood approach to parameter estimation.

Remark B.2.1.2 (Maximum likelihood and information criteria) A popular approach for parameter estimation of a model for a stochastic process (not necessarily a Markov process) is the *maximum likelihood* (ML) approach. The intuition behind this approach is to select the parameter values such that the observed time series $x_1, \ldots, x_T$ becomes most "plausible". In the discrete-valued case, with the model parameters summarized in the vector θ, the *likelihood function* is defined as the probability of observing what has already been observed, as a function of the unknown θ:

$$L(\theta) := P(X_T = x_T, \ldots, X_1 = x_1 \mid \theta).$$

The ML estimate of θ is defined to be a value $\hat{\theta}_{\mathrm{ML}}$ maximizing $L(\theta)$. In view of computational issues, it is common practice to determine $\hat{\theta}_{\mathrm{ML}}$, not by maximizing $L(\theta)$, but by maximizing the *logarithmic likelihood* function (log-likelihood) instead; that is, $\ell(\theta) := \ln(L(\theta))$. We denote the value of the maximized log-likelihood as ℓ_{max}.

In general, it is not clear if the ML estimators (uniquely) exist and if they are consistent. Even if they exist, they can often only be computed numerically by using appropriate optimization routines. The computation of the (log-)likelihood function itself is also sometimes not easy. An exception are homogeneous Markov chains, for which the joint distributions of segments from $(X_t)_{\mathbb{N}_0}$ are easily computed,

$$P(X_r = x_r, \ldots, X_s = x_s) = P(X_r = x_r) \cdot \prod_{t=r+1}^{s} p_{x_t | x_{t-1}}, \qquad r < s.$$

Similar arguments apply to pth-order Markov processes, for which we obtain

$$L(\theta) = P(X_1 = x_1, \ldots, X_p = x_p) \cdot \prod_{t=p+1}^{T} p(x_t \mid x_{t-1}, \ldots, x_{t-p}),$$
$$\ell(\theta) = \ln P(X_1 = x_1, \ldots, X_p = x_p) + \sum_{t=p+1}^{T} \ln p(x_t \mid x_{t-1}, \ldots, x_{t-p}),$$
$$\tag{B.5}$$

where $p(x \mid x_{-1}, \ldots, x_{-p})$ denotes the time-homogeneous pth-order conditional probabilities for all $x, x_{-1}, \ldots, x_{-p} \in S$. If the Markov process is stationary, the initial probability $P(X_1 = x_1, \ldots, X_p = x_p)$ can be obtained as a function of the conditional probabilities (B.4). To avoid this step, one often simply maximizes the *conditional (log-)likelihood*,

$$L(\theta \mid x_p, \ldots, x_1) = \prod_{t=p+1}^{T} p(x_t \mid x_{t-1}, \ldots, x_{t-p}),$$
$$\ell(\theta \mid x_p, \ldots, x_1) = \sum_{t=p+1}^{T} \ln p(x_t \mid x_{t-1}, \ldots, x_{t-p}),$$
$$\tag{B.6}$$

and refers to the resulting estimates as *conditional ML* estimates $\hat{\theta}_{\text{CML}}$. Results concerning the existence, consistency and asymptotic normality of the (C)ML estimators for discrete-valued Markov chains are available in Part I of the book by Billingsley (1961). Under appropriate conditions (Billingsley, 1961), $\sqrt{T-p}\,(\hat{\theta}_{\text{CML}} - \theta)$ is asymptotically normally distributed according to $N(\mathbf{0}, \mathbf{I}^{-1}(\theta))$, where $\mathbf{0}$ denotes the zero vector, and where $\mathbf{I}(\theta)$ denotes the *expected Fisher information* per observation. The latter is defined as follows. The negative Hessian of $\ell(\theta)$ is said to be the *observed Fisher information* $\mathbf{J}(\theta)$, which can be written as $\mathbf{J}(\theta) = \sum_{t=p+1}^{T} \mathbf{J}_t(\theta)$ with $\mathbf{J}_t(\theta)$ being the Hessian of $-\ln p(x_t \mid x_{t-1}, \ldots, x_{t-p})$; then $\mathbf{I}(\theta)$ is the expectation of $\mathbf{J}_t(\theta)$. In practice, the mean observed Fisher information $\frac{1}{T-p} \mathbf{J}(\theta)$ can be used to approximate $\mathbf{I}(\theta)$. Plugging in the obtained estimates $\hat{\theta}_{\text{CML}}$ in $(\mathbf{J}(\theta))^{-1}$ allows us, for instance, to approximate the asymptotic standard errors of the CML estimators.

The ML approach can be applied in combination with so-called *information criteria* for model selection (if several candidate models are available for the time series). On the one hand, it is desirable to select a model with the largest $\ell_{\max}$ possible. On the other hand, small models are more easily interpretable and also overfitting has to be avoided. The idea behind information criteria is to balance the goodness of fit against the model size by adding an appropriate penalty term to $\ell_{\max}$. Let n_{model} abbreviate the number of parameters of the given model, then *Akaike's information criterion* (AIC) and the

Bayesian information criterion (BIC) are given by

$$\text{AIC} = -2\,\ell_{\max} + 2\,n_{\text{model}}, \qquad \text{BIC} = -2\,\ell_{\max} + n_{\text{model}}\,\ln T, \qquad \text{(B.7)}$$

respectively. That model producing the lowest value of AIC or BIC, respectively, is selected from among all candidate models. With increasing T, the BIC more strongly penalizes the model size, so it tends to select smaller models than the AIC. If applied to estimate the order p of a finite Markov process, Katz (1981) showed that only the BIC is consistent while the AIC leads to an overestimation.

AIC and BIC can also be computed based on the CML approach. Since the number of terms in $\ell(\theta \mid x_p, \dots, x_1)$ from (B.6) varies with varying p, one may insert the factor $T/(T - p)$ before $\ell_{\max}$ in (B.7) to account for this distortion.

B.2.2 Stationary Markov Chains

Criteria for the stationarity of a Markov chain will depend on the cardinality of S; the existence of an invariant distribution (and hence of a stationary solution to the Markov condition) will be much easier to establish if S is finite (*finite Markov chain*).

So let us now introduce, step-by-step, the potentially relevant conditions for the existence of a stationary solution (Feller, 1968; Seneta, 1983). The Markov chain is said to be *irreducible* if for any $i, j \in S$, there exists some lag $h \in \mathbb{N}$ such that $p_{i|j}^{(h)} > 0$. Note that h may differ for varying i, j. But if there exists a unique $h \in \mathbb{N}$ such that all $p_{i|j}^{(h)} > 0$, then the Markov chain is said to be *primitive*.

For a *finite* Markov chain, the following conclusions can be drawn:

- An irreducible finite Markov chain possesses a *unique* stationary distribution $\boldsymbol{p}$, being the unique solution of the invariance equation (B.4).
- A primitive finite Markov chain is even *ergodic*; that is, the distributions $\boldsymbol{p}_t$ always converge to the stationary distribution $\boldsymbol{p}$ for $t \to \infty$, independent of the initial distribution $\boldsymbol{p}_0$.

Expressed in terms of the h-step-ahead transition probabilities $p_{i|j}^{(h)}$, *ergodicity* means $p_{i|j}^{(h)} \to p_i$ for $h \to \infty$; that is, the h-step-ahead forecasting distribution $\boldsymbol{P}^h$ converges to the marginal distribution $\boldsymbol{p}\,\boldsymbol{1}^\top$, where $\boldsymbol{1}$ denotes the vector of ones (concerning the rate of convergence, see Remark B.2.2.1 below). This characterization offers a way of numerically computing the stationary distribution (instead of solving the eigenvalue problem (B.4)), by considering $p_{i|j}^{(M)}$ with a sufficiently large M as an approximation for p_i.

Another approach to ergodicity is to look at the *period* of the Markov chain. An irreducible Markov chain has a unique period, and this period equals d if the greatest common divisor of those $h \in \mathbb{N}$, for which $p_{i|i}^{(h)} > 0$, equals d (for $i \in S$ chosen arbitrarily). For $d = 1$, the Markov chain is said to be *aperiodic*, a property that may be established by the following criterion: if at least one diagonal element $p_{i|i}$ of $\boldsymbol{P}$ is positive, then the irreducible Markov chain is aperiodic.

For a *finite* Markov chain, the following equivalence holds:

- irreducible and aperiodic ⇔ primitive (and these imply ergodicity).

Finally, in view of the mixing concepts discussed in Definition B.1.5 above, it is worth mentioning that an ergodic *finite* Markov chain is ϕ-mixing with geometrically decreasing weights; that is, weights $\phi_n = a \cdot \rho^n$ with $a > 0$ and $0 < \rho < 1$.

Remark B.2.2.1   (Perron–Frobenius theorem)   As summarized above, a primitive finite Markov chain is even ergodic; that is, $\mathbf{P}^h \to \boldsymbol{p}\, \mathbf{1}^\top$ for $h \to \infty$. The rate of convergence can be expressed by adapting the *Perron–Frobenius theorem*. This theorem applies to non-negative matrices $\mathbf{T}$ in general; that is, matrices $\mathbf{T}$ having non-negative entries $t_{ij} \geq 0$. The transition matrix $\mathbf{P}$ is a special case in which the column sums are equal to 1 (a so-called *stochastic matrix*). The theorem states that for a primitive non-negative matrix $\mathbf{T}$, there exists the so-called *Perron–Frobenius eigenvalue* λ_{PF}, which satisfies

- $\lambda_{\mathrm{PF}} > |\lambda|$ for any other eigenvalue $\lambda \neq \lambda_{\mathrm{PF}}$ ("largest eigenvalue");
- λ_{PF} takes a value between the minimum and the maximum of all row sums of $\mathbf{T}$, and, in the same way, also a value between the minimum and the maximum of all column sums of $\mathbf{T}$;
- λ_{PF} has geometric and algebraic multiplicity 1;
- with λ_{PF} can be associated strictly positive left and right eigenvectors.

In this case, let $\boldsymbol{v}$ and $\boldsymbol{w}$ be corresponding positive left and right eigenvectors such that $\boldsymbol{v}^\top \boldsymbol{w} = 1$. Denote the remaining distinct eigenvalues by $\lambda_2, \ldots, \lambda_r$, where $\lambda_{\mathrm{PF}} > |\lambda_2| \geq \ldots \geq |\lambda_r|$ and where λ_2 ("second largest eigenvalue") has maximal multiplicity among all eigenvalues with the same modulus as λ_2; denote this multiplicity of λ_2 by m_2. Then the behavior of $\mathbf{T}^h$ for growing h is as follows:

$$\mathbf{T}^h = \lambda_{\mathrm{PF}}^h \cdot \boldsymbol{w}\boldsymbol{v}^\top + O(h^{m_2-1} \cdot |\lambda_2|^h).$$

This general result is now applied to the case of a primitive transition matrix $\mathbf{P}$ with (unique) invariant distribution $\boldsymbol{p}$. Since then all column sums are equal to 1, it follows that the Perron–Frobenius eigenvalue equals $\lambda_{\mathrm{PF}} = 1$. Furthermore, because of (B.4) – that is, $\mathbf{P}\,\boldsymbol{p} = \boldsymbol{p}$, and $\mathbf{1}^\top \mathbf{P} = \mathbf{1}^\top$ (column sums equal to 1) – any right eigenvector for $\lambda_{\mathrm{PF}} = 1$ is a multiple of $\boldsymbol{p}$, and any left eigenvector a multiple of $\mathbf{1}$. So it follows that

$$\mathbf{P}^h = \boldsymbol{p}\mathbf{1}^\top + O(h^{m_2-1} \cdot |\lambda_2|^h);$$

that is, the rate of convergence of $\mathbf{P}^h \to \boldsymbol{p}\mathbf{1}^\top$ is determined by the absolute value of the second largest eigenvalue, which satisfies $|\lambda_2| < 1$.

Let us conclude this section about discrete-valued Markov chains by looking at the countably infinite case. Here, irreducibility alone does not guarantee the existence of an invariant distribution. Therefore, it is necessary to look at another characteristic of the states of a Markov chain: the *recurrence properties*. Let $r_{j|j}(n)$ denote the conditional probability that the Markov chain $(X_t)_{\mathbb{N}_0}$, given that it started in $X_0 = j$, returns to the state j for the first time at time $t = n$; that is,

$$r_{j|j}(n) := P(X_n = j, \ X_{n-1}, \dots, X_1 \neq j \mid X_0 = j).$$

Summing about all n, we get the probability $r_{j|j} := \sum_{n=1}^{\infty} r_{j|j}(n)$ that the Markov chain will once return to the state j. If this probability $r_{j|j}$ equals 1, then $r_{j|j}(1), r_{j|j}(2), \dots$ constitute a valid probability distribution, the *recurrence time distribution*. The corresponding mean, $\mu_j = \sum_{n=1}^{\infty} n \, r_{j|j}(n)$, is referred to as the *mean recurrence time*. Now, the following classification is done:

- The state j is *transient* if $r_{j|j} < 1$; that is, there is a positive probability of never returning to j at all.
- Otherwise, that is, if $r_{j|j} = 1$, it is said to be recurrent:
  - it is *positive recurrent* if its mean recurrence time is finite, $\mu_j < \infty$,
  - it is *null recurrent* if its recurrence time is infinite, $\mu_j = \infty$.

For an *irreducible* Markov chain, all states are of the same recurrence type, just as they also have a unique period; see above. The following conclusions can be drawn:

- If $(X_t)_{\mathbb{N}_0}$ is an irreducible, aperiodic and positive recurrent[1] Markov chain, then it possesses a unique stationary distribution, and it is also ergodic.
- If $(X_t)_{\mathbb{N}_0}$ is an irreducible and aperiodic Markov chain having an invariant distribution $\boldsymbol{p}$, then this distribution is unique, and $(X_t)_{\mathbb{N}_0}$ is also positive recurrent and hence ergodic, where $p_j = 1/\mu_j$ for all $j \in S$.

Finally, such a countably infinite Markov chain, being irreducible, aperiodic and stationary, is also α-mixing (Definition B.1.5); but in contrast to the finite case above, there is no further general assertion concerning the speed of convergence of the mixing weights (Bradley, 2005, Section 3).

B.3 ARMA Models: Definition and Properties

The remaining sections of Appendix B are about common models for continuous-valued time series (Box et al., 2015; Brockwell & Davis, 1991, 2016; Cryer & Chan, 2008; Falk et al., 2012; Shumway & Stoffer, 2011; Wei, 2006). For

---

1 Note that a finite Markov chain cannot have a null recurrent state, and it is also impossible that all states are transient. Hence an irreducible finite Markov chain is automatically positive recurrent.

such continuous-valued processes, the class of linear processes is of particular importance.

Background B.3.1 **(Linear processes)** A real-valued sequence $(a_u)_{\mathbb{Z}}$ of weights, the *filter*, is said to be *absolutely summable* if

$$\sum_{u=-\infty}^{\infty} |a_u| := \sum_{u=0}^{\infty} |a_u| + \sum_{u=1}^{\infty} |a_{-u}| < \infty.$$

If $(Z_t)_{\mathbb{Z}}$ is a weakly stationary process with mean μ_Z and autocovariance function $\gamma_Z(k)$, with $(a_u)_{\mathbb{Z}}$ being an absolutely summable filter, then the filtered process $(Y_t)_{\mathbb{Z}}$, defined by

$$Y_t := \sum_{u=-\infty}^{\infty} a_u Z_{t-u},$$

exists and is also weakly stationary, where

$$\mu_Y := E[Y_t] = \mu_Z \sum_{u=-\infty}^{\infty} a_u,$$

$$\gamma_Y(k) := Cov[Y_t, Y_{t-k}] = \sum_{u,v=-\infty}^{\infty} a_u\, a_v\, \gamma_Z(k - u + v).$$

In particular, if the process to be filtered is not only weakly stationary but square-integrable white noise $(\epsilon_t)_{\mathbb{Z}}$ (the *innovations*), then the *linear process*

$$X_t := \sum_{u=-\infty}^{\infty} a_u \epsilon_{t-u}$$

is also stationary with

$$\mu = \mu_\epsilon \sum_{u=-\infty}^{\infty} a_u, \qquad \gamma(k) = \sigma_\epsilon^2 \sum_{u=-\infty}^{\infty} a_u\, a_{u+k}. \tag{B.8}$$

Note that the autocovariance function is now (nearly) solely defined through the filter $(a_u)_{\mathbb{Z}}$. Therefore, it is essential to study the properties of such filters $(a_u)_{\mathbb{Z}}$. For this purpose, consider the Laurent series

$$A(z) := \sum_{u=-\infty}^{\infty} a_u\, z^u,$$

which is referred to as the *characteristic polynomial* of the filter $(a_u)_{\mathbb{Z}}$. Using the *backshift operator* B – that is, the operator defined by $BX_t := X_{t-1}$ – the application of the linear filter is expressed as

$$X_t = A(B)\, \epsilon_t.$$

The successive application of two filters $(a_u)_{\mathbb{Z}}$ and $(b_v)_{\mathbb{Z}}$ is expressed by multiplying the corresponding characteristic polynomials $A(z)$ and $B(z)$. We say that $(a_u)_{\mathbb{Z}}$ is *invertible* (and $(b_v)_{\mathbb{Z}}$ is its *inverse filter*) if $A(z) B(z) = 1$. So the weights of the inverse filter are obtained by expanding $A^{-1}(z) = 1/A(z)$.

An absolutely summable filter $(a_u)_{\mathbb{Z}}$ is said to be *causal* if $a_u = 0$ for all $u < 0$; that is, if $A(z) = \sum_{u=0}^{\infty} a_u z^u$ is a power series. The practical meaning of a causal filter is that X_t does not depend on future innovations. An important example of a causal filter is one having only finitely many non-zero weights,

$$A_p(z) := 1 - a_1 z - \ldots - a_p z^p. \tag{B.9}$$

$A_p(z)$ has an absolutely summable and causal inverse filter iff the p (possibly complex) roots of $A_p(z)$ are outside the unit circle; that is, iff they have an absolute value larger than 1.

Now, we are in a position to introduce the family of ARMA models, which is done in three steps. First, we consider the pure moving-average models of order q, where the current observation is defined as a weighted mean of the current observation and q past innovations. These models date back to works by G. U. Yule and E. E. Slutsky in the 1920s; see Nie & Wu (2013).

Definition B.3.2 (MA(q) model) Let $(\epsilon_t)_{\mathbb{Z}}$ be square-integrable white noise. Then $(X_t)_{\mathbb{Z}}$ defined by

$$X_t = \epsilon_t - \beta_1 \epsilon_{t-1} - \ldots - \beta_q \epsilon_{t-q} \qquad (\beta_q \neq 0)$$

$$= \beta(B) \cdot \epsilon_t \qquad \text{with} \quad \beta(z) = 1 - \beta_1 z - \ldots - \beta_q z^q$$

is said to be a *moving-average process* of order q, abbreviated as $MA(q)$ *process*.

The case $q = \infty$ ($MA(\infty)$ *process*) just corresponds to the causal linear process defined in Background B.3.1. The MA(q) process is strictly and weakly stationary with

$$\mu = \mu_\epsilon (1 - \beta_\bullet), \qquad \text{where} \quad \beta_\bullet := \sum_{i=1}^{q} \beta_i, \tag{B.10}$$

$$\gamma(k) = \begin{cases} 0 & \text{for } k > q, \\ \gamma(-k) & \text{for } k < 0, \\ \sigma_\epsilon^2 \sum_{u=0}^{q-k} \beta_u \beta_{u+k} & \text{for } 0 \leq k \leq q, \quad \text{where } \beta_0 := -1; \end{cases} \tag{B.11}$$

see (B.8). Note that the autocorrelation function vanishes after lag q; this property is used to identify the model order q of an MA model: one visually inspects the available correlogram for an (abrupt) drop towards zero; the largest lag with a (significantly) non-zero autocorrelation is used as an estimate of the model order.

Finally, it should be noted that result (B.9) for linear processes implies that the MA(q) process is *invertible* with a *causal* inverse iff the roots of $\beta(z)$ are outside the unit circle. Then $\epsilon_t = \sum_{u=0}^{\infty} \psi_u X_{t-u}$; that is, the innovation at time t can be recovered from the observations available at time t, where the coefficients ψ_u are obtained by expanding $\beta^{-1}(z)$ into a power series.

As the second step towards full ARMA models, we consider the autoregressive models of order p, where the current observation is a weighted mean of p past observations plus noise. So these processes are generated from their own pasts (hence "auto" regressive; see also the Markov models in Definition B.1.7). These models were established by G. U. Yule and G. T. Walker; see Nie & Wu (2013) for references.

Definition B.3.3 (AR(p) model) Let $(\epsilon_t)_{\mathbb{Z}}$ be square-integrable white noise. Then $(X_t)_{\mathbb{Z}}$ defined by

$$X_t = \alpha_1 X_{t-1} + \ldots + \alpha_p X_{t-p} + \epsilon_t \qquad (\alpha_p \neq 0)$$

$$\Leftrightarrow \alpha(B) \cdot X_t = \epsilon_t \qquad \text{with} \quad \alpha(z) = 1 - \alpha_1 z - \ldots - \alpha_p z^p$$

is said to be an *autoregressive process* of order p, abbreviated as *AR(p) process*.

In view of the above result (B.9), a (weakly) stationary and *causal* solution exists for the AR(p) recursion iff the roots of $\alpha(z)$ are outside the unit circle; then $X_t = \sum_{u=0}^{\infty} \theta_u \epsilon_{t-u}$ (a causal linear process), where the coefficients θ_u are obtained by expanding $\alpha^{-1}(z)$ into a power series.

If $(X_t)_{\mathbb{Z}}$ is a weakly stationary and causal AR(p) process, then its mean is computed as

$$\mu = E[X_t] = \frac{\mu_\epsilon}{1 - \alpha_\bullet}, \qquad \text{where} \quad \alpha_\bullet := \sum_{i=1}^{p} \alpha_i, \qquad (B.12)$$

while the autocorrelation function $\rho(k)$ is obtained by solving the so-called *Yule–Walker equations*:

$$\rho(k) = \sum_{i=1}^{p} \alpha_i \, \rho(|k - i|) \qquad \text{for } k = 1, 2, \ldots \qquad (B.13)$$

Now, we typically have $\rho(k) \neq 0$ for all k such that $\rho(k)$ cannot be (directly) applied to identify the model order of the autoregressive process, in contrast to the corresponding procedure for moving-average processes, as sketched below formula (B.11). However, transforming the (sample) autocorrelation function into the (sample) *partial* autocorrelation function first, an analogous procedure is possible.

Theorem B.1 **(Partial autocorrelation)** Let $r_k := (\rho(1), \ldots, \rho(k))^\top$ as well as

$$\mathbf{R}_k := (\rho(|i-j|))_{i,j=1,\ldots,k} = \begin{pmatrix} 1 & \rho(1) & \cdots & \rho(k-1) \\ \rho(1) & 1 & \ddots & \vdots \\ \vdots & \ddots & \ddots & \rho(1) \\ \rho(k-1) & \cdots & \rho(1) & 1 \end{pmatrix},$$

assume that $\mathbf{R}_k$ is invertible. Let $a_k \in \mathbb{R}^k$ be the unique solution of the equation

$$\mathbf{R}_k \, a_k = r_k, \qquad \text{that is,} \qquad a_k = \mathbf{R}_k^{-1} \, r_k. \tag{B.14}$$

Then $\rho_{\text{part}}(k) := a_{k,k}$ (last component of a_k) is referred to as the *partial autocorrelation* at lag k.

If $(X_t)_{\mathbb{Z}}$ is a weakly stationary and causal AR(p) process, then the partial autocorrelation function (PACF) satisfies

$$\rho_{\text{part}}(\text{p}) = \alpha_{\text{p}}, \qquad \rho_{\text{part}}(k) = 0 \quad \text{for } k > p.$$

Note that the PACF can be computed recursively according to

$$a_{k+1,k+1} = \frac{\rho(k+1) - \sum_{i=1}^k a_{k,i} \, \rho(k+1-i)}{1 - \sum_{i=1}^k a_{k,i} \, \rho(i)}, \tag{B.15}$$

$$a_{k+1,j} = a_{k,j} - a_{k+1,k+1} \, a_{k,k-j+1} \qquad \text{for } j = 1, \ldots, k.$$

The sample version of the PACF (SPACF) is derived in the same way as in (B.14) and (B.15), but with $\rho(\cdot)$ replaced by the corresponding sample autocorrelations, $\hat{\rho}(\cdot)$.

As the final step, we combine the AR model with the MA model to obtain the full ARMA model. This combination was suggested by A. M. Walker in 1950, and the embedding into the theory of linear processes (Background B.3.1) was initiated by H. Wold in the late 1930s (Nie & Wu, 2013).

Definition B.3.4 **(ARMA(p, q) model)** Let $(\epsilon_t)_{\mathbb{Z}}$ be square-integrable white noise. Then $(X_t)_{\mathbb{Z}}$, defined by ($\alpha_{\text{p}}, \beta_{\text{q}} \neq 0$):

$$X_t = \alpha_1 X_{t-1} + \ldots + \alpha_{\text{p}} X_{t-p} + \epsilon_t - \beta_1 \epsilon_{t-1} - \ldots - \beta_{\text{q}} \epsilon_{t-q}$$
$$\Leftrightarrow \alpha(\text{B}) X_t = \beta(\text{B}) \epsilon_t \quad \text{with} \quad \alpha(z) = 1 - \alpha_1 z - \ldots - \alpha_{\text{p}} z^{\text{p}}$$
$$\text{and} \quad \beta(z) = 1 - \beta_1 z - \ldots - \beta_{\text{q}} z^{\text{q}}$$

is said to be an *autoregressive moving-average process* of order (p, q), which is abbreviated as *ARMA*(p, q) *process*. To keep the order (p, q) minimal, we also require that $\alpha(z)$ and $\beta(z)$ have no common roots.

The pure AR and pure MA models are embedded in Definition B.3.4 as the cases q = 0 and p = 0, respectively.

With analogous arguments as above (result (B.9)), a (weakly) stationary and *causal* solution exists for the ARMA(p, q) recursion iff the roots of $\alpha(z)$ are outside the unit circle; then $X_t = \sum_{u=0}^{\infty} \theta_u \, \epsilon_{t-u}$ (a causal linear process), where the coefficients θ_u are obtained by expanding $\alpha^{-1}(z) \, \beta(z)$ into a power series. Similarly, invertibility requires the roots of $\beta(z)$ to be outside the unit circle.

The coefficients θ_u of the representation $X_t = \sum_{u=0}^{\infty} \theta_u \, \epsilon_{t-u}$ can also be computed recursively: setting $\theta_k := 0$ for $k < 0$, and $\beta_0 := -1$ as well as $\beta_k := 0$ for $k > q$, it holds that

$$\theta_k = \sum_{j=1}^{p} \alpha_j \, \theta_{k-j} - \beta_k. \tag{B.16}$$

If $(X_t)_{\mathbb{Z}}$ is a weakly stationary and causal ARMA(p, q) process, then its mean is computed as

$$\mu = E[X_t] = \frac{1 - \beta_{\bullet}}{1 - \alpha_{\bullet}} \, \mu_{\epsilon}, \tag{B.17}$$

while the autocorrelation function $\rho(k)$ is obtained after solving the *Yule–Walker equations*

$$\gamma(k) - \sum_{i=1}^{p} \alpha_i \, \gamma(|k - i|) = \begin{cases} -\sigma_{\epsilon}^2 \sum_{u=k}^{q} \beta_u \, \theta_{u-k} \\ \qquad \text{for } 0 \leq k < \max\{p, q + 1\}, \\ 0 \quad \text{for } k \geq \max\{p, q + 1\}. \end{cases} \tag{B.18}$$

B.4 Further Selected Models for Continuous-valued Time Series

The basic ARMA models (which themselves might be applied to stationary processes with a short memory) gave rise to innumerable modifications and extensions to deal with, for example, trend or seasonality, long memory, time-varying volatility or with multivariate observations. A compact survey of time series models related to ARMA models is provided by Holan et al. (2010). Here, two of these models appear to be particularly relevant: the famous GARCH models with their ability to generate conditional heteroskedasticity, and the VARMA models as a multivariate extension of ARMA models.

B.4.1 GARCH Models

Let us start with the ARCH models, which were developed by R. F. Engle III in 1979. In 2003, he received (one half of) the Nobel prize in economic sciences "for methods of analyzing economic time series with time-varying volatility

(ARCH)".[2] The ARCH models are motivated by a specific drawback of the AR models. If looking at the conditional mean and variance of an AR(p) process according to Definition B.3.3, then

$$E[X_t \mid X_{t-1}, \ldots] = \alpha_1 X_{t-1} + \ldots + \alpha_p X_{t-p} + \mu_\epsilon$$

varies in time according to the last p observations, while $V[X_t \mid X_{t-1}, \ldots] = \sigma_\epsilon^2$ is constant in time. But, especially for financial time series, it is common to observe clusters of large or low volatility, a phenomenon that cannot be reproduced by the AR models.

Definition B.4.1.1 (ARCH(p) model) Let $(\epsilon_t)_{\mathbb{Z}}$ be square-integrable white noise with $E[\epsilon_t] = 0$ and $V[\epsilon_t] = 1$. Then $(X_t)_{\mathbb{Z}}$ defined by

$$X_t = \sigma_t \cdot \epsilon_t, \qquad \text{where} \quad \sigma_t^2 = \beta_0 + \alpha_1 X_{t-1}^2 + \ldots + \alpha_p X_{t-p}^2$$

with $\beta_0, \alpha_p > 0$ and $\alpha_1, \ldots, \alpha_{p-1} \geq 0$, and where ϵ_t is required to be independent of $(X_s)_{s<t}$ (*causality*), is said to be an *autoregressive conditional heteroskedasticity process* of order p, abbreviated as *ARCH*(p) *process*.

As a result, we obtain the time-varying conditional variances

$$V[X_t \mid X_{t-1}, \ldots] = \sigma_t^2 = \beta_0 + \alpha_1 X_{t-1}^2 + \ldots + \alpha_p X_{t-p}^2. \tag{B.19}$$

So now, the AR(p)-like recursion is not applied to the observations but to their conditional variances. In contrast, the unconditional variance remains constant in time provided that the requirement $\sum_{j=1}^{p} \alpha_j < 1$ is satisfied. In fact, this condition (which is equivalent to requiring the roots of $\alpha(z) := 1 - \alpha_1 z - \ldots - \alpha_p z^p$ to be outside the unit circle, due to the non-negativity of the α_i) again guarantees the existence of a unique causal (weakly) stationary solution of the ARCH recursion in Definition B.4.1.1.

Looking at the autocorrelation structure, one might initially be surprised, since $\gamma(k) = 0$ for $k \neq 0$. So although an ARCH(p) process is obviously not serially independent by construction, it is serially uncorrelated. However, looking at the process of squared observations, autocorrelation becomes visible. If the weakly stationary and causal ARCH(p) process $(X_t)_{\mathbb{Z}}$ has existing fourth-order moments, then we can represent the squared process $(X_t^2)_{\mathbb{Z}}$ by an AR(p)-like recursion,

$$X_t^2 = \alpha_1 X_{t-1}^2 + \ldots + \alpha_p X_{t-p}^2 + v_t$$

with the $(v_t)_{\mathbb{Z}}$ being *weak* white noise having the mean $E[v_t] = \beta_0$. Therefore, the autocorrelation function of the squared process satisfies the Yule–Walker equations (B.13) above.

---

2 A lot of background information concerning ARCH models can be found in Engle's Nobel lecture at www.nobelprize.org/nobel_prizes/economic-sciences/laureates/2003/.

A few years after the introduction of the ARCH model, Engle's student T. Bollerslev proposed the *generalized* ARCH model, where the conditional variances not only depend on past observations but also on past conditional variances. So for a GARCH process of order (p, q), abbreviated as *GARCH*(p, q) *process*, the recursion

$$\sigma_t^2 = \beta_0 + \alpha_1 X_{t-1}^2 + \ldots + \alpha_p X_{t-p}^2 + \beta_1 \sigma_{t-1}^2 + \ldots + \beta_q \sigma_{t-q}^2 \qquad (B.20)$$

is required to be satisfied. The condition for the existence of a (weakly) stationary and causal solution now becomes $\alpha_1 + \ldots + \alpha_p + \beta_1 + \ldots + \beta_q < 1$.

B.4.2   VARMA Models

As aforementioned, one half of the 2003's Nobel prize in economic sciences was awarded to Engle; the other half was received by another statistician: C. W. J. Granger "for methods of analyzing economic time series with common trends (cointegration)", a topic related to multivariate time series. A multivariate (or vector-valued) time series $x_1, \ldots, x_T$ is obtained if multiple features are observed simultaneously over time. For modeling the underlying multivariate (or vector) process $(X_t)_{\mathbb{Z}}$ with its possible cross-correlations, again a large variety of models is available; see Lütkepohl (2005) for a comprehensive survey. But in view of the scope of the present book, we shall only consider the *vector autoregressive moving-average models* of order (p, q), abbreviated as *VARMA*(p, q) *models*.

Let $(X_t)_{\mathbb{Z}}$ be a process of d-dimensional random variables with range $\mathbb{R}^d$. Then the *mean vector* $\mu(t)$ and the *autocovariance matrix* $\Gamma_t(k)$ are defined by

$$\mu(t) := E[X_t] := (\ldots, E[X_{t,i}], \ldots)^{\top},$$

$$\Gamma_t(k) := E[(X_t - \mu(t))(X_{t-k} - \mu(t-k))^{\top}] \qquad (B.21)$$

$$:= \left(E[(X_{t,i} - \mu_i(t))(X_{t-k,j} - \mu_j(t-k))] \right)_{i,j=1,\ldots,d}.$$

If both expressions exist (square-integrable) and if they are constant in time t, then we refer to the process $(X_t)_{\mathbb{Z}}$ as being *weakly stationary* (as above), and we denote them as μ and $\Gamma(k)$ (without t) in this case. The *autocorrelation matrix* is then defined as

$$\mathbf{R}(k) = (\rho_{ij}(k))_{i,j=1,\ldots,d} \quad \text{with} \quad \rho_{ij}(k) = \frac{\gamma_{ij}(k)}{\sqrt{\gamma_{ii}(0)\gamma_{jj}(0)}}, \qquad (B.22)$$

where $\gamma_{ij}(k)$ denotes the (i, j)th element of $\Gamma(k)$. Note that the definition (B.21) implies that

$$\Gamma(k) = \Gamma(-k)^{\top}. \qquad (B.23)$$

Analogous to Example B.1.2, the multivariate process $(\epsilon_t)_{\mathbb{Z}}$ is said to be a *white noise* if the ϵ_t are i.i.d. If $(\epsilon_t)_{\mathbb{Z}}$ is also square-integrable with covariance matrix

$\boldsymbol{\Sigma}_\epsilon$, then the corresponding autocovariance matrix equals

$$\boldsymbol{\Gamma}_\epsilon(k) = \boldsymbol{\Sigma}_\epsilon \quad \text{for } k = 0, \quad \text{and} \quad \boldsymbol{\Gamma}_\epsilon(k) = \mathbf{O} \quad \text{otherwise,} \tag{B.24}$$

with $\mathbf{O}$ being the zero matrix.

Definition B.4.2.1 (VARMA(p, q) model) Let $(\epsilon_t)_{\mathbb{Z}}$ be square-integrable d-dimensional white noise. Then $(X_t)_{\mathbb{Z}}$ is said to be a *VARMA*(p, q) *process* if

$$X_t = \mathbf{A}_1 \, X_{t-1} + \dots + \mathbf{A}_{\mathrm{p}} \, X_{t-\mathrm{p}} + \epsilon_t - \mathbf{B}_1 \, \epsilon_{t-1} - \dots - \mathbf{B}_{\mathrm{q}} \, \epsilon_{t-\mathrm{q}}$$

with $\mathbf{A}_1, \dots, \mathbf{B}_{\mathrm{q}} \in \mathbb{R}^{d \times d}$, or equivalently if

$$\mathbf{A}(B) \, X_t = \mathbf{B}(B) \, \epsilon_t$$

with $\mathbf{A}(z) = \mathbf{I} - \mathbf{A}_1 \, z - \dots - \mathbf{A}_{\mathrm{p}} \, z^{\mathrm{p}}$ and $\mathbf{B}(z) = \mathbf{I} - \mathbf{B}_1 \, z - \dots - \mathbf{B}_{\mathrm{q}} \, z^{\mathrm{q}}$, where $\mathbf{I}$ denotes the identity matrix.

The criterion for the existence of a unique *causal and stationary* solution now becomes

$$\det(\mathbf{A}(z)) \neq 0 \quad \text{for } |z| \leq 1, \tag{B.25}$$

while the criterion for *invertibility* is given by

$$\det(\mathbf{B}(z)) \neq 0 \quad \text{for } |z| \leq 1. \tag{B.26}$$

The autocovariance matrices of a causal stationary VARMA(p, q) process are determined again by a set of *Yule–Walker equations* (by considering (B.23)), given by

$$\boldsymbol{\Gamma}(k) - \sum_{i=1}^{\mathrm{p}} \mathbf{A}_i \, \boldsymbol{\Gamma}(k - i) = \sum_{j=k}^{\mathrm{q}} \mathbf{B}_j \, \boldsymbol{\Sigma}_\epsilon \, \mathbf{C}_{j-k}^{\top} \quad \text{for } k = 0, 1, \dots,$$

where $\mathbf{C}_r = \mathbf{O}$ for $r < 0, \quad \mathbf{C}_0 = \mathbf{I}$

and $\mathbf{C}_r = \sum_{i=1}^{r} \mathbf{A}_i \, \mathbf{C}_{r-i} + \mathbf{B}_r$ for $r > 0$. $\tag{B.27}$

Up to this point, everything looks analogous to the corresponding results for the univariate ARMA(p, q) models; see Definition B.3.4 and below. Now, we have to discuss some important differences. First, it has to be mentioned that the VAR(1) model is of outstanding importance, since any d-dimensional VARMA(p, q) model can be translated into an $d(\mathrm{p} + \mathrm{q})$-dimensional VAR(1) model. This is done in an analogous way to the approach for Markov processes sketched below (B.1), by constructing the coefficient matrix $\mathbf{D} \in \mathbb{R}^{d(\mathrm{p}+\mathrm{q}) \times d(\mathrm{p}+\mathrm{q})}$ as follows:

$$\mathbf{D}_{11} = \begin{pmatrix} \mathbf{A}_1 & \cdots & \mathbf{A}_{\mathrm{p}-1} & \mathbf{A}_{\mathrm{p}} \\ \mathbf{I} & & \mathbf{O} & \mathbf{O} \\ & \ddots & & \vdots \\ \mathbf{O} & & \mathbf{I} & \mathbf{O} \end{pmatrix} \in \mathbb{R}^{d\mathrm{p} \times d\mathrm{p}}, \tag{B.28}$$

$$
\mathbf{D}_{12} = \begin{pmatrix} \mathbf{B}_1 & \cdots & \mathbf{B}_q \\ \mathbf{O} & \cdots & \mathbf{O} \\ \vdots & \ddots & \vdots \\ \mathbf{O} & \cdots & \mathbf{O} \end{pmatrix} \in \mathbb{R}^{dp \times dq}, \qquad \mathbf{D}_{21} = \mathbf{O} \in \mathbb{R}^{dq \times dp},
$$

$$
\mathbf{D}_{22} = \begin{pmatrix} \mathbf{O} & \cdots & \mathbf{O} & \mathbf{O} \\ \mathbf{I} & & \mathbf{O} & \mathbf{O} \\ & \ddots & & \vdots \\ \mathbf{O} & & \mathbf{I} & \mathbf{O} \end{pmatrix} \in \mathbb{R}^{dq \times dq}, \qquad \mathbf{D} = \begin{pmatrix} \mathbf{D}_{11} & \mathbf{D}_{12} \\ \mathbf{D}_{21} & \mathbf{D}_{22} \end{pmatrix}.
$$

With

$$
\boldsymbol{Y}_t := \begin{pmatrix} \boldsymbol{X}_t \\ \vdots \\ \boldsymbol{X}_{t-p+1} \\ \boldsymbol{\epsilon}_t \\ \vdots \\ \boldsymbol{\epsilon}_{t-q+1} \end{pmatrix}, \qquad \boldsymbol{U}_t := \begin{pmatrix} \boldsymbol{\epsilon}_t \\ \mathbf{0} \\ \vdots_{p-1} \\ \mathbf{0} \\ \boldsymbol{\epsilon}_t \\ \mathbf{0} \\ \vdots_{q-1} \\ \mathbf{0} \end{pmatrix},
$$

we have

$$
\boldsymbol{Y}_t = \mathbf{D}\, \boldsymbol{Y}_{t-1} + \boldsymbol{U}_t. \tag{B.29}
$$

Using this VAR(1)-representation, we may, for example, trace back the auto-correlation function of the general VARMA(p, q) model to that of the VAR(1) model. The latter is given by

$$
\boldsymbol{\Gamma}(k) = \mathbf{A}_1^k\, \boldsymbol{\Gamma}(0), \qquad \boldsymbol{\Gamma}(0) = \mathbf{A}_1\, \boldsymbol{\Gamma}(0)\, \mathbf{A}_1^\top + \boldsymbol{\Sigma}_\epsilon, \tag{B.30}
$$

see (B.27) with (p, q) = (1, 0). Furthermore, the causal solution of the VAR(1) model equals

$$
\boldsymbol{X}_t = \sum_{j=0}^{\infty} \mathbf{A}_1^j\, \boldsymbol{\epsilon}_{t-j}. \tag{B.31}
$$

As the second main difference to the univariate ARMA models, it has to be mentioned that sometimes there is not a unique representation of a VARMA process (counterexamples are easily constructed; see for example the one on p. 259 in Holan et al. (2010)). Therefore, in practice, one often restricts oneself to the purely autoregressive VAR(p) models in advance to avoid such *non-identifiability issues*. Further information is presented in the book by Lütkepohl (2005).

C

Computational Aspects

For illustrating the models and methods described in this book, a number of data examples and numerical examples have been presented. The corresponding computations would not have been possible without software support. The unique software solution used was the program R (R Core Team, 2016). This choice was mainly motivated by the fact that R is freely available to everyone and it also runs on every common operating system. It is widely used anyway. Other computational software packages could be used, for example Matlab or Mathematica. To allow the reader to easily do the computations on her/his own, and to modify and extend the computations, all R codes are provided on the companion website: www.wiley.com/go/weiss/discrete-valuedtimeseries as a password-protected zip file (Password: `DiscrValTS17`). The datasets that are freely available can be found there too. An overview of the R codes used for this book is given in Section C.2, while the datasets are listed in Section C.3.

To simplify code translation into other programming languages, the R codes were written by using only basic commands; that is, commands available in base R (and analogously in other common languages) without the need for special packages (with a few exceptions, as described in Section C.1 below). Another advantage of this approach is that "black boxes" are avoided to some degree, and the code can be better understood and related to the formulae used in this book. On the other hand, the presented R solutions will usually not be the most efficient ones, and the amount of programming code could be reduced by using appropriate R packages. The main problem with the latter approach, however, would be that ready solutions exist for only a small proportion of the models and methods discussed in this book (at least at the time of writing this book), so a unique presentation would not be possible in this way. To mention a few packages,

- some of the INGARCH and log-linear regression models from Sections 4 and 5.1 could also be fitted by using the R package "tscount" described in Liboschik et al. (2016)

An Introduction to Discrete-Valued Time Series, First Edition. Christian H. Weiss.
© 2018 John Wiley & Sons Ltd. Published 2018 by John Wiley & Sons Ltd.
Companion website: www.wiley.com/go/weiss/discrete-valuedtimeseries

- some GLARMA models (Section 5.1) are available through the R package "glarma" (Dunsmuir & Scott, 2015)
- methods for hidden-Markov models (Section 5.2) are offered by the R package "HiddenMarkov" (Harte, 2016), among others
- for the monitoring of discrete-valued processes, the R package "surveillance" (Höhle, 2007; Salmon et al., 2016) could be used.

C.1   Some Comments about the Use of R

Since the total amount of R code used for this book is very large (Section C.2), it makes no sense to discuss these codes line-by-line. Instead, some general comments will be given that will help readers better understand the code, and allow them to write their own code for models and methods not covered in this book. These comments are quite specific and assume that the reader already has a solid knowledge of R. If this is not yet the case, the reader may consult an introductory book on R, such as Gardener (2012). See also the R project's website for further book recommendations.[1]

Computation and Simulation for Count Models

The pmf of the count distributions discussed in Appendix A are readily implemented in R in only a few basic cases, such as the Poisson or (negative) binomial distribution. In other cases, the formula for the pmf has to be programmed by hand (if an explicit formula is available), or numeric pmf values might be obtained by a numeric series expansion of the pgf. Such a numerical series expansion can be done with standard methods offered in Matlab or Mathematica, while an extra package is required for R, such as the R package "pracma" by Borchers (2016). For visualizing a discrete pmf, one may use `plot(...,type="h", ...)`. The cdf corresponding to a given pmf vector is obtained by applying `cumsum`.

An analogous distinction applies to the simulation of the corresponding i.i.d. random variables: if a certain distribution is not readily available, one has to use a general command for taking random samples with replacement. In R, this is done by

```
sample(0:upper, reps, replace=TRUE, prob=pmf)
```

with *upper* being either the upper limit of the distribution's range if this is finite, or a sufficiently large value otherwise. A vector of the corresponding numerical values of the pmf is provided as *pmf*, and *reps* is the number of i.i.d. replications.

---

1 https://www.r-project.org/.

Stationary Marginal Distribution of Markov Chains

The computation of the stationary marginal distribution of a Markov chain is required for several examples in this book (for example, Example 2.1.3.5), and it makes use of the invariance equation (B.4). According to this equation, the unique non-negative eigenvector with component sum 1 corresponding to the eigenvalue 1 of the transition matrix is required. In R, this can be computed numerically by using `eigen(..., symmetric=FALSE)$vectors[,1]`; see also the discussion of the command `eigen` below. A matrix inversion is possible using `solve`; matrix multiplication is done in R by `%*%`. The visual representation of the transition matrices in Example 7.1.1 by weighted directed graphs was done by using the R package "qgraph" by Epskamp et al. (2012).

Simulation of Count Processes

The simulation of count processes uses the above methods for simulating count distributions. For a stationary Markov process, one needs the stationary marginal distribution (see also the discussion above) for the initial observation; the remaining observations are generated according to the respective conditional distribution (Remark 2.6.4). For some models, such as the INAR(1) model (Example 2.1.2.1), one can directly utilize the model recursion and simulate each of its parts separately. In other cases (such as the Poisson INARCH(1) model, as in Remark 4.1.7), one simulates the conditional distribution as a whole. An alternative way of initializing the simulation (more generally of ergodic processes) is to work with a prerun (Remarks 2.6.4 and 4.1.7, Examples 5.1.2 and 5.1.3).

Numerical Maximum Likelihood Estimation

For the numerical computation of ML estimates, the log-likelihood function $\ell(\theta)$ has to be maximized by using an appropriate optimization routine. In R, such routines always search for a local minimum, so in fact, the negative log-likelihood $-\ell(\theta)$ has to be minimized in θ. R offers several different routines for numerical optimization, where the decision to use a particular routine is often determined by a given parameter constraint. In many cases, the constraints can be expressed as box constraints (for example, in Section 2.5); that is, $\theta \in [l, u]$ must be satisfied. Then one may use

```
optim(..., method="L-BFGS-B", lower=low, upper=up, ...)
```

with *low* being the vector l of the lower box limits, and *up* the vector u of upper limits. The argument `hessian=TRUE` is optional, but obtaining the Hessian of $-\ell$ is attractive in view of approximating the standard errors; see Remark B.2.1.2. A possible alternative to `optim` is `nlminb`.

If more general inequality constraints of the form $\mathbf{A}\,\boldsymbol{\theta} \geq \boldsymbol{b}$ are given (for example, in Example 7.2.4), one can use `constrOptim(..., ui=A, ci=b, ...)`. Since box constraints are particular inequality constraints, `constrOptim` could also be used in the case described above, but usually, it requires more computing time than `optim` or `nlminb`. Furthermore, the hessian of $-\ell$ can only be computed if the gradient of $-\ell$ is also delivered to `constrOptim`.

Finally, there are sometimes no constraints on the model parameters (that is, the full set of real numbers is allowed for each model parameter), as in Example 7.4.5. In this case, one can again use `optim` with a `method` for unconstrained optimization; alternatively, one can use `nlm`.

Categorical Random Variables

Categorical random variables can be simulated with the help of the command `rmultinom(..., size=1, ...)` for multinomially distributed random variables. As a result, a binary column vector (that is, a $(d+1) \times 1$-matrix) is generated, which corresponds to the binarization of the categorical random variable (Example A.3.3). This is changed into a numerical coding by the numbers $0, \ldots, d$ with the help of matrix multiplication; that is, by adding `c(0:d) %*%` before the binarized vector. Given the numerical coding, one can replace these numbers with certain strings $s_0, \ldots, s_d$ describing the categorical states, and the translation is done the other way round by applying `match` together with a vector of these strings. The numerical coding, in turn, is easily changed back into a binarization, which is the most useful representation for frequency computations. As an example, the binarization was used for creating the rate evolution graphs in Example 6.1.2 (by also using `apply(..., cumsum)` and then `matplot`), and for computing the measures of serial dependence in Example 6.3.1.

Eigenvalues and Sparse Matrices

The eigenvalues computed with `eigen` are sorted in descending order according to their absolute value, so the jth largest eigenvalue and its corresponding eigenvector are obtained by adding `$values[j]` and `$vectors[,j]`, respectively.

In contrast to some other computational software solutions, R requires an additional package for creating sparse matrices and vectors: the R package "matrix" by Bates & Mächler (2016). To use these sparse matrix techniques (as in Example 8.3.2.1), one has to specify the non-zero entries by creating a list of their values plus two lists with the coordinates of these entries. Then the matrix is created by the `sparseMatrix` command. The common matrix operations (such as `%*%`) and commands (such as `solve`) can be applied to the sparse matrices too. To facilitate the computation of the dominant eigenvector (the one belonging to the largest eigenvalue), a further R package might be used, namely "SQUAREM" by Varadhan (2016), and there the command `squarem`.

C.2 List of R Codes

Section 2.0: `Counts.r`.
Computation and simulation for count models.
Example 2.1.2.1: `PINAR1paths.r`.
Simulation of Poisson INAR(1) process.
Example 2.1.3.5: `NB-INAR1.r`.
Computations for NB-INAR(1) model.
Section 2.5: `INAR1fit.r`.
Fitting INAR(1) models.
Example 2.6.2: `PINAR1forecast.r`.
Forecasting Poisson INAR(1) process.
Remark 2.6.3: `INAR1fit.r`.
Forecasting INAR(1) process.
Example 3.1.6: `INARpfit.r`.
Fitting INAR(p) models.
Remark 3.1.7: `PINAR1nonstatpaths.r`.
Simulation of extensions of INAR(1) process.
Example 3.2.1: `INAR1fit.r`.
Fitting NB-RCINAR(1) model.
Example 3.3.4: `BAR1fit.r`.
Fitting (beta-)binomial AR(1) model.
Example 3.4.2: `BINAR1fit.r`.
Fitting BINAR(1) models.
Example 4.1.5: `INGARCH11fit.r`.
Fitting Poisson INGARCH(1, 1) model.
Remark 4.1.7: `PINARCH1paths.r`.
Simulation of Poisson INARCH(1) process.
Example 4.1.8: `INARCH1fit.r`.
Fitting Poisson INARCH(1) model.
Example 4.2.4: `INGARCH11fit.r`.
Fitting INGARCH(1, 1) models.
Example 4.2.6: `BINARCH1fit.r`.
Fitting binomial INARCH(1) model.
Example 5.1.2: `GLARMAsim.r`.
Simulation of Poisson GLARMA(0,1) process.
Example 5.1.3: `LogLinAutoRsim.r`.
Simulation of log-linear Poisson autoregression.
Example 5.1.6: `MargRegression.r`.
Fitting marginal seasonal log-linear model.
Example 5.1.7: `CondRegression.r`.
Fitting conditional seasonal log-linear model.
Example 5.2.2: `PoissonHMMprop.r`.
Computations for Poisson HMM.

Example 5.2.5: `PoissonHMMfit.r`.
Fitting Poisson HMM.
Section 5.3: `DAR1paths.r`.
Simulation of Poisson DAR(1) process.
Example 6.1.1: `plot_ordinal_ts.r`.
Plotting an ordinal time series.
Example 6.1.2: `plot_nominal_ts.r`.
Plotting a rate evolution graph.
Remark 6.1.3: `dependence:nominal_ts.r`.
Plotting a spectral envelope.
Example 6.2.2: `plot_ordinal_ts.r, plot_nominal_ts.r`.
Measuring categorical dispersion.
Example 6.3.1: `dependence:nominal_ts.r`.
Measuring categorical serial dependence.
Example 7.1.1: `MTDp.r`.
Computations for Markov chains.
Example 7.1.2: `MTDp.r`.
Computations for MTD models.
Example 7.2.4: `modelfit_nominal_ts.r`.
Fitting DAR and MTD models.
Example 7.3.2: `CatHMMprop.r`.
Computations for categorical HMM.
Example 7.3.4: `CatHMMfit.r`.
Fitting categorical HMM.
Example 7.4.5: `CatRegression.r`.
Fitting autoregressive logit model.
Example 7.4.6: `OrdLogitAR1paths.r`.
Computations for ordinal regression model.
Example 7.4.8: `OrdLogitAR1paths.r`.
Computations and simulation for ordinal autoregressive logit model.
Example 8.2.1.1: `c_chart_iid.r`.
c chart and ARL computations for i.i.d. data.
Example 8.2.1.2: `c_chart_iid_est.r`.
c chart with estimated parameters.
Example 8.2.2.3: `c_chart_INAR1.r`.
c chart and ARL computations for INAR(1) data.
Example 8.3.1.1: `CUSUM_iid.r`.
CUSUM chart and ARL computations for i.i.d. data.
Example 8.3.2.1: `CUSUM_INAR1.r`.
CUSUM chart and ARL computations for INAR(1) data.
Example 8.3.3.1: `EWMA_iid.r`.
ARL computation for EWMA chart for i.i.d. data.

Example 8.3.3.2: `EWMA_INAR1.r`.
EWMA chart and ARL computations for INAR(1) data.
Example 9.1.1.2: `SampleMon_bin.r`.
np chart, CUSUM chart and ARL computations for i.i.d. samples data.
Example 9.1.2.2: `SampleMon_cat.r`.
Pearson and Gini chart for i.i.d. samples data, ARL simulation.
Example 9.2.1.1: `runs_chart_iid.r`.
ATS computation for runs chart.
Example 9.2.1.2: `BinCUSUM_iid.r`.
CUSUM chart and ARL computations for i.i.d. binary data.
Example 9.2.2.1: `CatCUSUM_iid.r`.
CUSUM chart and ARL computations for i.i.d. categorical data.

C.3   List of Datasets

Download counts: Downloads.txt.
   Source: Weiß (2008a). (Section 2.5, Remark 2.6.3, Examples 3.2.1 and 5.2.5)
Rig counts: OffshoreRigcountsAlaska.txt.
   Source: Baker Hughes. (Example 2.6.2)
Counts of gold particles: goldparticle380.txt.
   Source: Westgren (1916). (Example 3.1.6)
Price stability counts: PriceStability.txt.
   Source: Weiß & Kim (2014). (Example 3.3.4)
Counts of road accidents: Accidents.txt.
   Source: Pedeli & Karlis (2011). (Example 3.4.2)
Transactions counts: EricssonB_Jul2.txt.
   Source: Brännäs & Quoreshi (2010). (Examples 4.1.5 and 4.2.4)
Strikes counts: Strikes.txt.
   Source: U S Bureau of Labor Statistics. (Example 4.1.8)
Counts of hantavirus infections: Hanta.txt.
   Source: Robert-Koch-Institut (2016). (Example 4.2.6)
Counts of Legionella infections: LegionnairesDisease_02-08.txt.
   Source: Robert-Koch-Institut (2016). (Example 5.1.6)
Counts of cryptosporidiosis infect.: Cryptosporidiosis_02-08.txt.
   Source: Robert-Koch-Institut (2016). (Example 5.1.7)
EEG sleep states: InfantEEGsleepstates.txt.
   Source: Stoffer et al. (2000). (Examples 6.1.1, 6.2.2 and 6.3.1)
Song of wood pewee: WoodPeweeSong.txt.
   Source: Craig (1943), Raftery & Tavaré (1994). (Examples 6.1.2, 6.2.2, 6.3.1)
DNA sequence of bovine leukemia virus: Bovine.txt.
   Source: NCBI. (Examples 7.2.4, 7.3.4 and 7.4.5)
IP counts: IPs.txt.
   Source: Weiß (2007). (Examples 8.2.2.3, 8.3.2.1 and 8.3.3.2)
Medical diagnoses: FattyLiver.txt, FattyLiver2.txt.
   Source: Weiß & Atzmüller (2010). (Examples 9.1.1.2 and 9.2.1.2)

References

Adnaik, S.B., Gadre, M.P., Rattihalli, R.N. (2015) Single attribute control charts for a Markovian-dependent process. *Communications in Statistics—Theory and Methods* **44**(17), 3723–3737.

Agosto, A., Cavaliere, G., Kristensen, D., Rahbek, A. (2016) Modeling corporate defaults: Poisson autoregressions with exogenous covariates (PARX). *Journal of Empirical Finance* **38**(B), 640–663.

Aigner, W., Miksch, S., Schumann, H., Tominski, C. (2011) *Visualization of Time-oriented Data*. Springer-Verlag, London.

Aitchison, J. (1986) *The Statistical Analysis of Compositional Data*. Chapman and Hall, New York.

Albers, W., Kallenberg, W.C.M. (2004) Are estimated control charts in control? *Statistics* **38**(1), 67–79.

Al-Osh, M.A., Alzaid, A.A. (1987) First-order integer-valued autoregressive (INAR(1)) process. *Journal of Time Series Analysis* **8**(3), 261–275.

Al-Osh, M.A., Alzaid, A.A. (1988) Integer-valued moving average (INMA) process. *Statistical Papers* **29**(1), 281–300.

Alwan, L.C., Roberts, H.V. (1995) The problem of misplaced control limits. *Journal of the Royal Statistical Society, Series C* **44**(3), 269–278.

Alzaid, A.A., Al-Osh, M.A. (1988) First-order integer-valued autoregressive process: distributional and regression properties. *Statistica Neerlandica* **42**(1), 53–61.

Alzaid, A.A., Al-Osh, M.A. (1990) An integer-valued pth-order autoregressive structure (INAR(p)) process. *Journal of Applied Probability* **27**(2), 314–324.

Alzaid, A.A., Al-Osh, M.A. (1993) Generalized Poisson ARMA processes. *Annals of the Institute of Statistical Mathematics* **45**(2), 223–232.

Andersson, J., Karlis, D. (2014) A parametric time series model with covariates for integers in $\mathbb{Z}$. *Statistical Modelling* **14**(2), 135–156.

Andreassen, C.M. (2013) *Models and Inference for Correlated Count Data*. PhD thesis, Aarhus University.

Azzalini, A. (1994) Logistic regression and other discrete data models for serially correlated observations. *Journal of the Italian Statistical Society* **3**(2), 169–179.

An Introduction to Discrete-Valued Time Series, First Edition. Christian H. Weiss.
© 2018 John Wiley & Sons Ltd. Published 2018 by John Wiley & Sons Ltd.
Companion website: www.wiley.com/go/weiss/discrete-valuedtimeseries

Bates, D., Mächler, M. (2016) *Matrix: Sparse and Dense Matrix Classes and Methods*. R package, version 1.2-7.1. https://cran.r-project.org/web/packages/Matrix/

Baum, L.E., Petrie, T. (1966) Statistical inference for probabilistic functions of finite state Markov chains. *The Annals of Mathematical Statistics* **37**(6), 1554–1563.

Benjamin, M.A., Rigby, R.A., Stasinopoulos D.M. (2003) Generalized autoregressive moving average models. *Journal of the American Statistical Association* **98**(1), 214–223.

Berchtold, A. (1999) The double chain Markov model. *Communications in Statistics—Theory and Methods* **28**(11), 2569–2589.

Berchtold, A. (2002) High-order extensions of the double chain Markov model. *Stochastic Models* **18**(2), 193–227.

Berchtold, A., Raftery, A.E. (2001) The mixture transition distribution model for high-order Markov chains and non-Gaussian time series. *Statistical Science* **17**(3), 328–356.

Bhat, U.N., Lal, R. (1990) Attribute control charts for Markov dependent production processes. *IIE Transactions* **22**(2), 181–188.

Billingsley, P. (1961) *Statistical Inference for Markov Processes*. University of Chicago Press, Chicago.

Billingsley, P. (1999) *Convergence of Probability Measures*. 2nd edition, John Wiley & Sons, New York.

Biswas, A., Song, P.X.-K. (2009) Discrete-valued ARMA processes. *Statistics and Probability Letters* **79**(17), 1884–1889.

Blatterman, D.K., Champ, C.W. (1992) A Shewhart control chart under 100 % inspection for Markov dependent attribute data. *Proceedings of the 23rd Annual Modeling and Simulation Conference*, 1769–1774.

Borchers, H.W. (2016) *pracma: Practical Numerical Math Functions*. R package, version 1.9.3. https://cran.r-project.org/web/packages/pracma/

Borror, C.M., Champ, C.W., Ridgon, S.E. (1998) Poisson EWMA control charts. *Journal of Quality Technology* **30**(4), 352–361.

Bourke, P.D. (1991) Detecting a shift in fraction nonconforming using run-length control charts with 100 % inspection. *Journal of Quality Technology* **23**(3), 225–238.

Box, G.E.P., Jenkins, G.M. (1970) *Time Series Analysis: Forecasting and Control*. 1st edition, Holden-Day, San Francisco.

Box, G.E.P., Jenkins, G.M., Reinsel, G.C., Ljung, G.M. (2015) *Time Series Analysis: Forecasting and Control*. 5th edition, John Wiley & Sons, Inc.

Bradley, R.C. (2005) Basic properties of strong mixing conditions: a survey and some open questions. *Probability Surveys* **2**, 107–144.

Brännäs, K., Hall, A. (2001) Estimation in integer-valued moving average models. *Applied Stochastic Models in Business and Industry* **17**(3), 277–291.

Brännäs, K., Quoreshi, A.M.M.S. (2010) Integer-valued moving average modelling of the number of transactions in stocks. *Applied Financial Economics* **20**(18), 1429–1440.

Braun, W.J. (1999) Run length distributions for estimated attributes charts. *Metrika* **50**(2), 121–129.

Brenčič, M., Kononova, N.K., Vreča, P. (2015) Relation between isotopic composition of precipitation and atmospheric circulation patterns. *Journal of Hydrology* **529**(3), 1422–1432.

Brockwell, P.J., Davis, R.A. (1991) *Time Series: Theory and Methods*. 2nd edition, Springer-Verlag, New York.

Brockwell, P.J., Davis, R.A. (2016) *Introduction to Time Series and Forecasting*. 3rd edition, Springer, Switzerland.

Brook, D., Evans, D.A. (1972) An approach to the probability distribution of CUSUM run length. *Biometrika* **59**(3), 539–549.

Bu, R., McCabe, B., Hadri, K. (2008) Maximum likelihood estimation of higher-order integer-valued autoregressive processes. *Journal of Time Series Analysis* **29**(6), 973–994.

Bühlmann, P., Wyner, A.J. (1999) Variable length Markov chains. *Annals of Statistics* **27**(2), 480–513.

Bulla, J., Berzel, A. (2008) Computational issues in parameter estimation for stationary hidden Markov models. *Computational Statistics* **23**(1), 1–18.

Cameron, A.C., Trivedi, P.K. (2013) *Regression Analysis of Count Data*. 2nd edition, Cambridge University Press, New York.

del Castillo, J., Pérez-Casany, M. (1998) Weighted Poisson distributions for overdispersion and underdispersion situations. *Annals of the Institute of Statistical Mathematics* **50**(3), 567–585.

Chan, K.S., Ledolter, J. (1995) Monte Carlo EM estimation for time series models involving counts. *Journal of the American Statistical Association* **90**(1), 242–252.

Chang, T.J., Kavvas, M.L., Delleur, J.W. (1984) Daily precipitation modeling by discrete autoregressive moving average processes. *Water Resources Research* **20**(5), 565–580.

Christou, V., Fokianos, K. (2014) Quasi-likelihood inference for negative binomial time series models. *Journal of Time Series Analysis* **35**(1), 55–78.

Christou, V., Fokianos, K. (2015) On count time series prediction. *Journal of Statistical Computation and Simulation* **85**(2), 357–373.

Churchill, G.A. (1989) Stochastic models for heterogeneous DNA sequences. *Bulletin of Mathematical Biology* **51**(1), 79–94.

Cox, D.R. (1975) Partial likelihood. *Biometrika* **62**(2), 269–276.

Cox, D.R. (1981) Statistical analysis of time series: some recent developments. *Scandinavian Journal of Statistics* **8**(2), 93–115.

Craig, W. (1943) *The Song of the Wood Pewee, Myiochanes virens Linnaeus: A Study of Bird Music.* New York State Museum Bulletin No. 334, University of the State of New York, Albany.

Cryer, J.D., Chan, K.-S. (2008) *Time Series Analysis—With Applications in R.* 2nd edition, Springer-Verlag, New York.

Cui, Y., Lund, R. (2009) A new look at time series of counts. *Biometrika* **96**(4), 781–792.

Cui, Y., Lund, R. (2010) Inference in binomial AR(1) models. *Statistics and Probability Letters* **80**(23–24), 1985–1990.

Cui, Y., Wu, R. (2016) On conditional maximum likelihood estimation for INGARCH(p,q) models. *Statistics and Probability Letters* **118**, 1–7.

Cutting, D., Kupiec, J., Pedersen, J., Sibun, P. (1992) A practical part-of-speech tagger. *ANLC'92 Proceedings of the third conference on applied natural language processing*, 133–140.

Czado, C., Gneiting, T., Held, L. (2009) Predictive model assessment for count data. *Biometrics* **65**(4), 1254–1261.

David, H.A., 1985. Bias of S^2 under dependence. *The American Statistician* **39**(3), 201.

Davis, R.A., Dunsmuir, W.T.M., Wang, Y. (2000) On autocorrelation in a Poisson regression model. *Biometrika* **87**(3), 491–505.

Davis, R.A., Dunsmuir, W.T.M., Streett, S.B. (2003) Observation-driven models for Poisson counts. *Biometrika* **90**(4), 777–790.

Davis, R.A., Holan, S.H., Lund, R., Ravishanker, N. (eds.) (2016) *Handbook of Discrete-Valued Time Series.* Chapman & Hall/CRC Press, Boca Raton.

Davis, R.A., Liu, H. (2016) Theory and inference for a class of nonlinear models with application to time series of counts. *Statistica Sinica* **26**(4), 1673–1707.

Davis, R.A., Wu, R. (2009) A negative binomial model for time series of counts. *Biometrika* **96**(3), 735–749.

Davoodi, M., Niaki, S.T.A., Torkamani, E.A. (2015) A maximum likelihood approach to estimate the change point of multistage Poisson count processes. *International Journal of Advanced Manufacturing Technology* **77**(5–8), 1443–1464.

Dehnert, M., Helm, W.E., Hütt,M.-Th. (2003) A discrete autoregressive process as a model for short-range correlations in DNA sequences. *Physica A* **327**(3–4), 535–553.

Deligonul, Z.S., Mergen, A.E. (1987) Dependence bias in conventional *p*-charts and its correction with an approximate lot quality distribution. *Journal of Applied Statistics* **14**(1), 75–81.

Delleur, J.W., Chang, T.J., Kavvas, M.L. (1989) Simulation models of sequences of dry and wet days. *Journal of Irrigation and Drainage Engineering* **115**(3), 344–357.

Doukhan, P., Latour, A., Oraichi, D. (2006) A simple integer-valued bilinear time series model. *Advances in Applied Probability* **38**(2), 559–578.

Doukhan, P., Fokianos, K., Li, X. (2012) On weak dependence conditions: the case of discrete valued processes. *Statistics and Probability Letters* **82**(11), 1941–1948.

Doukhan, P., Fokianos, K., Li, X. (2013) Corrigendum to "On weak dependence conditions: the case of discrete valued processes". *Statistics and Probability Letters* **83**(2), 674–675.

Drost, F.C., van den Akker, R., Werker, B.J.M. (2009) Efficient estimation of auto-regression parameters and innovation distributions for semiparametric integer-valued AR(p) models. *Journal of the Royal Statistical Society, Series B* **71**(2), 467–485.

Du, J.-G., Li, Y. (1991) The integer-valued autoregressive (INAR(p)) model. *Journal of Time Series Analysis* **12**(2), 129–142.

Dunsmuir, W.T.M., Scott, D.J. (2015) The glarma package for observation-driven time series regression of counts. *Journal of Statistical Software* **67**(7), 1–36.

Duran, R.I., Albin, S.L. (2009) Monitoring and accurately interpreting service processes with transactions that are classified in multiple categories. *IIE Transactions* **42**(2), 136–145.

Elton, C., Nicholson, M. (1942) The ten-year cycle in numbers of the lynx in Canada. *Journal of Animal Ecology* **11**(2), 215–244.

Emiliano, P.C., Vivanco, M.J.F., de Menezes, F.S. (2014) Information criteria: How do they behave in different models? *Computational Statistics and Data Analysis* **69**, 141–153.

Ephraim, Y., Merhav, N. (2002) Hidden Markov processes. *IEEE Transactions on Information Theory* **48**(6), 1518–1569.

Epskamp, S., Cramer, A.O.J., Waldorp, L.J., Schmittmann, V.D., Borsboom, D. (2012) qgraph: Network visualizations of relationships in psychometric data. *Journal of Statistical Software* **48**(4), 1–18.

Falk, M., Marohn, F., Michel, R., Hofmann, D., Macke, M., Spachmann, C., Englert, S. (2012) *A First Course on Time Series Analysis.* Open source book. http://statistik.mathematik.uni-wuerzburg.de/timeseries/

Fahrmeir, L., Kaufmann, H. (1987) Regression models for non-stationary categorical time series. *Journal of Time Series Analysis* **8**(2), 147–160.

Fahrmeir, L., Tutz, G. (2001) *Multivariate Statistical Modelling based on Generalized Linear Models.* 2nd edition, Springer-Verlag, New York.

Feller, W. (1968) *An Introduction to Probability Theory and its Applications—Volume I.* 3rd edition, John Wiley & Sons, New York.

Ferland, R., Latour, A., Oraichi, D. (2006) Integer-valued GARCH processes. *Journal of Time Series Analysis* **27**(6), 923–942.

Fokianos, K. (2011) Some recent progress in count time series. *Statistics* **45**(1), 49–58.

Fokianos, K., Kedem, B. (2003) Regression theory for categorical time series. *Statistical Science* **18**(3), 357–376.

Fokianos, K., Kedem, B. (2004) Partial likelihood inference for time series following generalized linear models. *Journal of Time Series Analysis* **25**(2), 173–197.

Fokianos, K., Rahbek, A., Tjøstheim, D. (2009) Poisson autoregression. *Journal of the American Statistical Association* **104**(4), 1430–1439.

Fokianos, K., Tjøstheim, D. (2011) Log-linear Poisson autoregression. *Journal of Multivariate Analysis* **102**, 563–578.

Fokianos, K., Tjøstheim, D. (2012) Nonlinear Poisson autoregression. *Annals of the Institute of Statistical Mathematics* **64**(6), 1205–1225.

Franke, J., Kirch, C., Kamgaing, J.T. (2012) Changepoints in times series of counts. *Journal of Time Series Analysis* **33**(5), 757–770.

Franke, J., Subba Rao, T. (1993) Multivariate first-order integer-valued autoregression. *Technical report* No. 95, Universität Kaiserslautern.

Freeland, R.K., McCabe, B.P.M. (2004a) Analysis of low count time series data by Poisson autoregression. *Journal of Time Series Analysis* **25**(5), 701–722.

Freeland, R.K., McCabe, B.P.M. (2004b) Forecasting discrete valued low count time series. *International Journal of Forecasting* **20**(3), 427–434.

Freeland, R.K., McCabe, B.P.M. (2005) Asymptotic properties of CLS estimators in the Poisson AR(1) model. *Statistics and Probability Letters* **73**(2), 147–153.

Gan, F.F. (1990a) Monitoring Poisson observations using modified exponentially weighted moving average control charts. *Communications in Statistics—Simulation and Computation* **19**(1), 103–124.

Gan, F.F. (1990b) Monitoring observations generated from a binomial distribution using modified exponentially weighted moving average control chart. *Journal of Statistical Computation and Simulation* **37**(1–2), 45–60.

Gan, F.F. (1993) An optimal design of CUSUM control charts for binomial counts. *Journal of Applied Statistics* **20**(4), 445–460.

Gandy, A., Kvaløy, J.T. (2013) Guaranteed conditional performance of control charts via bootstrap methods. *Scandinavian Journal of Statistics* **40**(4), 647–668.

Gardener, M. (2012) *Beginning R: The Statistical Programming Language.* John Wiley & Sons, Inc., Indianapolis.

Genest, C., Nešlehová, J. (2007) A primer on copulas for count data. *Astin Bulletin* **37**(2), 475–515.

Gonçalves, E., Mendes-Lopes, N., Silva, F. (2015) Infinitely divisible distributions in integer-valued GARCH models. *Journal of Time Series Analysis* **36**(4), 503–527.

Grunwald, G., Hyndman, R.J., Tedesco, L., Tweedie, R.L. (2000) Non-Gaussian conditional linear AR(1) models. *Australian and New Zealand Journal of Statistics* **42**(4), 479–495.

Hagmark, P.-E. (2009) A new concept for count distributions. *Statistics and Probability Letters* **79**(8), 1120–1124.

Harte, D. (2016) *HiddenMarkov: Hidden Markov Models*. R package, version 1.8-7. https://cran.r-project.org/web/packages/HiddenMarkov/

Harvey, A.C., Fernandes, C. (1989) Time series models for count or qualitative observations. *Journal of Business & Economic Statistics* **7**(4), 407–417.

Hawkins, D.M., Olwell, D.H. (1998) *Cumulative Sum Charts and Charting for Quality Improvement*, Springer-Verlag, New York.

Heathcote, C.R. (1966) Corrections and comments of the paper "A branching process allowing immigration". *Journal of the Royal Statistical Society, Series B* **28**(1), 213–217.

Heinen, A. (2003) Modelling time series count data: an autoregressive conditional Poisson model. *CORE Discussion Paper* 2003/62, University of Louvain, Belgium.

Heinen, A., Rengifo, E. (2003) Multivariate modelling of time series count data: an autoregressive conditional Poisson model. *CORE Discussion Paper* 2003/25, University of Louvain, Belgium.

Held, L., Höhle, M., Hofmann, M. (2005) A statistical framework for the analysis of multivariate infectious disease surveillance counts. *Statistical Modelling* **5**(3), 187–199.

Held, L., Hofmann, M., Höhle, M. (2006) A two-component model for counts of infectious diseases. *Biostatistics* **7**(3), 422–437.

Heyman, D.P., Tabatabai, A., Lakshman, T.V. (1992) Statistical analysis and simulation study of video teleconference traffic in ATM networks. *IEEE Transactions on Circuits and Systems for Video Technology* **2**(1), 49–59.

Heyman, D.P., Lakshman, T.V. (1996) Source models for VBR broadcast-video traffic. *IEEE/ACM Transactions on Networking* **4**(1), 40–48.

Heyman, P., Vaheri, A., Lundkvist, Å., Avsic-Zupanc, T. (2009) Hantavirus infections in Europe: from virus carriers to a major public-health problem. *Expert Review of Anti-infective Therapy* **7**(2), 205–217.

Höhle, M. (2007) Surveillance: an R package for the monitoring of infectious diseases. *Computational Statistics* **22**(4), 571–582.

Höhle, M. (2010) Online change-point detection in categorical time series. In Kneib & Tutz (eds): Statistical Modelling and Regression Structures, *Festschrift in Honour of Ludwig Fahrmeir*, Physica-Verlag, Heidelberg, 377–397.

Höhle, M., Paul, M. (2008) Count data regression charts for the monitoring of surveillance time series. *Computational Statistics and Data Analysis* **52**(9), 4357–4368.

Holan, S.H., Lund, R., Davis, G. (2010) The ARMA alphabet soup: a tour of ARMA model variants. *Statistics Surveys* **4**, 232–274.

Horn, S.D. (1977) Goodness-of-fit tests for discrete data: a review and an application to a health impairment scale. *Biometrics* **33**(1), 237–248.

Hudecová, Š., Hušková, M., Meintanis, S. (2015) Detection of changes in INAR models. In Steland et al. (eds), *Stochastic Models, Statistics and Their Applications*, Springer Proceedings in Mathematics & Statistics 122, 11–18.

Ibragimov, I. (1962). Some limit theorems for stationary processes. *Theory of Probability & Its Applications* **7**(4), 349–382.

Jacobs, P.A., Lewis, P.A.W. (1978a) Discrete time series generated by mixtures. I: correlational and runs properties. *Journal of the Royal Statistical Society, Series B* **40**(1), 94–105.

Jacobs, P.A., Lewis, P.A.W. (1978b) Discrete time series generated by mixtures. II: asymptotic properties. *Journal of the Royal Statistical Society, Series B* **40**(2), 222–228.

Jacobs, P.A., Lewis, P.A.W. (1978c) Discrete time series generated by mixtures. III: autoregressive processes (DAR(p)). *Naval Postgraduate School Technical Report* NPS55-78-022.

Jacobs, P.A., Lewis, P.A.W. (1983) Stationary discrete autoregressive-moving average time series generated by mixtures. *Journal of Time Series Analysis* **4**(1), 19–36.

Jazi, M.A., Jones, G., Lai, C.-D. (2012) First-order integer valued AR processes with zero inflated Poisson innovations. *Journal of Time Series Analysis* **33**(6), 954–963.

Jensen, W.A., Jones-Farmer, L.A., Champ, C.W., Woodall, W.H. (2006) Effects of parameter estimation on control chart properties: a literature review. *Journal of Quality Technology* **38**(4), 349–364.

Joe, H. (1996) Time series models with univariate margins in the convolution-closed infinitely divisible class. *Journal of Applied Probability* **33**(3), 664–677.

Joe, H. (1997) *Multivariate Models and Dependence Concepts*. Chapman & Hall/CRC Press, Boca Raton.

Joe, H. (2016) Markov models for count time series. In Davis et al. (eds): *Handbook of Discrete-Valued Time Series*, Chapman & Hall/CRC Press, Boca Raton, 29–49.

Johnson, N.L., Kemp, A.W., Kotz, S. (2005) *Univariate Discrete Distributions*. 3rd edition, John Wiley & Sons, Hoboken, New Jersey.

Johnson, N.L., Kotz, S., Balakrishnan, N. (1997) *Discrete Multivariate Distributions*. John Wiley & Sons, Hoboken, New Jersey.

Jørgensen, B., Lundbye-Christensen, S., Song, P.X.-K., Sun, L. (1999) A state space model for multivariate longitudinal count data. *Biometrika* **86**(1), 169–181.

Jung, R.C., Ronning, G., Tremayne, A.R. (2005) Estimation in conditional first order autoregression with discrete support. *Statistical Papers* **46**(2), 195–224.

Jung, R.C., Tremayne, A.R. (2003) Testing for serial dependence in time series models of counts. *Journal of Time Series Analysis* **24**(1), 65–84.

Jung, R.C., Tremayne, A.R. (2006) Coherent forecasting in integer time series models. *International Journal of Forecasting* **22**(2), 223–238.

Jung, R.C., Tremayne, A.R. (2011a) Convolution-closed models for count time series with applications. *Journal of Time Series Analysis* **32**(3), 268–280.

Jung, R.C., Tremayne, A.R. (2011b) Useful models for time series of counts or simply wrong ones? *AStA Advances in Statistical Analysis* **95**(1), 59–91.

Jung, R.C., Liesenfeld, R., Richard, J.-F. (2011) Dynamic factor models for multivariate count data: an application to stock-market trading activity. *Journal of Business & Economic Statistics* **29**(1), 73–85.

Kang, J., Lee, S. (2014) Parameter change test for Poisson autoregressive models. *Scandinavian Journal of Statistics* **41**(4), 1136–1152.

Kang, J., Song, J. (2015) Robust parameter change test for Poisson autoregressive models. *Statistics and Probability Letters* **104**, 14–21.

Kanter, M. (1975) Autoregression for discrete processes mod 2. *Journal of Applied Probability* **12**(2), 371–375.

Karlis, D., Pedeli, X. (2013) Flexible bivariate INAR(1) processes using copulas. *Communications in Statistics—Theory and Methods* **42**(4), 723–740.

Karlis, D. (2016) Models for multivariate count time series. In Davis et al. (eds): *Handbook of Discrete-Valued Time Series*, Chapman & Hall/CRC Press, Boca Raton, 407–424.

Katz, R.W. (1981) On some criteria for estimating the order of a Markov chain. *Technometrics* **23**(3), 243–249.

Kedem, B. (1980) *Binary Time Series*. Marcel Dekker, New York.

Kedem, B., Fokianos, K. (2002) *Regression Models for Time Series Analysis*. John Wiley & Sons, Hoboken, New Jersey.

Kemp, C.D. (1967) 'Stuttering-Poisson' distributions. *Journal of the Statistical and Social Inquiry Society of Ireland* **21**(5), 151–157.

Kenett, R.S., Pollak, M. (1996) Data-analytic aspects of the Shiryayev–Roberts control chart: surveillance of a non-homogeneous Poisson process. *Journal of Applied Statistics* **23**(1), 125–137.

Kenett, R.S., Pollak, M. (2012) On assessing the performance of sequential procedures for detecting a change. *Quality and Reliability Engineering International* **28**(5), 500–507.

Kiesl, H. (2003) *Ordinale Streuungsmaße—Theoretische Fundierung und statistische Anwendungen* (in German). Josef Eul Verlag, Lohmar, Cologne.

Kim, H.-Y., Park, Y. (2008) A non-stationary integer-valued autoregressive model. *Statistical Papers* **49**(3), 485–502.

Kirch, C., Kamgaing, J.T. (2015) On the use of estimating functions in monitoring time series for change points. *Journal of Statistical Planning and Inference* **161**, 25–49.

Kirch, C., Kamgaing, J.T. (2016) Detection of change points in discrete-valued time series. In Davis et al. (eds): *Handbook of Discrete-Valued Time Series*, Chapman & Hall/CRC Press, Boca Raton, 219–244.

Klein, J.L. (1997) *Statistical Visions in Time: A History of Time Series Analysis, 1662–1938*. Cambridge University Press, New York.

Klimko, L.A., Nelson, P.I. (1978) On conditional least squares estimation for stochastic processes. *Annals of Statistics* **6**(3), 629–642.

Knoth, S. (2006) The art of evaluating monitoring schemes — How to measure the performance of control charts? In Lenz & Wilrich (eds): *Frontiers in Statistical Quality Control 8*, Physica-Verlag, Heidelberg, 74–99.

Kocherlakota, S., Kocherlakota, K. (1992) *Bivariate Discrete Distributions*. Marcel Dekker, New York.

Kotz, S., Balakrishnan, N., Johnson, N.L. (2000) *Continuous Multivariate Distributions—Volume 1: Methods and Applications*. 2nd edition, John Wiley & Sons, New York.

Krogh, A., Brown, M., Mian, I.S., Sjölander, K., Haussler, D. (1994) Hidden Markov models in computational biology — applications to protein modeling. *Journal of Molecular Biology* **235**(5), 1501–1531.

Kulasekera, K.B., Tonkyn, D.W. (1992) A new distribution with applications to survival dispersal and dispersion. *Communications in Statistics—Simulation and Computation* **21**(2), 499–518.

Lai, C.D., Xie, M., Govindaraju, K. (2000) Study of a Markov model for a high-quality dependent process. *Journal of Applied Statistics* **27**(4), 461–473.

Latour, A. (1997) The multivariate GINAR(p) process. *Advances in Applied Probability* **29**(1), 228–248.

Latour, A. (1998) Existence and stochastic structure of a non-negative integer-valued autoregressive process. *Journal of Time Series Analysis* **19**(4), 439–455.

Lazaris, A., Koutsakis, P. (2010) Modeling multiplexed traffic from H.264/AVC videoconference streams. *Computer Communications* **33**(10), 1235–1242.

Liboschik, T., Fokianos, K., Fried, R. (2016) `tscount`: An R package for analysis of count time series following generalized linear models. To appear in *Journal of Statistical Software*.

Liu, H. (2012) *Some Models for Time Series of Counts*. PhD thesis, Columbia University.

Lorden, G. (1971) Procedures for reacting to a change in distribution. *Annals of Mathematical Statistics* **42**(6), 1897–1908.

Lütkepohl, H. (2005) *New Introduction to Multiple Time Series Analysis*. Springer-Verlag, New York.

MacDonald, I.L. (2014) Numerical maximisation of likelihood: a neglected alternative to EM? *International Statistical Review* **82**(2), 296–308.

Makhoul, J., Starner, T., Schwartz, R., Chou, G. (1994) On-line cursive handwriting recognition using hidden Markov models and statistical grammar. *HLT'94 Proceedings of the workshop on Human Language Technology*, Plainsboro, NJ, 432–436.

Marcucci, M. (1985) Monitoring multinomial processes. *Journal of Quality Technology* **17**(2), 86–91.

McKenzie, E. (1981) Extending the correlation structure of exponential autoregressive-moving-average processes. *Journal of Applied Probability* **18**(1), 181–189.

McKenzie, E. (1985) Some simple models for discrete variate time series. *Water Resources Bulletin* **21**(4), 645–650.

McKenzie, E. (1988) Some ARMA models for dependent sequences of Poisson counts. *Advances in Applied Probability* **20**(4), 822–835.

McKenzie, E. (2003) Discrete variate time series. In Rao & Shanbhag (eds): *Handbook of Statistics 21*, Elsevier, Amsterdam, 573–606.

Meintanis, S.G., Karlis, D. (2014) Validation tests for the innovation distribution in INAR time series models. *Computational Statistics* **29**(5), 1221–1241.

Mills, T.C. (2011) *The Foundations of Modern Time Series Analysis*. Palgrave Macmillan UK.

Mills, T.M., Seneta, E. (1991) Independence of partial autocorrelations for a classical immigration branching process. *Stochastic Processes and their Applications* **37**(2), 275–279.

Mohammadipour, M., Boylan, J.E. (2012) Forecast horizon aggregation in integer autoregressive moving average (INARMA) models. *Omega* **40**(6), 703–712.

Möller, T.A. (2016) Self-exciting threshold models for time series of counts with a finite range. *Stochastic Models* **32**(1), 77–98.

Möller, T.A., Silva, M.E., Weiß, C.H., Scotto, M.G., Pereira, I. (2016) Self-exciting threshold binomial autoregressive processes. *AStA Advances in Statistical Analysis* **100**(4), 369–400.

Monteiro, M., Scotto, M.G., Pereira, I. (2010) Integer-valued autoregressive processes with periodic structure. *Journal of Statistical Planning and Inference* **140**(6), 1529–1541.

Monteiro, M., Scotto, M.G., Pereira, I. (2012) Integer-valued self-exciting threshold autoregressive processes. *Communications in Statistics—Theory and Methods* **41**(15), 2717–2737.

Montgomery, D.C. (2009) *Introduction to Statistical Quality Control*. 6th edition, John Wiley & Sons, New York.

Moriña, D., Puig, P., Ríos, J., Vilella, A., Trilla, A. (2011) A statistical model for hospital admissions caused by seasonal diseases. *Statistics in Medicine* **30**(26), 3125–3136.

Mousavi, S., Reynolds, M.R. Jr. (2009) A CUSUM chart for monitoring a proportion with autocorrelated binary observations. *Journal of Quality Technology* **41**(4), 401–414.

Moysiadis, T., Fokianos, K. (2014) On binary and categorical time series models with feedback. *Journal of Multivariate Analysis* **131**, 209–228.

Mukhopadhyay, A.R. (2008) Multivariate attribute control chart using Mahalanobis D^2 statistic. *Journal of Applied Statistics* **35**(4), 421–429.

Natarajan, P., Lu, Z., Schwartz, R., Bazzi, I., Makhoul, J. (2001) Multilingual machine printed OCR. In Bunke & Caelli (eds): *Hidden Markov Models: Applications in Computer Vision*, World Scientific Publishing Company, 43–63.

Neumann, M.H. (2011) Absolute regularity and ergodicity of Poisson count processes. *Bernoulli* **17**(4), 1268–1284.

Ng, K.W., Tian, G.-L., Tang, M.-L. (2011) *Dirichlet and Related Distributions—Theory, Methods and Applications*. John Wiley & Sons, Chichester.

Nie, S.-Y., Wu, X.-Q. (2013) A historical study about the developing process of the classical linear time series models. In Yin et al. (eds): *Proceedings of The Eighth International Conference on Bio-Inspired Computing: Theories and Applications (BIC-TA), 2013*, Springer-Verlag, Berlin, Heidelberg, 425–433.

Page, E. (1954) Continuous inspection schemes. *Biometrika* **41**(1), 100–115.

Pakes, A.G. (1971) Branching processes with immigration. *Journal of Applied Probability* **8**(1), 32–42.

Paulino, S., Morais, M.C., Knoth, S. (2016) An ARL-unbiased c-chart. *Quality and Reliability Engineering International* **32**(8), 2847–2858.

Pawlowsky-Glahn, V., Buccianti, A. (eds) (2011) *Compositional Data Analysis—Theory and Practice*, John Wiley & Sons, Chichester.

Pedeli, X., Karlis, D. (2011) A bivariate INAR(1) process with application. *Statistical Modelling* **11**(4), 325–349.

Pedeli, X., Karlis, D. (2013a) Some properties of multivariate INAR(1) processes. *Computational Statistics and Data Analysis* **67**, 213–225.

Pedeli, X., Karlis, D. (2013b) On composite likelihood estimation of a multivariate INAR(1) model. *Journal of Time Series Analysis* **34**(2), 206–220.

Pegram, G.G.S. (1980) An autoregressive model for multilag Markov chains. *Journal of Applied Probability* **17**(2), 350–362.

Pevehouse, J.C., Brozek, J.D. (2008) Time-series analysis. In Box-Steffensmeier et al. (eds): *The Oxford Handbook of Political Methodology*, Oxford University Press.

Pickands III, J., Stine, R.A., 1997. Estimation for an $M/G/\infty$ queue with incomplete information. *Biometrika* **84**(2), 295–308.

Puig, P., Valero, J. (2006) Count data distributions: some characterizations with applications. *Journal of the American Statistical Association* **101**(1), 332–340.

Puig, P., Valero, J. (2007) Characterization of count data distributions involving additivity and binomial subsampling. *Bernoulli* **13**(2), 544–555.

Rabiner, L.R. (1989) A tutorial on hidden Markov models and selected applications in speech recognition. *Proceedings of the IEEE* **77**(2), 257–286.

Raftery, A.E. (1985) A model for high-order Markov chains. *Journal of the Royal Statistical Society, Series B* **47**(3), 528–539.

Raftery, A.E., Tavaré, S. (1994) Estimation and modelling repeated patterns in high order Markov chains with the mixture transition distribution model. *Applied Statistics* **43**(1), 179–199.

R Core Team (2016) *R: A Language and Environment for Statistical Computing*. R Foundation for Statistical Computing, Vienna, Austria. https://www.r-project.org

Reynolds, M.R. Jr., Stoumbos, Z.G. (1999) A CUSUM chart for monitoring a proportion when inspecting continuously. *Journal of Quality Technology* **31**(1), 87–108.

Ribler, R.L. (1997) *Visualizing Categorical Time Series Data with Applications to Computer and Communications Network Traces.* PhD thesis, Virginia Polytechnic Institute and State University.

Ristić, M.M., Bakouch, H.S., Nastić, A.S. (2009) A new geometric first-order integer-valued autoregressive (NGINAR(1)) process. *Journal of Statistical Planning and Inference* **139**(7), 2218–2226.

Robert-Koch-Institut (2016) SurvStat@RKI 2.0. https://survstat.rki.de/

Roberts, S.W. (1959) Control chart tests based on geometric moving averages. *Technometrics* **1**(3), 239–250.

Romano, J.P., Thombs, L.A. (1996) Inference for autocorrelations under weak assumptions. *Journal of the American Statistical Association* **91**(2), 590–600.

Ron, D., Singer, Y., Tishby, N. (1996) The power of amnesia: learning probabilistic automata with variable memory length. *Machine Learning* **25**(2), 117–149.

Ryan, A.G., Wells, L.J., Woodall, W.H. (2011) Methods for monitoring multiple proportions when inspecting continuously. *Journal of Quality Technology* **43**(3), 237–248.

Rydberg, T.H., Shephard, N. (2000) BIN models for trade-by-trade data. Modelling the number of trades in a fixed interval of time. *Econometric Society World Congress 2000*, Contributed Papers No 0740, Econometric Society.

Salmon, M., Schumacher, D., Höhle, M. (2016) Monitoring count time series in R: aberration detection in public health surveillance. *Journal of Statistical Software* **70**(10), 35 pages.

Schweer, S. (2015) Time-reversibility of integer-valued autoregressive processes of order p. In Steland et al. (eds): *Stochastic Models, Statistics and Their Applications*, Springer Proceedings in Mathematics & Statistics 122, Springer-Verlag, Heidelberg, 397–405.

Schweer, S., Weiß, C.H. (2014) Compound Poisson INAR(1) processes: stochastic properties and testing for overdispersion. *Computational Statistics and Data Analysis* **77**, 267–284.

Scotto, M.G., Weiß, C.H., Silva, M.E., Pereira, I. (2014) Bivariate binomial autoregressive models. *Journal of Multivariate Analysis* **125**, 233–251.

Scotto, M.G., Weiß, C.H., Gouveia, S. (2015) Thinning-based models in the analysis of integer-valued time series: a review. *Statistical Modelling* **15**(6), 590–618.

Seneta, E. (1983) *Non-negative Matrices and Markov Chains.* 2nd edition, Springer Verlag, New York.

Shewhart, W.A. (1926) Quality control charts. *Bell System Technical Journal* **5**(4), 593–603.

Shewhart, W.A. (1931) *Economic Control of Quality of Manufactured Product.* D. van Nostrand Company, Toronto, Canada.

Shmueli, G., Minka, T.P., Kadane, J.B., Borle, S., Boatwright, P. (2005) A useful distribution for fitting discrete data: revival of the Conway-Maxwell-Poisson distribution. *Journal of the Royal Statistical Society, Series C* **54**(1), 127–142.

Shumway, R.H., Stoffer, D.S. (2011) *Time Series Analysis and Its Applications.* 3rd edition, Springer-Verlag, New York.

Silva, M.E., Oliveira, V.L. (2005) Difference equations for the higher order moments and cumulants of the INAR(p) model. *Journal of Time Series Analysis* **26**(1), 17–36.

Slud, E., Kedem, B. (1994) Partial likelihood analysis of logistic regression and autoregression. *Statistica Sinica* **4**(1), 89–106.

Sparks, R.S., Keighley, T., Muscatello, D. (2010) Early warning CUSUM plans for surveillance of negative binomial daily disease counts. *Journal of Applied Statistics* **37**(11), 1911–1929.

Steutel, F.W., van Harn, K. (1979) Discrete analogues of self-decomposability and stability. *Annals of Probability* **7**(5), 893–899.

Stoffer, D.S. (1987) Walsh–Fourier analysis of discrete-valued time series. *Journal of Time Series Analysis* **8**(4), 449–467.

Stoffer, D.S., Tyler, D.E., McDougall, A.J. (1993) Spectral analysis for categorical time series: scaling and the spectral envelope. *Biometrika* **80**(3), 611–622.

Stoffer, D.S., Tyler, D.E., Wendt, D.A. (2000) The spectral envelope and its applications. *Statistical Science* **15**(3), 224–253.

Sutradhar, B.C. (2008) On forecasting counts. *Journal of Forecasting* **27**(2), 109–129.

Tanwir, S., Perros, H.G. (2014) *VBR Video Traffic Models*, ISTE Ltd, London and John Wiley & Sons, Hoboken.

Taylor, W.R. (1986) The classification of amino acid conservation. *Journal of Theoretical Biology* **119**(2), 205–218.

Testik, M.C. (2007) Conditional and marginal performance of the Poisson CUSUM control chart with parameter estimation. *International Journal of Production Research* **45**(23), 5621–5638.

Thavaneswaran, A., Ravishanker, N. (2016) Estimating equation approaches for integer-valued time series models. In Davis et al. (eds): *Handbook of Discrete-Valued Time Series*, Chapman & Hall/CRC Press, Boca Raton, 145–163.

Thede, S.M., Harper, M.P. (1999) A second-order hidden Markov model for part-of-speech tagging. *ACL'99 Proceedings of the 37th annual meeting of the Association for Computational Linguistics on Computational Linguistics*, 175–182.

Tjøstheim, D. (2012) Some recent theory for autoregressive count time series. *TEST* **21**(3), 413–438.

Torkamani, E.A., Niaki, S.T.A., Aminnayeri, M., Davoodi, M. (2014) Estimating the change point of correlated Poisson count processes. *Quality Engineering* **26**(2), 182–195.

Tsay, R.S. (1992) Model checking via parametric bootstraps in time series analysis. *Journal of the Royal Statistical Society, Series C* **41**(1), 1–15.

Tsay, R.S. (2000) Time series and forecasting: brief history and future research. *Journal of the American Statistical Association* **95**(2), 638–643.

Tsiamyrtzis, P., Karlis, D. (2004) Strategies for efficient computation of multivariate Poisson probabilities. *Communications in Statistics—Simulation and Computation* **33**(2), 271–292.

Turkman, K.F., Scotto, M.G., de Zea Bermudez, P. (2014) *Non-Linear Time Series: Extreme Events and Integer Value Problems.* Springer-Verlag, Switzerland.

Varadhan, R. (2016) *SQUAREM: Squared Extrapolation Methods for Accelerating EM-Like Monotone Algorithms.* R package, version 2016.8-2. https://cran.r-project.org/web/packages/SQUAREM/

Venkataraman, K.N. (1982) A time series approach to the study of the simple subcritical Galton-Watson process with immigration. *Advances in Applied Probability* **14**(1), 1–20.

Wang, Y.H., Yang, Z. (1995) On a Markov multinomial distribution. *Mathematical Scientist* **20**(1), 40–49.

Wang, C., Liu, H., Yao, J.-F., Davis, R.A., Li, W.K. (2014) Self-excited threshold Poisson autoregression. *Journal of the American Statistical Association* **109**(2), 777–787.

Wei, W.W.S. (2006) *Time Series Analysis: Univariate and Multivariate Methods.* 2nd edition, Pearson Education, Boston.

Weiß, C.H. (2007) Controlling correlated processes of Poisson counts. *Quality and Reliability Engineering International* **23**(6), 741–754.

Weiß, C.H. (2008a) Thinning operations for modelling time series of counts — a survey. *AStA Advances in Statistical Analysis* **92**(3), 319–341.

Weiß, C.H. (2008b) Serial dependence and regression of Poisson INARMA models. *Journal of Statistical Planning and Inference* **138**(10), 2975–2990.

Weiß, C.H. (2008c) The combined INAR(p) models for time series of counts. *Statistics and Probability Letters* **78**(13), 1817–1822.

Weiß, C.H. (2008d) Visual analysis of categorical time series. *Statistical Methodology* **5**(1), 56–71.

Weiß, C.H. (2009a) Monitoring correlated processes with binomial marginals. *Journal of Applied Statistics* **36**(4), 399–414.

Weiß, C.H. (2009b) A new class of autoregressive models for time series of binomial counts. *Communications in Statistics—Theory and Methods* **38**(4), 447–460.

Weiß, C.H. (2009c) Modelling time series of counts with overdispersion. *Statistical Methods and Applications* **18**(4), 507–519.

Weiß, C.H. (2009d) Properties of a class of binary ARMA models. *Statistics* **43**(2), 131–138.

Weiß, C.H. (2009e) EWMA monitoring of correlated processes of Poisson counts. *Quality Technology and Quantitative Management* **6**(2), 137–153.

Weiß, C.H. (2009f) Group inspection of dependent binary processes. *Quality and Reliability Engineering International* **25**(2), 151–165.

Weiß, C.H. (2010a) INARCH(1) processes: higher-order moments and jumps. *Statistics and Probability Letters* **80**(23–24), 1771–1780.

Weiß, C.H. (2010b) The INARCH(1) model for overdispersed time series of counts. *Communications in Statistics—Simulation and Computation* **39**(6), 1269–1291.

Weiß, C.H. (2011a) Empirical measures of signed serial dependence in categorical time series. *Journal of Statistical Computation and Simulation* **81**(4), 411–429.

Weiß, C.H. (2011b) The Markov chain approach for performance evaluation of control charts — a tutorial. In Werther (ed.): *Process Control: Problems, Techniques and Applications*, Nova Science Publishers, New York, 205–228.

Weiß, C.H. (2011c) Detecting mean increases in Poisson INAR(1) processes with EWMA control charts. *Journal of Applied Statistics* **38**(2), 383–398.

Weiß, C.H. (2012) Continuously monitoring categorical processes. *Quality Technology and Quantitative Management* **9**(2), 171–188.

Weiß, C.H. (2013a) Integer-valued autoregressive models for counts showing underdispersion. *Journal of Applied Statistics* **40**(9), 1931–1948.

Weiß, C.H. (2013b) Serial dependence of NDARMA processes. *Computational Statistics and Data Analysis* **68**, 213–238.

Weiß, C.H. (2013c) Monitoring k-th order runs in binary processes. *Computational Statistics* **28**(2), 541–563.

Weiß, C.H. (2015a) A Poisson INAR(1) model with serially dependent innovations. *Metrika* **78**(7), 829–851.

Weiß, C.H. (2015b) SPC methods for time-dependent processes of counts — a literature review. *Cogent Mathematics* **2**(1), 1111116.

Weiß, C.H., Atzmüller, M. (2010) EWMA control charts for monitoring binary processes with applications to medical diagnosis data. *Quality and Reliability Engineering International* **26**(8), 795–805.

Weiß, C.H., Göb, R. (2008) Measuring serial dependence in categorical time series. *AStA Advances in Statistical Analysis* **92**(1), 71–89.

Weiß, C.H., Kim, H.-Y. (2013) Binomial AR(1) processes: moments, cumulants, and estimation. *Statistics* **47**(3), 494–510.

Weiß, C.H., Kim, H.-Y. (2014) Diagnosing and modeling extra-binomial variation for time-dependent counts. *Applied Stochastic Models in Business and Industry* **30**(5), 588–608.

Weiß, C.H., Pollett, P.K. (2012) Chain binomial models and binomial autoregressive processes. *Biometrics* **68**(3), 815–824.

Weiß, C.H., Pollett, P.K. (2014) Binomial autoregressive processes with density dependent thinning. *Journal of Time Series Analysis* **35**(2), 115–132.

Weiß, C.H., Schweer, S. (2015) Detecting overdispersion in INARCH(1) processes. *Statistica Neerlandica* **69**(3), 281–297.

Weiß, C.H., Schweer, S. (2016) Bias corrections for moment estimators in Poisson INAR(1) and INARCH(1) processes. *Statistics and Probability Letters* **112**, 124–130.

Weiß, C.H., Testik, M.C. (2009) CUSUM monitoring of first-order integer-valued autoregressive processes of Poisson counts. *Journal of Quality Technology* **41**(4), 389–400.

Weiß, C.H., Testik, M.C. (2012) Detection of abrupt changes in count data time series: cumulative sum derivations for INARCH(1) models. *Journal of Quality Technology* **44**(3), 249–264.

Westgren, A. (1916) Die Veränderungsgeschwindigkeit der lokalen Teilchenkonzentration in kolloiden Systemen. *Arkiv för Matematik, Astronomi och Fysik* **11**(14), 1–24.

Wong, W.H. (1986) Theory of partial likelihood. *Annals of Statistics* **14**(1), 88–123.

Wood, R.A., McInish, T.H., Ord, J.K. (1985) An investigation of transactions data for NYSE stocks. *The Journal of Finance* **40**(3), 723–739.

Woodall, W.H. (1997) Control charts based on attribute data: bibliography and review. *Journal of Quality Technology* **29**(2), 172–183.

Woodall, W.H., Montgomery, D.C. (2014) Some current directions in the theory and application of statistical process monitoring. *Journal of Quality Technology* **46**(1), 78–94.

Xie, M., Goh, N., Kuralmani, V. (2000) On optimal setting of control limits for geometric chart. *International Journal of Reliability, Quality and Safety Engineering* **7**(1), 17–25.

Xu, H.-Y., Xie, M., Goh, T.N., Fu, X. (2012) A model for integer-valued time series with conditional overdispersion. *Computational Statistics and Data Analysis* **56**(12), 4229–4242.

Yang, M., Zamba, G.K.D., Cavanaugh, J.E. (2013) Markov regression models for count time series with excess zeros: a partial likelihood approach. *Statistical Methodology* **14**, 26–38.

Ye, N., Masum, S., Chen, Q., Vilbert, S. (2002) Multivariate statistical analysis of audit trails for host-based intrusion detection. *IEEE Transactions on Computers* **51**(7), 810–820.

Yeh, A.B., McGrath, R.N., Sembower, M.A., Shen, Q. (2008) EWMA control charts for monitoring high-yield processes based on non-transformed observations. *International Journal of Production Research* **46**(20), 5679–5699.

Yontay, P., Weiß, C.H., Testik, M.C., Bayindir, Z.P. (2013) A two-sided CUSUM chart for first-order integer-valued autoregressive processes of Poisson counts. *Quality and Reliability Engineering International* **29**(1), 33–42.

Zeger, S.L. (1988) A regression model for time series of counts. *Biometrika* **75**(4), 621–629.

Zeger, S.L., Qaqish, B. (1988) Markov regression models for time series: a quasi-likelihood approach. *Biometrics* **44**(4), 1019–1031.

Zheng, H., Basawa, I.V., Datta, S. (2006) Inference for pth-order random coefficient integer-valued autoregressive processes. *Journal of Time Series Analysis* **27**(3), 411–440.

Zheng, H., Basawa, I.V., Datta, S. (2007) First-order random coefficient integer-valued autoregressive processes. *Journal of Statistical Planning and Inference* **173**(1), 212–229.

Zhu, F. (2011) A negative binomial integer-valued GARCH model. *Journal of Time Series Analysis* **32**(1), 54–67.

Zhu, F. (2012a) Modeling overdispersed or underdispersed count data with generalized Poisson integer-valued GARCH models. *Journal of Mathematical Analysis and Applications* **389**(1), 58–71.

Zhu, F. (2012b) Zero-inflated Poisson and negative binomial integer-valued GARCH models. *Journal of Statistical Planning and Inference* **142**(4), 826–839.

Zhu, F. (2012c) Modeling time series of counts with COM-Poisson INGARCH models. *Mathematical and Computer Modelling* **56**(9–10), 191–203.

Zhu, R., Joe, H. (2003) A new type of discrete self-decomposability and its application to continuous-time Markov processes for modeling count data time series. *Stochastic Models* **19**(2), 235–254.

Zhu, R., Joe, H. (2006) Modelling count data time series with Markov processes based on binomial thinning. *Journal of Time Series Analysis* **27**(5), 725–738.

Zucchini, W., MacDonald, I.L. (2009) *Hidden Markov Models for Time Series: An Introduction Using R*. Chapman & Hall/CRC, London.

List of Acronyms

An Introduction to Discrete-Valued Time Series, First Edition. Christian H. Weiss.
© 2018 John Wiley & Sons Ltd. Published 2018 by John Wiley & Sons Ltd.
Companion website: www.wiley.com/go/weiss/discrete-valuedtimeseries

List of Notations

$\circ$, 17, 65
$\bullet$, 54, 65
$*$, 54
$\odot$, 58
$\otimes$, 70
$\mathbf{0}$, 27, 235
$\mathbf{1}$, 236

$\mathbb{1}$, 33, 189

$\mathfrak{A}$, 229
$a_{k,j}$, 242
$(\alpha_n)_{\mathbb{N}}$, 232
ARL, 167, 172
$\text{ARL}^{(\tau)}$, 172
$\text{ARL}^{(\infty)}$, 172
$A(z)$, 239
$\alpha(z)$, 241
$\mathbf{A}(z)$, 246

$\mathbf{B}$, 239
$\beta(z)$, 240
$\mathbf{B}(z)$, 246

c, 142
cgf(z), 12
κ_j, 13
$\kappa_{(j)}$, 13
$Corr[\cdot,\cdot]$, 230

$Cov[\cdot,\cdot]$, 230
C_t^+, 178
C_t^-, 178

d, 121, 225
$\delta_{k,l}$, 220

$E[\cdot]$, 230
$\mathbf{e}$, 121, 225
$e^{(h)}$, 38
e_t, 32

$\overline{F}$, 33
fcgf(z), 12
f_k, 12
$(\phi_n)_{\mathbb{N}}$, 231

$B(x,y)$, 222
$\Gamma(x)$, 217
$\gamma(k)$, 230
$\mathbf{\Gamma}(k)$, 245
$\hat{\gamma}(k)$, 231
$\gamma_t(k)$, 230
$\mathbf{\Gamma}_t(k)$, 245

I, 13
I_{Bin}, 15
$\mathbf{I}$, 66, 173, 246

$I_n(z)$, 221
I_{zero}, 14

$\kappa(k)$, 130
$\kappa_{\text{part}}(k)$, 141
$\hat{\kappa}(k)$, 130
$\hat{\kappa}_{\text{part}}(k)$, 130

L, 165
$\text{Li}_\nu(z)$, 219
λ_{PF}, 237
$L(\boldsymbol{\theta})$, 234
$\ell(\boldsymbol{\theta})$, 234
$\ell_{\max}$, 234
LR, 180
ℓR, 181
$\lambda(\omega)$, 125
$\hat{\lambda}(\omega)$, 125

μ, 12, 230
$\boldsymbol{\mu}$, 245
mgf(z), 12
μ_n, 12
$\overline{\mu}_n$, 12
$\overline{\mu}_{(n)}$, 12
M_t, 74, 96
$\mu(t)$, 230
$\mathbf{M}_t$, 93
$\boldsymbol{\mu}(t)$, 245

An Introduction to Discrete-Valued Time Series, First Edition. Christian H. Weiss.
© 2018 John Wiley & Sons Ltd. Published 2018 by John Wiley & Sons Ltd.
Companion website: www.wiley.com/go/weiss/discrete-valuedtimeseries

Index

An Introduction to Discrete-Valued Time Series, First Edition. Christian H. Weiss.
© 2018 John Wiley & Sons Ltd. Published 2018 by John Wiley & Sons Ltd.
Companion website: www.wiley.com/go/weiss/discrete-valuedtimeseries